Principles of Inverter Circuits

Principles of Inverter Circuits

B. D. Bedford

Senior Electrical Engineer,
Advanced Technology Laboratories,
General Electric Company,
Schenectady, New York

R. G. Hoft

Formerly, Manager, Converter Circuits Engineering,
Advanced Technology Laboratories,
General Electric Company,
Schenectady, New York

(Currently, Doctoral Candidate, Electrical Engineering,
Iowa State University of Science and Technology, Ames, Iowa)

John Wiley & Sons, Inc., New York · London · Sydney

Library of Congress Catalog Card Number: 64-20078
Printed in the United States of America

Contributing Authors

J. D. HARNDEN, JR.

Project Engineer,
Advanced Technology Laboratories,
Schenectady, New York

W. McMURRAY

Electrical Engineer,
Advanced Technology Laboratories,
Schenectady, New York

R. E. MORGAN

Electronics Engineer,
Advanced Technology Laboratories,
Schenectady, New York

D. P. SHATTUCK

Supervisor of Physics and
Applied Mathematics Laboratory,
Locomotive and Car Equipment Department,
General Electric Company,
Erie, Pennsylvania

F. G. TURNBULL

Electronics Engineer
Advanced Technology Laboratories,
General Electric Company,
Schenectady, New York

Preface

It is the purpose of this book to explain the fundamental principles of inverter circuits. The book is written for the circuit-development engineer and the graduate student in electrical engineering.

The *Bibliography on Electronic Power Converters*, published by the American Institute of Electrical Engineers in February, 1950, contains a chronological list of references beginning with the year 1903. The first inverter paper listed in this bibliography was published in 1925. In subsequent years, a number of additional inverter technical articles were written, and some inverter equipments were developed. These equipments involved the application of the controlled electronic valve of that era—the grid-controlled, gas-filled tube. Only a few commercial equipments were produced because of the limitations of the available valves. Mercury tubes, with holding arcs, and hot cathode thyratron tubes are difficult to start when only d-c power is available. In addition, these controlled-rectifier devices have considerable anode-cathode voltage drop when conducting and, hence, are not efficient in low voltage circuits. Mainly for these reasons, the inverter technology remained relatively dormant until a new device appeared on the scene.

The transistor was announced by Bell Laboratories in 1948. This discovery was the first of a whole new generation of controlled semiconductor devices. Initially, most effort was directed toward producing improved devices for application in relatively low-power electronic circuits. The silicon-controlled rectifier (SCR) was introduced by the General Electric Company in 1957. This provided "gate control" in a semiconductor of considerable power-handling capacity and, as a result, there has been a tremendous rebirth of interest in the inverter technology. Many semiconductor inverter equipments are commercially available. These equipments are extremely efficient, compact, and highly reliable d-c to a-c power converters.

In the normal rectifier for conversion of alternating current to direct current, the transfer of current from one valve to the next occurs naturally and automatically. The inverter requires control to definitely establish the conducting period of each valve during every cycle of operation. SCR inverters, additionally, require a means of interrupting the flow of

vii

current through each valve. A "commutating circuit" provides this current interruption. The term commutation is used to describe the transfer of current from one valve to the next in rectifier and inverter circuits.

This book primarily deals with SCR inverters, which require gate control and a commutating means. A principal portion of the book is devoted to discussions of alternate commutating circuits, as these are the key to the reliable inversion process. SCR devices are presently available in much higher power ratings than devices that have the ability to interrupt the flow of current; that is, power transistors or turn-off SCR's. This appears to be an inherent characteristic of all power-switching devices. Therefore, commutating circuits are expected to be required in most inverters with substantial power ratings.

The first chapter includes a description of the operation and characteristics of the silicon-controlled rectifier, as this is necessary background for the circuit engineer. Chapter 2 explains some of the more basic electronic oscillators, which can deliver a-c power. Inverters that use switching elements with the ability to interrupt the flow of current are also discussed in this chapter. Chapters 3 through 6 describe the more well-known means of commutation in inverters. These techniques have applications in certain situations, and form an excellent background for understanding the newer inverter commutation techniques discussed in Chapter 7. The impulse-commutated inverters, described in this chapter, currently have the most widespread application, as they operate reliably over wide ranges of load and load-power factor. These impulse-commutation circuits utilize the advantageous features of modern semiconductor-controlled rectifiers.

Simple inverters, generally, do not provide control of the output voltage, and produce output-voltage waveforms with considerable harmonic content. Techniques for controlling the voltage and improving the waveform are discussed in Chapters 8 and 9.

Chapter 10 describes several of the more basic d-c to d-c converter circuits. These circuits use the commutation principles discussed in the previous chapters. The final chapter contains a description of a number of different practical equipments, which use the approaches discussed throughout the book.

We are deeply indebted to the contributors who prepared the initial material for Chapters 1, 7, 9, 10, and 11. In addition, we make grateful acknowledgement to Dr. H. F. Storm for his valuable suggestions, F. E. Gentry for his comments on Chapter 1, G. F. Wright for devising the analog computer simulation in Chapter 4, C. W. Flairty for his contributions to Chapter 8, F. W. Gutzwiller and N. Mapham for reviewing the entire draft, and to our wives for their patience and understanding during all of the time that we have devoted to this book.

B. D. BEDFORD
R. G. HOFT

Note. This book has been prepared without giving particular consideration to the possible applicability of patents or patent applications to the material presented. The total patent situation should be checked in connection with any manufacture, use, or sale of the devices, circuits, or techniques described.

Contents

xi

Nomenclature and Definitions

The symbols and terms are defined in the text as they are used. A few of the more important ones are repeated below as an aid in becoming familiar with the "language" of this book.

Commutation (*in inverters*). Transfer of current from one electric valve to the next to conduct in sequence and the recovery of forward-blocking capability by the originally conducting valve.

Commutating time. Time interval required for complete commutation, which includes a period of negative voltage on the originally conducting valve to enable it to regain its forward-blocking capability. The complete negative voltage period provided by an inverter circuit is often called the "commutating angle."

e	Instantaneous voltage.
e_{a-o}	Instantaneous potential of point a with respect to o, where positive voltage occurs when a is positive with respect to o.
E	Average value of d-c voltage or half-cycle average value of an a-c voltage.
E_d	Average value of d-c supply voltage.
E_e	Effective or rms voltage.
E_m	Maximum or crest voltage.
$E_{m,1}$	Maximum value of fundamental component.
$E_{m,n}$	Maximum value of nth harmonic component.
i	Instantaneous current.
I	Average value of d-c current or half-cycle average value of an a-c current.
I_d	Average value of d-c supply current.
I_e	Effective or rms current.
I_m	Maximum or crest current.
$I_{m,1}$	Maximum value of fundamental component.
$I_{m,n}$	Maximum value of nth harmonic component.
SCR	Silicon-controlled rectifier device; also called electric valve, controlled rectifier, and solid-state switch.

Turn-off time. Time interval required for an SCR to regain its forward-blocking capability immediately after forward conduction.

Turn-on time. Time interval between initiation of gate turn-on signal and the reduction of the SCR forward voltage to 10% of its blocking value.

(In circuit diagrams, polarity dots are used on transformer and reactor windings to indicate the positive terminal for the induced voltage produced by a positive flux change. Arrows are used to indicate the positive direction of current flow or positive polarity of voltage.)

Chapter One

Controlled-Rectifier Device Theory and Characteristics

by F. G. Turnbull

A basic knowledge of the operation and characteristics of the devices or components to be used is necessary background for the advanced circuit engineer. This chapter includes a brief discussion of semiconductor theory. The basic theory is then extended to explain the principles of operation of rectifiers, transistors, and controlled-rectifier devices. After these theoretical discussions, the remainder of the chapter is devoted to the physical characteristics and ratings of controlled rectifiers. Because semiconductor theory is the subject of numerous textbooks, only the broad principles of device operation are included here. A rather extensive list of references is provided for those desiring a depth treatment of the theory of *p-n-p-n* device operation.

1.1 MODEL OF AN ATOM

The atom may be considered to consist of a central, relatively heavy, positively charged core surrounded by the proper number of electrons to produce a neutral charge on the whole structure. The electrons act as if they were particles having a definite mass and obeying Newton's laws of motion and Coulomb's law of attraction of oppositely charged bodies. A second concept is to consider the electrons as if they were electromagnetic waves. Consideration of the electron as both a wave and a particle is called the "wave-particle duality."

The elementary model of the hydrogen atom consists of a nucleus with one positive charge and a single orbital electron with a negative charge, as shown in Figure 1.1. If the orbit of this single electron is circular, there are two equal and opposite forces acting on this particle. The first force is due to Coulomb's law of attraction of oppositely charged bodies. The

1

second force acting on the electron is due to the orbital velocity of the rotating particle. The energy possessed by the rotating electron is composed of two parts—a potential energy and a kinetic energy. The potential energy is due to the electron position in the electric field of the nucleus, and the kinetic energy results from the orbital velocity of the electron about the nucleus.

The foregoing description of the hydrogen atom has been simplified, since it was assumed that the electron traveled in a circular orbit around the nucleus. A more exact solution is to assume that the orbit of the electron is elliptical. The principles of quantum mechanics are used to define the energy of an electron in an elliptical orbit. The principal quantum number, n, defines the major axis of the ellipse and specifies the main "shell"; the orbital quantum number, ℓ, defines the minor axis of the ellipse, and subdivides the main shell into "subshells." A third quantum number is required to describe the orbit of an electron in order to take into account the three-dimensional property of the atom. The elliptical orbit can be in any plane with respect to any other orbit. This quantum number, called the magnetic orbital quantum number, m_ℓ, defines the plane of the orbit. The electrons that are revolving around the nucleus are also spinning about their own axes. The spin of two electrons can be in the same direction or in opposite directions, resulting in the need for a fourth quantum number. This last quantum number, called the magnetic spin quantum number, m_s, specifies the direction of electron spin. These four quantum numbers completely specify the path of the electron. The Pauli exclusion principle states that no two electrons can have the same set of four quantum numbers.

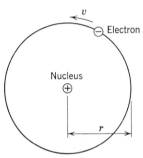

Figure 1.1   Model of a hydrogen atom.

In the materials used for semiconductors, each atom contains a number of electrons traversing different orbits about the nucleus. The electrons that are of paramount importance to the operation of semiconductors are the valence electrons or those that are located farthest away from the central nucleus. These valence electrons can take part in electrical conduction because they need acquire only a small amount of energy to enable them to go to the next vacant higher energy level, and possibly escape from the attraction of the central nucleus altogether. An electron at a lower energy level must acquire enough energy to go all the way up to the valence band because all of the intervening levels are occupied. When the electron leaves its orbital path and is not attached to any nucleus, the atom is left with a net positive charge and is called an ion. The positively charged ion is rigidly held in the crystal structure and is not available for electrical conduction.

For a single atom, the principles of quantum mechanics require the electrons to be at discrete energy levels. In the case of a solid, there are many atoms closely bound together, and there can be interaction between one nucleus and its neighbors. The permissible energy levels, especially those farthest from the nucleus, are extremely numerous and overlap with those of the adjacent atoms. Thus, the valence electrons of the atoms in a solid may be considered to have a continuous band of permissible energy levels.

1.2 THE INTRINSIC SEMICONDUCTOR

The two elements of most practical interest as semiconductors are germanium and silicon. These two elements occur in the fourth column of the periodic table, and each atom has four valence electrons. An additional four electrons are required to completely fill the valence energy subshell. The crystal structure of both of these two elements is a diamond cubic, as shown in Figure 1.2. Figure 1.2(a) shows the germanium or silicon atoms located at each corner of the cube and in the center of each cube face. The four additional germanium or silicon atoms in the crystal structure are shown separately in Figure 1.2(b) for clarity. Figure 1.2(c) shows the bonds between one of the inner germanium or silicon atoms and its four nearest neighbors. All other atoms in the crystal are similarly bonded together. Since each atom has four equidistant neighbors and requires four additional electrons to completely fill its valence subshell, the valence electrons of a given atom and its four nearest neighbors are shared by one another. This electron sharing is shown in Figure 1.3 by the simple two-dimensional model of the germanium or silicon crystal indicating only the nucleii and their valence electrons. The completion of the valence energy subshell by electron sharing produces electron pair bonds and the crystal formed is called a covalent solid. The perfect germanium or silicon crystal with no impurity atoms present is called an intrinsic semiconductor. In these intrinsic semiconductors, the valence energy level band is completely filled with electrons. The necessary number of valence electrons exist, and these electrons are all bonded to atoms in the crystal. At a temperature of absolute zero, there are no free electrons in the crystal and, hence, no electrical conduction is possible. The conduction energy level band is in the next range of permissible energy levels above the valence band, corresponding to the next set of higher quantum numbers. There is a forbidden band of energy levels between the top of the valence band

and the bottom of the conduction band. This band gap is illustrated in Figure 1.4. The width of the gap at 300°K is 0.72 ev for germanium and 1.1 ev for silicon.[3]

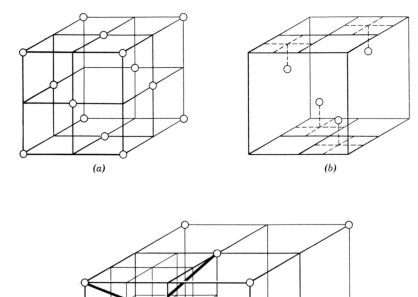

(a)   (b)

(c)

Figure 1.2 (a) Germanium or silicon crystal structure showing atoms at corners and faces. (b) Germanium or silicon crystal structure showing atoms in the body of the cube. (c) Germanium or silicon crystal structure showing the four nearest neighbor bonds.

Electrons can be established in the conduction band or holes in the valence band of an intrinsic semiconductor by the application of heat, light, nuclear radiation, injection from a doped material, or electric field. By the addition of energy, some of the valence electrons acquire sufficient

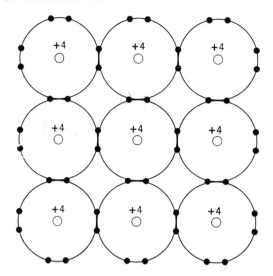

Figure 1.3   Germanium or silicon atoms, sharing valence electrons.

energy to break their bonds, thus creating electron-hole pairs which can now move freely through the crystal. These are available for electrical conduction. As the amount of energy is increased, a larger number of electron-hole pairs are created. For every electron bond broken, one free electron and one free hole are formed. Thus, the number of free holes in the valence band and free electrons in the conduction band in an intrinsic semiconductor are equal.

1.3   THE EXTRINSIC SEMICONDUCTOR

The previous section considered a crystal composed of only germanium or silicon atoms in perfect array. The addition of impurities or "doping" the intrinsic semiconductor crystal markedly changes the electrical prop-erties of the material, and this gives

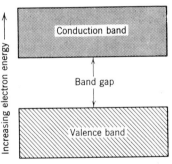

Figure 1.4   Energy band model for a semiconductor.

rise to useful characteristics in semiconductor devices. Atoms of elements in the third and fifth columns of the periodic table can occupy the place of a germanium or silicon atom in the crystal substitutionally; that is, without changing the lattice structure. Elements in the third column of the periodic table have only three electrons in their valence subshells, and

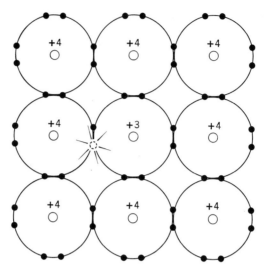

Figure 1.5   Germanium or silicon atoms with a group-three impurity atom.

are called acceptor impurities. If one of these atoms is located in a germanium or silicon crystal, there is a shortage of one electron to form the four electron-pair bonds. The four nearest neighbors to the acceptor impurity atom only have seven electrons to share, as shown in Figure 1.5. Since one of the nucleii has only three electrons in its valence band, the electron-pair bonds are incomplete. The resultant structure is lacking one

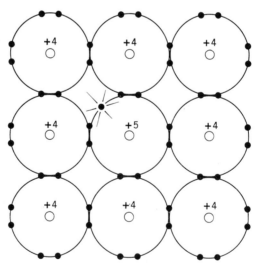

Figure 1.6   Germanium or silicon atoms with a group-five impurity atom.

electron to form a perfect crystal. This deficiency of an electron is called a hole. If an electron from a neighboring atom fills the deficiency in the valence band, it will cause the deficiency or hole to move in the opposite direction through the crystal. Examples of these group-three atoms are aluminum, gallium, and indium. At absolute zero, there are no free electrons in the germanium or silicon crystal with this impurity present, and the electron-pair deficiency cannot be remedied. These group-three

Table 1.1 Donor and acceptor impurity levels

		Germanium	Silicon
		Difference between acceptor or donor impurity energy level and valence or conduction band in electron volts	
Group-3 acceptors			
Boron	B	0.010	0.045
Aluminum	Al	0.010	0.057
Gallium	Ga	0.011	0.065
Indium	In	0.011	0.16
Group-5 donors			
Phosphorus	P	0.012	0.044
Arsenic	As	0.013	0.049
Antimony	Sb	0.010	0.039

elements are called acceptors because they can accept free electrons from the semiconductor crystal. The doping of an intrinsic semiconductor material with acceptor impurities forms a *p*-type semiconductor.

Elements in the fifth column of the periodic table—for example, arsenic, antimony, and phosphorus—can be substituted into the crystal structure. These impurity atoms have five electrons in their subshell and are called donor impurities. Therefore, there is one extra electron left over after the four nearest germanium or silicon atoms have shared the required eight electrons to completely fill their valence bands, as shown in Figure 1.6. At a temperature of absolute zero, this extra electron is not free to move in response to electric fields, as it remains fixed to the donor atom. Impurity atoms in this category are called donors because they donate an extra electron to the crystal. When donor impurities are added to an intrinsic material, an *n*-type semiconductor is formed.

The energy levels associated with the common impurity atoms are close to the bottom of the conduction band in the case of acceptors. For a donor, the additional electron is loosely held by the impurity atom. Therefore, only a small amount of energy is required to enable the additional electron to go into the conduction band. Since the acceptor hole is close to the top of the valence band, only a small amount of energy is required to cause a valence electron to fill an acceptor hole. The donor electrons in the conduction band or the holes produced in the valence band are both available for electric conduction. Table 1.1 gives the location of common donor and acceptor impurity levels.[5]

1.4 FERMI LEVEL

An important concept in semiconductor operation is the Fermi function[3]

$$f(W) = \frac{1}{\epsilon^{(W-W_f)/kT} + 1} \tag{1.1}$$

where W = energy in electron volts
W_f = Fermi level in electron volts
k = Boltzmann's constant in joules/$^\circ$K
T = temperature in $^\circ$K

The Fermi function, $f(W)$, gives the probability of finding an energy level, W, occupied with an electron or hole. The energy is measured with respect to the Fermi level, W_f. The Fermi level W_f, is that energy level where the Fermi function is equal to $\frac{1}{2}$. A second important concept is the distribution function, $g(W)$:[3]

$$g(W) = \frac{8\sqrt{2}\pi(m_e)^{3/2}}{h^3}(W - W_c)^{1/2} \tag{1.2}$$

where m_e = effective mass of an electron in kilograms
h = Planck's constant in joule-seconds

This function gives the number of permissible energy states expressed as a function of energy, and measured from the bottom of the conduction band, W_c. Between any two given energy levels in the conduction band the electron density is given by the integral of the product of the Fermi function and the distribution function[3]

$$n = \int_{W_1}^{W_2} f(W)g(W)\,dW \tag{1.3}$$

To find the number of electrons per cubic centimeter in the entire conduction band, equation (1.3) is integrated from W_c to infinity. The approximate result of this integration at room temperature is[3]

$$n = 2\left(\frac{2\pi m_e kT}{h^2}\right)^{3/2} \epsilon^{-(W_c - W_f)/kT} \tag{1.4}$$

A similar analysis yields the following expression for the density of free holes in the valence band.[3]

$$p = 2\left(\frac{2\pi m_h kT}{h^2}\right)^{3/2} \epsilon^{-(W_f - W_v)/kT} \tag{1.5}$$

where m_h = effective mass of a hole in kilograms
 W_v = energy level at the top of the valence band in electron volts

At room temperature, equations (1.4) and (1.5) give $n = p = 2.5 \times 10^{13}$ electrons or holes per cubic centimeter for intrinsic germanium and $n = p = 1.5 \times 10^{10}$ electrons or holes per cubic centimeter for intrinsic silicon.[4]

Free electrons and holes are produced in equal numbers in an intrinsic semiconductor. Therefore, equations (1.4) and (1.5) may be equated and solved for the Fermi level, giving the following:

$$W_f = \frac{W_v + W_c}{2} + \frac{3kT}{4} \ln\left(\frac{m_h}{m_e}\right) \tag{1.6}$$

At a temperature of absolute zero, the Fermi level is located halfway between the top of the valence band and the bottom of the conduction band. If the effective masses of holes and electrons are equal, the second term of equation (1.6) is again zero, and the Fermi level is halfway between the valence and conduction bands. Actually, there is some difference in the effective masses of holes and electrons and, thus, the Fermi level in an intrinsic semiconductor increases slightly with increasing temperature. Since the value of the Fermi function is 0.5 at the Fermi level, one half of the total number of carriers are located above and the other half are located below the Fermi level. Therefore, the Fermi level defines the average energy level in the system. When two semiconductor materials are brought together to form a junction, under equilibrium conditions, the Fermi level is continuous across the junction so that the average energy on each side of the junction is the same. The application of an external voltage to a semiconductor junction does not constitute an equilibrium condition. In this case, the Fermi level between the two sides of the junction is displaced by an amount proportional to the applied voltage.

The Fermi level for an extrinsic semiconductor may be determined in a similar manner. For an n-type semiconductor, the Fermi level is

$$W_f = W_c - kT \ln 2 \frac{\left(\frac{2\pi m_e kT}{h^2}\right)^{3/2}}{N_d} \tag{1.7}$$

where N_d = density of donor atoms. Equation (1.7) was derived by assuming that the total density of electrons in the conduction band is equal to the density of donor atoms. This means that the number of free electrons produced by the donor impurities is large compared to the free electrons in the intrinsic material. In addition, all of the donor atoms are assumed to be ionized so that their free electrons are in the conduction band.

For a p-type semiconductor, the Fermi level is

$$W_f = W_v + kT \ln 2 \frac{\left(\frac{2\pi m_h kT}{h^2}\right)^{3/2}}{N_a} \tag{1.8}$$

where N_a = density of acceptor atoms. This equation assumes that the density of holes in the valence band from the ionized impurity atoms is much larger than the hole density in the intrinsic material.

1.5   CONDUCTION PROCESSES

In semiconductor crystals, there are two important conduction processes. The first method of conduction is by diffusion. If there is a difference in the density of electrons or holes between two parts of a semiconductor crystal, the charged particles will flow from the higher density region toward the lower density region. The second method of conduction is the motion of charged particles acting under the influence of an electric field. The charged particles are undergoing constant motion due to collisions produced by the thermal energy of the particles. However, there is no average velocity in any one direction unless there is an applied electric field. The conductivity resulting from an applied electric field is

$$\sigma = q(n\mu_e + p\mu_n) \tag{1.9}$$

where σ = conductivity in ohm^{-1} cm^{-1}

q = charge on an electron or hole in coulombs

n = number of electrons per cubic centimeter

μ_e = electron mobility in cm^2/volt-second

p = number of holes per cubic centimeter

μ_n = hole mobility in cm^2/volt-second

The conductivity at room temperature is $\sigma = 0.02$ ohm^{-1} cm^{-1} for intrinsic germanium, and $\sigma = 4.3 \times 10^{-6}$ ohm^{-1} cm^{-1} for intrinsic silicon.[4]

1.6   *p-n* JUNCTION

The diode rectifier has a single *p-n* junction. If a semiconductor material is "doped" with an acceptor impurity and joined to a material doped with a donor impurity so as to form a perfect continuation of the crystal lattice, a *p-n* junction will result. The actual fabrication of a *p-n* junction is an extremely complex process, extensively covered in previous literature.[6-8]

When the two sides of a junction are joined, some of the excess electrons on the *n* side will diffuse into the *p* side. Also, some of the excess holes on the *p* side will diffuse into the *n* side. The result of these two diffusions of charged particles is to form a small electrostatic potential across the junction. The atoms of the semiconductor on the *n* side near the junction are now positive ions, having lost an electron. The atoms of the semiconductor on the *p* side near the junction are now negative ions, having gained an electron. This process is very rapid but stops as soon as the magnitude of the potential barrier across the junction prevents any additional charge flow. The magnitude of the potential barrier satisfies the condition that the Fermi level is continuous across the junction.

A very small number of majority carriers in the *n* region have enough energy to overcome the opposing electric field at the junction and cross over into the *p* side. After an electron is in the *p* region, it can combine with one of the majority carriers, holes, in a process called recombination. Likewise, some of the holes on the *p* side acquire enough energy to overcome the opposing field at the junction and cross over to recombine with an electron on the *n* side. The sum of these two charge flows is called the recombination current.[9]

As the temperature of the crystal is increased, some of the electron bonds of the semiconductor crystal are broken, thereby producing a free electron and a free hole. The free electrons in the *p* region are the minority carriers and are relatively few in number. However, the potential barrier at the junction is in the direction to attract these minority carriers. When an electron from the *p* region drifts to the barrier, it is accelerated across the barrier to the n region, where it becomes one of the majority carriers. Likewise, free holes, produced by the breaking of an electron bond, are minority carriers in the n region, and are accelerated across the junction into the p region by the electric field. This charge flow by the minority carriers is called the thermal current, since this component of current is dependent on temperature. As there is no external circuit to carry current, the thermal and recombination currents must be equal in magnitude and of opposite direction, in order that the total junction current equal zero.

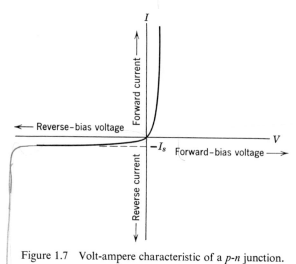

Figure 1.7   Volt-ampere characteristic of a p-n junction.

The fixed charges located on both sides of a p-n junction constitute a voltage difference and an effective capacitance. This is called the depletion-layer capacitance. The width of the depletion layer and the effective capacitance are a function of the junction-bias voltage.[3] A reverse-bias voltage on a p-n junction is positive on the n side and negative on the p side. Since the resistivity of the bulk of the semiconductor material is low, the entire external bias voltage appears at the junction, increasing the width of the depletion layer and the height of the potential barrier. The majority carriers that possess sufficient energy to overcome the increased voltage gradient are greatly reduced in number. Therefore, the recombination current decreases rapidly as the magnitude of the reverse-bias voltage increases. The charge flow due to minority carriers, thermal current, is not reduced by the application of a reverse bias, because the potential gradient at the junction accelerates all minority carriers that appear at the junction.

Therefore, the magnitude of this thermal current is independent of the magnitude of the external reverse-bias voltage. As this voltage increases in magnitude, the total junction current, which is the sum of the thermal and recombination currents, increases from zero and reaches a constant value called the reverse saturation current. See Figure 1.7.[1]

As the reverse-bias voltage is increased, so is the accelerating field of the junction to minority carrier flow. At some value of reverse potential, the minority carriers acquire so much energy that they collide with the fixed atoms of the crystal and dislodge additional minority carriers. These dislodged carriers add to the total current. This "high-field" emission is somewhat analogous to secondary emission in vacuum tubes. At this value of reverse voltage, the current through the device increases from the value of saturation current by many orders of magnitude. This constant voltage characteristic of a reverse-biased *p-n* junction is the basis for Zener diodes.

With an external forward-bias voltage applied to a *p-n* junction, all of the bias voltage again appears at the junction because of the low resistivity of the bulk material. This polarity of external bias voltage decreases the height of the potential barrier. Again, the magnitude of the minority carrier flow, thermal current, is unaffected by the decrease in accelerating potential across the junction. There will still be a net accelerating voltage to attract any minority carriers that appear at the junction. However, the decrease in height of the potential barrier means that many more of the majority carriers have sufficient energy to overcome the retarding electric field, and flow in the external circuit. The total current in a *p-n* junction is given by[1]

$$I_t = I_s(\epsilon^{qV/kT} - 1) \tag{1.10}$$

where I_t = total junction current in amperes

I_s = reverse saturation current in amperes

V = applied voltage in volts

V is positive for a forward-bias voltage and negative for a reverse-bias voltage. Equation (1.10) is plotted in Figure 1.7. This curve shows that the reverse current approaches $-I_s$ when $V < 0$, and the forward current is approximately $I_s\epsilon^{qV/kT}$ if $V > 0$.

1.7 *n-p-n* TRANSISTOR

The *n-p-n* transistor contains two *p-n* junctions in a single structure (see Figure 1.8). With no external bias voltages applied, the majority carriers diffuse until there is sufficient "built-in" voltage at each junction to oppose any further diffusion. A very small additional percentage of the majority carriers will have enough energy to overcome the retarding electric field and enter the *p* region. These majority-carrier electrons from both *n* regions become minority carriers in the central *p* region, and they recombine with some of the majority carriers; that is, holes. When the crystal is at room temperature, some of the electron bonds are broken, causing free electrons and free holes to exist in all three regions. In the collector and emitter, the free holes are minority carriers, and any that drift to the base-region junctions are accelerated by the electric fields into the base region. This minority carrier or thermal current just equals the majority carrier or recombination current so that the sum of the two at each junction is zero.

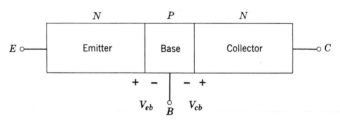

Figure 1.8   *n-p-n* Transistor.

Assume that the emitter-to-base junction is forward biased and that the collector-to-base junction is reverse biased. The positive voltage of the base with respect to the emitter appears as a reduction in the built-in voltage barrier. This reduced retarding voltage means that many more of the majority carriers, electrons, from the emitter can cross over into the base region. The magnitude of the retarding force for the majority carriers, holes, in the base region is also decreased by the forward-bias voltage. In spite of this fact, the number of holes that cross over to the emitter is kept small since the base region has a high resistivity compared to the emitter. The reverse bias on the collector-to-base junction increases the potential barrier to the majority carriers in both the collector and the base. The emitter has injected a large number of electrons into the base region. Since there are no electric fields present in the bulk of the base region, the electrons at the emitter-to-base junction will diffuse through the base material. The collector-to-base junction is biased such that any of the injected electrons that reach this junction are accelerated across and, then, flow out to the external circuit. In an ideal transistor, all of the electrons injected into the base would diffuse to the collector junction and be "collected" in the collector. However, in practical transistors, some of the electrons will recombine with the base-region holes. The ratio of the collected current to the total injected current from the emitter is called the current gain or α. In all junction transistors, α is less than unity but normally greater than 0.9. In order to reduce the amount of recombination in the base region, the base region is made very thin so that a greater portion of the injected electrons will diffuse to the collector.

The equations for the collector and base current flow are similar to equation (1.10) but each is composed of two parts—one due to majority-carrier flow caused by the forward-bias voltage, and the second due to the portion of the injected current collected at the collector junction.

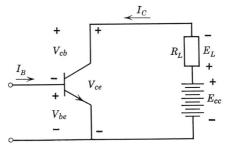

Figure 1.9 Common emitter *n-p-n* transistor switch.

Assuming the common emitter connection shown in Figure 1.9, the equations for the currents are[2,10]

$$I_e = \alpha_n I_{es}(\epsilon^{qV_{be}/kT} - 1) - I_{cs}(\epsilon^{-qV_{cb}/kT} - 1) \tag{1.11}$$

$$I_b = (1 - \alpha_n)I_{es}(\epsilon^{qV_{be}/kT} - 1) + (1 - \alpha_i)I_{cs}(\epsilon^{-qV_{cb}/kT} - 1) \tag{1.12}$$

where I_c = collector current in amperes
α_n = normal current gain
I_{es} = emitter saturation current in amperes
I_{cs} = collector saturation current in amperes
I_b = base current in amperes
α_i = inverted current gain; that is, with collector and emitter interchanged

There are three normal regions of operation for transistors. In Region I, called the "cutoff region," both junctions are reverse biased and the current flowing from collector to emitter is small. The substitution of the conditions that $V_{cb} \gg 0$ and $V_{be} \ll 0$ into equations (1.11) and (1.12) results in the following expression for the collector current:[2]

$$I_c = (1 - \alpha_i) I_{cs} \tag{1.13}$$

The value of collector current given by equation (1.13) is very small, and the load voltage, $I_c R_L$, is also very small. Under these conditions, most of the collector bias battery voltage, E_{cc} in Figure 1.9, appears on the transistor from collector to emitter.

A very narrow base width is required to obtain a value of α as near to unity as possible. Since both junctions are reverse biased in the cutoff region, the width of the depletion layers from both junctions will increase. During conditions of reverse bias, the accelerating potentials to minority-carrier flow are increased. The minority carriers crossing the junctions acquire enough energy that collisions with the crystal structure will dislodge additional minority and majority carriers. This is somewhat similar to secondary emission of electrons in a vacuum tube. The value of collector-to-emitter voltage that causes this "avalanche" effect is the maximum rating of the transistor, and determines the maximum value of collector voltage. If the depletion layers become so wide that they touch in the base region, the transistor is destroyed if the circuit allows large values of collector current to flow. The requirement for a narrow base width reduces the value of the "punch-through" voltage. Power transistors are presently manufactured that have voltage ratings of several hundred volts, although the majority are rated at less than 100 v.

In Region II, called the "linear region," the emitter-to-base junction is forward biased, and the collector-to-base junction is reverse biased. The substitution of these conditions into equations (1.11) and (1.12) results in the following value of collector current:[2]

$$I_c = \frac{\alpha_n I_b + I_{co}}{1 - \alpha_n} \qquad (1.14)$$

where

$$I_{co} = (1 - \alpha_n \alpha_i) I_{cs}$$

Neglecting the I_{co} term, the collector current is proportional to the base current and, hence, the load voltage, $I_c R_L$, in Figure 1.9 is proportional to the base current. The ratio $\alpha_n/(1 - \alpha_n)$ is called the grounded emitter current gain, and is given the symbol β. If $\alpha_n = 0.975$, then $\beta = 39$, indicating that the base current need be only 2.5 per cent of the collector current. The reason that Region II is called the linear region of operation is that the collector current (and, hence, the load voltage) is linearly proportional to the base current.

In normal operation, the base current I_b is a d-c bias current, upon which the signal current is superimposed. If the signal current is a high-frequency signal, at some value of frequency the collector current will not be able to follow the instantaneous value of the base signal current. The limiting parameter in determining this maximum frequency is the transit time for the injected electrons to diffuse through the base material. A very narrow base width decreases the transit time and results in a faster transistor. The frequency at which α is down 3 db from its low-frequency value is called the α-cutoff frequency. At present, this cutoff frequency is as high as 2000 megacycles for signal transistors. Higher voltage transistors and power transistors presently have α-cutoff frequencies of approximately 30 kilocycles to as great as 10 megacycles.

In Region III, referred to as the "saturation region," both junctions are forward biased; V_{be} is positive, and V_{cb} is negative. The substitution of these two constraints into equations (1.11) and (1.12) results in the following expression for V_{ce}:[2]

$$V_{ce} = \frac{kT}{q}(1 - \alpha_i) \qquad (1.15)$$

Equation (1.15) assumes that the base current is sufficient to cause the transistor to be well into the saturated region. The value of kT/q is 26 mv at $300°K$. Therefore, the theoretical voltage drop from the collector to the emitter is less than 26 mv. The actual saturated voltage drop is considerably greater because of the resistance of the collector-to-emitter bulk material. The collector current under these conditions is limited by the value of the load resistance.

The common emitter transistor approximates an ideal switch. The "open" leakage current from equation (1.13) is negligible compared with the "closed" current, E_{cc}/R_L; and the "closed" voltage drop given by equation (1.15) is usually negligible compared to the "open" voltage, E_{cc}. In the transistor inverter circuits, discussed in Chapter 2, the transistors are switched rapidly from Region I to Region III. Neglecting base currents, the power dissipated in the transistor is equal to $(1 - \alpha_i)(I_{cs}E_{cc})$ in Region I and $(kT/q)(1 - \alpha_i)(E_{cc}/R_L)$ in Region III, both of which are small compared to the maximum load power, E_{cc}^2/R_L. The transistor is switched through Region II as rapidly as possible to reduce the "switching" losses that occur during the transition from Region I to Region III.

The p-n-p transistor has the same three operating regions, and it operates in a similar fashion to the n-p-n transistor.

1.8 p-n-p-n CONTROLLED RECTIFIER—NO BIAS VOLTAGE

The p-n-p-n structure[5,11–18] is a three-junction, four-layer semiconductor device. When two leads are attached to the device, it is called a "four-layer diode."[17] The structure of the controlled rectifier and its electrical symbols are shown in Figure 1.10. In other devices external leads are attached to each of the four layers. During the rest of this section, discussion will be limited to the three-terminal controlled-rectifier configuration. In this device the end p region is called the anode, the end n region is called the cathode, and the p region in the center is called the gate.[25]

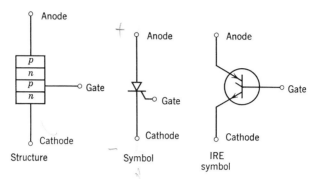

Figure 1.10   *p-n-p-n* Controlled rectifier.

Figure 1.11 shows the construction technique and materials used in a typical 25-amp silicon-controlled rectifier (SCR). Figure 1.12 shows a series of silicon-controlled rectifiers with rms current ratings of 1.6 to 235 amp.[11]

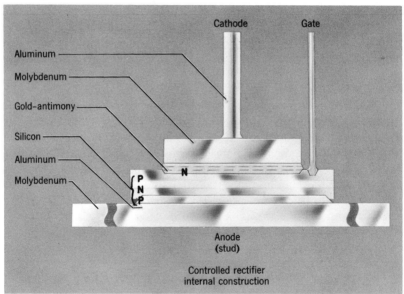

Figure 1.11   *p-n-p-n* Controlled rectifier structure.

The *p-n-p-n* structure is a bistable switching device. It normally has no linear region of operation, such as Region II for the transistor. Thus the SCR operates either in the saturated region or in the cutoff region. With a reverse voltage applied to its anode-to-cathode terminals (that is, positive voltage on the cathode), the device will not conduct any appreciable current. When the anode is made positive with respect to the cathode, again no appreciable current flows. However, when the anode is positive with respect to the cathode, and a small positive voltage pulse is applied across

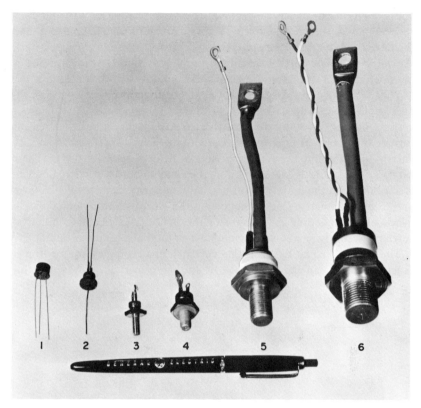

Figure 1.12 *p-n-p-n* Controlled rectifier, physical size. The illustration shows General Electric Company silicon-controlled rectifiers (SCR).

1. 1.6 amp rms   4. 25 amp rms
2. 2.7 amp rms   5. 110 amp rms
3. 7.0 amp rms   6. 235 amp rms

the gate-to-cathode terminals, the device switches "on." The current flow from anode to cathode is limited only by the external load impedance. Once the device is conducting appreciable forward current from anode to cathode, the gate-to-cathode signal is no longer required to maintain the device in its "on" condition. When the device has switched to its on condition, the forward current can be extinguished by increasing the circuit impedance to reduce the current below a specified value, making the cathode terminal positive with respect to the anode terminal or, in relatively low current controlled rectifiers, by applying a negative current pulse to the gate terminal.[26]

The controlled rectifier does have a linear region of operation that is occasionally used at very low power levels. If the anode to cathode is reverse biased, the anode current is proportional to the gate current, and the device exhibits the same properties as a transistor. In power circuits, the value of anode current that flows during conditions of reverse bias is considered "leakage current" and is usually neglected.

The electron energy diagram of a controlled rectifier with no bias voltages applied to its external terminals is shown in Figure 1.13. The majority carriers in each layer diffuse until there is a built-in voltage that retards further diffusion. However, some of the majority carriers have enough energy to surmount the retarding electric field at each junction. These carriers, once through the junction, become minority carriers and can recombine with the majority carriers. The minority carriers in each layer are accelerated across each junction by the electrostatic potential established at the junctions. The sum of the majority-carrier and minority-carrier currents is zero at each junction when there is no external circuit.[23,24]

The general equations for the currents and voltages in the three-junction device can be determined by the same method as was used for

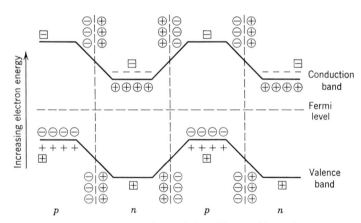

Figure 1.13   *p-n-p-n* Controlled rectifier, no bias voltages.

⊕ or ⊖ = fixed ions
⊞ or ⊟ = thermally generated carriers
+ or − = carriers produced by acceptor or
            donor impurities

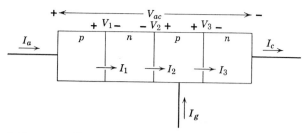

Figure 1.14   *p-n-p-n* Controlled rectifier, definitions of currents and voltages.

the one- and two-junction devices. The currents and voltages at each junction are defined in Figure 1.14. The currents I_1, I_2, and I_3, flowing across each junction, are composed of several components. The equations for these currents are as follows.[13,15,16]

$$I_1 = I_{s1}(\epsilon^{\beta V_1} - 1) - \alpha_{1i}I_{s2}(\epsilon^{\beta V_2} - 1) \tag{1.16}$$

$$I_2 = \alpha_{1n}I_{s1}(\epsilon^{\beta V_1} - 1) - I_{s2}(\epsilon^{\beta V_2} - 1) - \alpha_{2n}I_{s3}(\epsilon^{\beta V_3} - 1) \tag{1.17}$$

$$I_3 = -\alpha_{2i}I_{s2}(\epsilon^{\beta V_2} - 1) + I_{s3}(\epsilon^{\beta V_3} - 1) \tag{1.18}$$

where $\beta = q/kT$ in volts^{-1}

α_{1n} = normal alpha with junction one emitting and junction two collecting

α_{1i} = inverted alpha with junction two emitting and junction one collecting

α_{2n} = normal alpha with junction three emitting and junction two collecting

α_{2i} = inverted alpha with junction two emitting and junction three collecting

I_{s1} = the saturation current for junction one in amperes

I_{s2} = the saturation current for junction two in amperes

I_{s3} = the saturation current for junction three in amperes

Equation (1.16) has four terms. The first term is due to the majority carriers, that is, holes, crossing junction one; the second term is due to the minority carriers crossing junction one; the third term is due to the holes injected at junction two diffusing through the *n* region and crossing junction one; and the last term is due to minority carriers from junction two diffusing through the *n* region and crossing junction one. The current that crosses junction three is also composed of four terms; the current flowing across junction two is composed of six terms because of current flow from both junctions one and three. The following equations can be derived from an inspection of Figure 1.14.

$$I_a = I_1 = I_2 \qquad (1.19)$$

$$I_c = I_3 \qquad (1.20)$$

$$I_g = I_c - I_a \qquad (1.21)$$

The positive directions of I_a, I_c, and I_g are for the normal direction of current flow.

1.9 *p-n-p-n* CONTROLLED RECTIFIER—REGION A

This region of operation occurs when the cathode is positive with respect to the anode. This polarity of anode-to-cathode voltage reverse biases junctions one and three and forward biases junction two. From our previous discussion of reverse-biased *p-n* junctions, the controlled rectifier in Region A allows only a small leakage current to flow from cathode to anode. Junctions one and three share the applied reverse voltage in proportion to their "reverse resistances." Assuming that the junction-current gains (alphas) are much less than unity, equations (1.16), (1.17), and (1.19) may be combined to produce the following approximate expression for the anode current:

$$I_a, \text{ Region A} = -[I_{s1} + I_g(\alpha_{2n}\,\alpha_{1n})] \qquad (1.22)$$

Equation (1.22) shows that the anode current is negative and equal to the reverse saturation current of junction one plus a fraction of the gate current. The increasing of the reverse leakage current by the addition of gate current can increase the junction heating and cause thermal runaway. Since the junction saturation current is dependent on temperature, any rise in junction temperature will increase the saturation current and further increase the junction heating. For this reason, some device specifications limit the permissible gate voltage during conditions of reverse bias.[11]

As the magnitude of the anode-to-cathode reverse voltage is increased, the widths of the depletion layers at junctions one and three also increase. In normal device construction, junction one blocks most of the anode-to-cathode voltage and, hence, its depletion layer is relatively wide. In order that high voltage units can be constructed, the width of the center *n* region is made large so that the depletion layer from junction one does not cross over to junction two resulting in a "punch-through." Anode-to-cathode reverse-voltage ratings approaching 1500 v are the maximum presently available. The amount of reverse-leakage current, with no gate current, ranges from microamperes to several milliamperes in present devices, depending upon the rated anode current and anode-to-cathode voltage rating.[11]

1.10 p-n-p-n CONTROLLED RECTIFIER—REGION B

In this region of operation, the anode is positive with respect to the cathode, and the device is in the "blocking" condition. With this polarity of anode-cathode voltage, junctions one and three are forward biased and junction two is reverse biased. Since one of the p-n junctions is reverse biased, the anode current is very small. Assuming that the junction-current gains (alphas) are much less than unity, equations (1.16) through (1.21) may be combined to produce the following approximate expression for the anode current:

$$I_a, \text{ Region B} = I_{s2} + I_g\alpha_{2n} \qquad (1.23)$$

Equation (1.23) shows that the anode current during Region-B operation is positive, as defined in Figure 1.14, and is equal to the saturation current of junction two plus a fraction of the gate current. Since the term involving the gate current is positive, the increasing of the gate current increases the anode current during this mode of operation. The value of forward-leakage current ranges from microamperes to several milliamperes in presently available devices, depending on their anode current and voltage ratings.[11]

1.11 p-n-p-n CONTROLLED RECTIFIER—REGION C

There are four methods of switching the controlled rectifier from Region B to Region C. As the forward voltage from anode to cathode is increased, the width of the depletion layer at junction two is increased, and so is the accelerating voltage for minority carriers crossing junction two. When these carriers are accelerated across junction two, they collide with the fixed atoms of the crystal structure and dislodge additional minority carriers. As the anode-cathode forward voltage is further increased, the minority carriers dislodge many additional carriers during their transit across junction two, resulting in an avalanche breakdown of the junction. At this point, junction two becomes forward biased.[26] Since all three junctions are now forward biased, the anode current is limited only by the external load impedance. The controlled rectifier has switched from Region B, characterized by high voltage across the device and a low forward-leakage current, to Region C, corresponding to low voltage drop across the device and large forward current. The anode-to-cathode voltage at which the device switches from Region B to Region C is called the forward-breakover voltage. For practical units, the forward-breakover voltage is equal to or greater than the reverse-voltage rating. This anode-cathode voltage breakdown is the normal means of switching in a four-layer diode and a controlled rectifier with no gate current.

The gate lead on the controlled rectifier allows minority carriers to be injected into the gate region. This injection of additional minority carriers into the gate region is a second method of switching the device from Region B to Region C. Switching takes place with the anode-to-cathode potential less than the forward-breakover voltage. The injection of more carriers (that is, a larger gate current) causes the controlled rectifier to switch at lower values of anode-to-cathode voltage. For sufficiently large values of gate current, the device does not exhibit Region-B operation, but switches to Region C as soon as the anode is positive with respect to the cathode. The normal method of gating controlled rectifiers is to supply sufficient gate current at the desired instant of time to cause the device to switch to Region C at any value of forward voltage.

A third method of causing a controlled rectifier to switch from Region B to Region C is to illuminate the gate-to-cathode junction. The light can provide sufficient energy to break electron bonds in the semiconductor. The minority carriers thus formed in the gate region cause the controlled rectifier to switch "on" and conduct load current.

The fourth method by which the controlled rectifier may be turned on is to rapidly increase the anode-to-cathode forward voltage. This "dv/dt" "turn-on" is believed to be caused by anode-to-gate and gate-to-cathode capacitances. Rapid changes in the anode-cathode voltage produce transient gate current causing the device to turn on.

The time required for the device to switch from Region B to Region C is in the order of 1 to 3 microseconds, provided load circuit reactance does not limit the buildup of anode current. This time, called the turn-on time, can be increased or decreased somewhat by modifying the gate current and voltage waveshapes.

When operating in Region C, all three junctions in the controlled rectifier are forward biased, and the anode current must be limited by the external load impedance. With all of the junctions forward biased, equations (1.16) through (1.21) may be combined to produce the following approximate expression for the anode-to-cathode voltage:

$$V_{a\text{-}c} = \frac{1}{\beta} \ln \frac{I_a}{I_s} \left(\frac{1}{\alpha_{1n} + \alpha_{2n} - 1} \right) \qquad (1.24)$$

The forward voltage drop of a single forward-biased rectifier can be derived from equation (1.10) to give the following expression:

$$V_{a-c} = \frac{1}{\beta} \ln \left(\frac{I_a}{I_s} \right) \tag{1.25}$$

The sum of α_{1n} plus α_{2n} is approximately two when the controlled rectifier is conducting a large anode current. Thus, the controlled-rectifier forward-voltage drop, given by equation (1.24), approaches the voltage drop of a single forward-biased rectifier. The voltage drop across a conducting controlled rectifier is proportional to the natural log of the anode current. Therefore, the voltage drop increases only slightly for an increase in anode current of several orders of magnitude.

Once the controlled rectifier has switched to Region C, gate current is no longer required to keep the device operating in this region. In the case of the transistor, some of the injected carriers recombined in the base region, thereby requiring continuous base current to insure that the device stayed in saturation. With the sum of the alphas greater than unity in the controlled rectifier, any carriers lost due to recombination in the two center regions are replenished.

1.12   *p-n-p-n* CONTROLLED RECTIFIER TURNOFF

There are three means of stopping the flow of anode current in a controlled rectifier. First, the anode current can be reduced below a minimum value called the holding current. The value of holding current is reached when the sum of the alphas is again less than unity and the device switches back to Region-B operation. Rated anode current is about 2500 times larger than the holding current for a 25-amp controlled rectifier.[11] In this case, the voltage across the unit does not reverse, and the anode is maintained at a positive potential with respect to the cathode. The anode current can be reduced by opening a line switch, increasing the load resistance, or by shunting part of the load current through a transistor or other switch placed in parallel with the controlled rectifier.

The second method is similar to the first except that anode current is interrupted by reversing the anode-to-cathode voltage. If the anode is made negative with respect to the cathode, the device will switch to Region-A operation. This method is used in alternating-voltage circuits where the circuit voltage reverses every half-cycle. It is also used with commutating circuits where a previously charged capacitor is connected in parallel with the controlled rectifier. The polarity of the capacitor voltage is such that the anode of the controlled rectifier is forced negative with respect to the cathode.

The third method, used in low-current gate turn-off controlled rectifiers, is to increase the value of holding current by supplying a negative gate current.[26] When the increased holding current exceeds the load current, the device switches into Region-B operation.

The time required for the device to switch from Region-C operation to Region-B operation is called the "turn-off time."[11] This is defined as the time interval required for the controlled rectifier to regain its forward-blocking capability after forward conduction. While the controlled rectifier is operating in Region C, with the sum of the alphas greater than unity, all four regions of the device contain numerous moving charges. When the anode-to-cathode voltage is made negative to stop the flow of anode current, the moving charges are able to flow in the opposite direction. This reversal of charge flow may result in a large reverse anode current through the device. The reverse current builds up at a rate determined by the external circuit parameters. During this condition, junction two is still forward biased. The reverse current flows until most of the carriers at junctions one and three have been removed. At this time, junctions one and three revert to a blocking state, and the anode current approaches zero. The device can now hold off a reverse voltage because junctions one and three are reverse biased. However, junction two is still forward biased and contains numerous charges. These carriers cannot flow to the external circuit; therefore, they recombine at a rate independent of the external circuitry. The controlled rectifier cannot block a forward voltage until the excess carriers at junction two have recombined.

The turn-off time increases with increasing junction temperature because the "trapped" charges require a longer time in which to recombine at higher temperatures. The amount of forward current determines the amount of trapped charge near junction two. Therefore, large values of anode current prior to turning off, increase the turn-off time. The external circuitry should be arranged so as to allow a reverse current to flow through the controlled rectifier. This reverse current slightly decreases the turn-off time because it allows junctions one and three to become reverse biased in a shorter time than would be required if no reverse current were allowed to flow. The rate of rise of the forward voltage cannot be greater than a specified maximum or the device will again switch into Region-C operation because of dv/dt turn-on. With a negative gate-to-cathode voltage, the rate of rise of forward voltage can be increased without the device switching "on." The turn-off time of present controlled rectifiers is 10 to 20 microseconds, and the allowable dv/dt is 20 to 200 v per microsecond.[11]

The first and third methods of stopping the flow of anode current in the controlled rectifier are generally not used in modern power-conversion equipment. In such equipment it is not feasible to increase the load impedance by several orders of magnitude or divert the load current to a switch in parallel with the controlled rectifier, in order to reduce the anode current below the holding current. The third method of stopping the anode current is limited to low current devices. The maximum anode

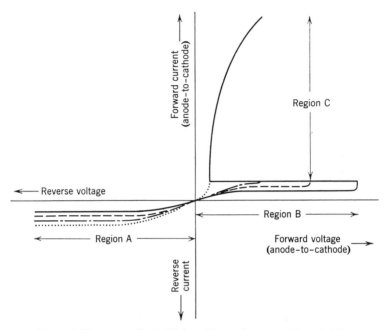

Figure 1.15   *p-n-p-n* Controlled rectifier, volt-ampere characteristic.

$$I_{G1} < I_{G2} < I_{G3} < I_{G4}$$

current of presently available gate turn-off controlled rectifiers is several amperes.

The second method of stopping the flow of anode current, applying a negative anode-to-cathode voltage to the device, is the most widely used method in power equipment. In alternating-voltage systems, the reversal of the supply voltage every half cycle provides a reliable means for the application of a reverse anode-to-cathode voltage to the controlled recti-fier. In this type of operation, the controlled rectifier is usually reverse biased for an appreciable portion of the negative half cycle of supply voltage. At power frequencies, equal to or less than 400 cps, the duration of negative voltage from anode to cathode is more than sufficient to allow the trapped charge at junction two to recombine. The reapplication of forward voltage is along a sine wave, thereby minimizing the problem of dv/dt turn-on of the controlled rectifier.

When the supply voltage is dc, or where stoppage of the anode current is desired more frequently than every half-cycle of the supply frequency, additional components must be added to the circuit to force the anode of the controlled rectifier negative with respect to the cathode, thereby stopping the flow of anode current. These additional components are the "commutation" circuit. Since the size and the weight of these extra components is proportional to the length of time the anode-cathode voltage must be negative, it is desirable to have controlled-rectifier devices with the shortest possible turn-off time. Additional components also generally must be added to the circuit to decrease the rate of reapplication of forward voltage to prevent dv/dt turn-on of the controlled rectifier.

The three regions of operation of the controlled-rectifier device are shown in Figure 1.15. This figure is a steady-state characteristic, and does not indicate the transient conditions that exist during turn-on and turn-off.

1.13   PHYSICAL CHARACTERISTICS AND RATINGS OF p-n-p-n CONTROLLED-RECTIFIER DEVICES

Power Losses

The power losses in a controlled rectifier can be divided into five parts.

(a) Forward conduction loss due to load current (Region C).
(b) Forward power loss due to forward leakage (Region B).
(c) Reverse power loss due to reverse leakage (Region A).
(d) Gate power losses.
(e) Turn-on losses during switching from Region B to Region C.

The forward volt-ampere characteristic of the controlled rectifier for Region-C operation can be approximated by either of the following equations. [19,20]

$$e = A + B \ln (i) + Ci \qquad (1.26)$$

$$e = V_0 + R_F i \qquad (1.27)$$

where e = instantaneous anode to cathode voltage
  i = instantaneous anode current
  $A, B, C, V_0,$ and R_F are constants

Equation (1.26) is more exact than equation (1.27) by including a term proportional to the natural logarithm of the instantaneous forward current; it is, however, more difficult to solve for the average forward power loss. The average power loss over a given time interval is the integral of the product of voltage and current over the specified time interval divided by the time interval, as given below.

$$P_{av} = \frac{1}{T} \int_0^T ei \, dt \qquad (1.28)$$

Combining equations (1.27) and (1.28) produces the following:[20]

$$P_{av} = \frac{I_{max}^2 \pi R_F}{2 \sin^4 (\Psi/2)} \left(\frac{\Psi}{2} - \frac{\sin 2\Psi}{4} \right) \qquad (1.29)$$

where I_{max} = peak value of current that would flow if the conduction angle were greater than 90 electrical degrees
Ψ = conduction angle

A much more complicated calculation of the average power, using equation (1.26), gives the following:[19]

$$P_{av} = I_{av}[A + B(\ln DI_{av} - E) + FCI_{av}] \qquad (1.30)$$

Values of the constants D, E, and F for various waveforms and conduction angles are given in Reference 19. Manufacturers' specification sheets for controlled rectifiers generally give data on the average forward power dissipation as a function of average forward current for several conduction angles. When this information is available, it is not necessary to use equation (1.29) or (1.30) to calculate the approximate forward conduction loss.

The forward power loss due to the average forward leakage current multiplied by the average forward voltage can be determined from equation (1.23) with zero gate current. Most manufacturers' specification sheets give the value of forward "leakage" current as a function of forward voltage for each voltage rating controlled rectifier at its maximum junction temperature. With this data, an estimate of the losses during this mode of operation can be calculated.

The reverse power loss is equal to the reverse leakage current, given by equation (1.22), multiplied by the reverse voltage from anode to cathode. Most specifications plot the value of leakage current at maximum junction temperature for several voltage ratings as a function of reverse voltage. With this data, a value of reverse power loss can be calculated.

The gate losses are due to the energy supplied to the gate to cathode of the controlled rectifier to cause it to switch from Region-B to Region-C operation. In circuits that apply a pulse of gate current to "turn-on" the controlled rectifier, the average gate power loss can be very low. Larger average gate power is required in circuits that require a continuous gate-to-cathode signal to insure that the controlled rectifier will conduct at all required times. Most manufacturers specify the maximum average gate power.

The losses due to switching are caused by the finite time required for the anode-to-cathode voltage to decrease and the anode current to increase. An approximate formula to evaluate the loss due to switching on and off at 400 cps and below is[11]

$$P = \frac{V_m I_m}{4.6} \qquad (1.31)$$

where V_m = anode-to-cathode voltage prior to switching
I_m = anode current after switching

At higher switching frequencies, the controlled rectifier must be further derated, as a finite time is required for conduction to spread from the gate connection across the entire junction. This spreading of conduction from the gate connection generally limits the maximum tolerable di/dt during turn-on.

The five contributions to loss in the controlled rectifier are added together to determine the total power losses. Where the supply frequency or switching rate is 400 cps or less, the forward power loss in Region C is usually the largest of the five individual power losses. When the supply frequency or switching rate is above 400 cycles, the switching losses become increasingly important and, at frequencies above two or three kilocycles, additional circuitry generally is used to reduce the switching losses.

Once the value of total power losses is known, the controlled-rectifier stud temperature can be calculated, knowing the thermal resistance in °C per watt of the particular controlled rectifier. Manufacturers' specification sheets normally give the steady-state junction to stud thermal resistance. Some device specification sheets also give the transient thermal resistance, which is useful in determining the junction temperature for intermittent duty applications or overload conditions.

Once the stud temperature and the maximum ambient temperature are known, the size of the heat sink required can be calculated, knowing the thermal resistance of the heat sink from stud to ambient. This value of thermal resistance can be calculated for several types of flat-plate heat sinks both with free convection or forced air cooling.[11]

Voltage Ratings

The peak reverse voltage (PRV) rating is the maximum repetitive reverse voltage that can be applied from anode to cathode. Some device manufacturers also indicate a transient PRV rating in excess of the repetitive PRV. If this transient PRV or the repetitive PRV rating is exceeded, the controlled rectifier may go into a condition of avalanche breakdown and, if the anode current is not limited, the unit will be damaged.

The forward breakover voltage (V_{BO}) is the minimum value of anode-to-cathode voltage that will cause the controlled rectifier to switch from Region-B to Region-C operation with no gate signal. The V_{BO} rating is generally measured at the maximum junction temperature with the gate open-circuited or with a fixed resistor connected from gate to cathode. The forward breakover voltage is a function of the rate of reapplication of forward voltage (dv/dt). The higher the dv/dt, the lower the V_{BO} value.

The peak forward voltage, PFV, is greater than the V_{BO} rating. If the anode-to-cathode voltage increases above the V_{BO} rating due to a voltage transient, the controlled rectifier may switch "on." This non-destructive breakover during voltage transients is an important asset for the controlled rectifier. However, the forward breakover voltage increases with lower junction temperature. In addition, the actual breakover voltage is generally somewhat higher than the V_{BO} rating. When voltages greater than the PFV rating are required to switch the controlled rectifier on, the unit may be damaged.

Series Operation

When controlled rectifiers are to be operated in series, care must be taken to insure that each unit equally shares the total forward and reverse voltage. The voltage across each cell should be equal, both during steady-state operation and transient conditions. The steady-state voltage equalization can be accomplished by a resistor or Zener diode in parallel with each controlled rectifier in a series string. For transient voltage equalization, a low value of noninductive resistance is used in series with a capacitor, all in parallel with each controlled rectifier.

If several series-controlled rectifiers are to be turned on by gate signals, it may be necessary to add circuit elements to delay the buildup of anode current until all of the series units have turned on and their anode-to-cathode voltages have dropped to the forward voltage drop in Region C. This prevents a voltage spike from appearing on the controlled rectifier with the longest turn-on time.

In some applications, only one of the controlled rectifiers in series need be turned on by an external signal into its gate; the other units in series are "slave fired" or triggered by the decay in anode-to-cathode voltage of this controlled rectifier. However, in most cases, each controlled rectifier in a series string requires an isolated gate-to-cathode turn-on signal to provide the most reliable turn-on with a minimum of detrimental transients.

Parallel Operation

When controlled rectifiers are to be operated in parallel, they should equally share the total anode current. The forward voltage drop of a controlled rectifier is nearly independent of load current in Region C; therefore, wide variations in load current division are possible. If matched cells are used, their anode current ratings must be reduced, depending on the cell type and the number of cells in parallel.[11] In order to keep the junction temperature of each cell approximately equal, all paralleled cells should be mounted on the same heat sink. The gate-to-cathode signal should remain on all of the paralleled units until after the anode current has built up beyond the holding current. After one of the parallel cells has decreased the anode-to-cathode voltage on the other parallel cells to a few volts, a gate signal is still required to insure that the remaining units will turn on and conduct load current. If precision is required or un-matched cells are used, balancing resistors or reactors can be used. The value of resistance should be large enough so that the voltage drop on the balancing resistor is comparable to the anode-to-cathode voltage drop of the controlled rectifiers. Current-balancing reactors are connected in the anode circuit of two controlled rectifiers in such a way as to cause the ampere-turns of the two circuits to be equalized.

1.14   COMPARISON OF TRANSISTORS AND CONTROLLED RECTI-FIERS AS SWITCHES

In power conversion equipment, the following factors are generally most important to the circuit designer.[11,21]

(1) The maximum reverse voltage that the device can block.
(2) The maximum forward voltage that the device can block.
(3) The forward current rating of the device.
(4) The forward voltage drop at rated forward current.

(5) The characteristics of the device in regard to turn-on power, turn-off power, and switching speed.

The controlled-rectifier device is capable of withstanding higher values of reverse voltage and forward blocking voltage because of the wider center n layer. The transistor, in order to minimize recombination in the base region and the corresponding reduction in α, requires a narrow base width resulting in lower values of blocking voltage.

The maximum value of forward voltage that the controlled rectifier can block is generally given with the gate lead open, that is, with no gate bias voltage. The maximum forward voltage that a transistor can block is usually given with the emitter-to-base junction reverse biased by a separate bias supply. The controlled-rectifier device does not require a reverse voltage from gate to cathode when blocking rated forward anode-to-cathode voltage, while the transistor requires a reverse bias from base to emitter while blocking rated forward collector-to-emitter voltage.

The forward current rating of the controlled rectifier is greater than that of an equivalent transistor because the forward current density is generally more uniform at the junctions. However, the forward voltage drop of a controlled rectifier is generally larger than that of modern power transistors.

The turn-on and turn-off characteristics of the devices are the major differences. The ratio of rated anode current to gate turn-on current in a controlled rectifier is approximately 3000.[22] The ratio of collector current to base current in presently available power transistors, typically, may be an order of magnitude less than 3000. The controlled rectifier can be switched on with a current pulse into the gate terminal. Therefore, the *average* turn-on current gain can be much greater than 3000 if only a several-microsecond wide current pulse is used to switch the controlled rectifier. Base current is required during the entire period of time the power transistor is on. Any decrease in base current will cause the transistor to operate in the linear region, thereby causing higher power dissipation in the transistor. Therefore, the turn-on circuit in a controlled-rectifier power circuit can have a much smaller power rating than that of an equivalent transistor drive circuit, and can also be of an intermittent duration.

In a transistor, the base current maintains the device in saturation. Therefore, the transistor can be turned off by the removal or reversal of the base current. The higher current-controlled rectifiers, once conducting, cannot be turned off from the gate circuit. The controlled rectifier is turned off by the action of a commutating circuit, as discussed in the subsequent chapters of this book. The turn-off time of a typical controlled-rectifier device is approximately 10 to 20 microseconds.

34   PRINCIPLES OF INVERTER CIRCUITS

REFERENCES

1. T. L. Martin, Jr., *Physical Basis for Electrical Engineering*, Prentice-Hall, Englewood Cliffs, N.J., 1957.
2. R. B. Hurley, *Junction Transistor Electronics*, John Wiley & Sons, New York, 1958.
3. R. D. Middlebrook, *An Introduction to Junction Transistor Theory*, John Wiley & Sons, New York, 1957.
4. E. M. Conwell, "Properties of Silicon and Germanium II," *Proceedings of the IRE*, Vol. 46, November 1952, pp. 1281–1300.
5. A. K. Jonscher, *Principles of Semiconductor Device Operation*, John Wiley & Sons, New York, 1960.
6. H. E. Bridgers, J. H. Scaff, and J. N. Shive, *Transistor Technology*, Vol. 1, Van Nostrand, Princeton, N. J., 1958.
7. F. J. Biondi, *Transistor Technology*, Vols. 2 and 3, Van Nostrand, Princeton, N.J., 1958.
8. R. W. Aldrich and N. Holonyak, Jr., "Silicon–Controlled Rectifiers from Oxide-Masked Diffused Structures," *AIEE Transactions*, Vol. 77, Part 1, 1958, pp. 952–954.
9. L. P. Hunter *et al.*, *Handbook of Semiconductor Electronics*, McGraw-Hill, New York, 1956.
10. J. J. Ebers and J. L. Moll, "Large Signal Behavior of Junction Transistors," *Proceedings of the IRE*, Vol. 42, December 1954, pp. 1761–1772.
11. F. W. Gutzwiller *et al.*, *Silicon Controlled Rectifier Manual*, Second Edition, General Electric Company, Auburn, N.Y., 1961.
12. J. J. Ebers, "Four Terminal p–n–p–n Transistors," *Proceedings of the IRE*, Vol. 40, November 1952, pp. 1361–1364.
13. J. L. Moss, M. Tanebaum, J. M. Goldey, and N. Holonyak, "p–n–p–n Transistor Switches," *Proceedings of the IRE*, Vol. 44, September 1956, pp. 1174–1182.
14. A. K. Jonscher, "p–n–p–n Switching Diodes," *Journal of Electronics and Control*, London, Vol. 3, December 1957, pp. 573–586.
15. I. M. MacKintosh, "The Electrical Characteristics of Silicon p–n–p–n Triodes," *Proceedings of the IRE*, Vol. 46, June 1958, pp. 1229–1235.
16. R. W. Aldrich and N. Holonyak, Jr., "Multiterminal p–n–p–n Switches," *Proceedings of the IRE*, Vol. 46, June 1958, pp. 1236–1239.
17. W. Shockley and J. F. Gibbons, "Introduction to the Four-Layer Diode," *Semiconductor Products*, Vol. 11, No. 1, January–February 1958, pp. 9–13.
18. J. M. Goldy, "p–n–p–n Switches—Diode and Triodes," *Control Engineering*, New York, Vol. 7, No. 10, October 1960, pp. 101–104.
19. C. D. Mohler, "Digital Computer Calculation of Rectifier and Silicon Controlled Rectifier Ratings," AIEE Conference Paper CP 62–433, New York, January 30, 1962.
20. J. I. Missen, "The Power Rating of Semiconductor Rectifiers and Silicon Controlled Rectifiers," *Direct Current*, London, Vol. 6, No. 5, August 1961, pp. 134–149.
21. H. W. Henkels and F. S. Stein, "Comparison of n–p–n Transistors and n–p–n–p Devices as Twenty Ampere Switches," *IRE Transactions on Electron Devices*, Vol. ED–6, No. 1, January 1960, pp. 39–45.
22. J. D. Harnden, Jr., "Properties of the Silicon Controlled Rectifier—a Survey," AIEE Conference Paper CP 61–332, New York, February 2, 1961.
23. J. N. Shive, *The Properties, Physics, and Design of Semiconductor Devices*, Van Nostrand, Princeton, N.J., 1959.

24. W. C. Dunlap, Jr., *An Introduction to Semiconductors*, John Wiley & Sons, New York, 1957.
25. AIEE Rectifier Device Working Group of the AIEE Semiconductor Rectifiers Committee, "Proposed Definitions for Semiconductor Switches," AIEE Transactions Paper 61–127, New York, January 29–February 3, 1961.
26. H. F. Storm, "Introduction to Turn-Off Silicon Controlled Rectifiers," IEEE Transactions Paper 63–321, New York, January 27–February 1, 1963.

Chapter Two

Introduction to Inverter Principles

A rectifier circuit is used to convert ac to dc, while an inverter accomplishes the reverse process. In an inverter, the electric valves must have the ability to hold off forward voltage and the cyclic conducting period of each valve must be controllable. Vacuum-tube amplifiers and oscillators are examples of circuits which produce ac from a d-c source where the current through the valves is proportionally controlled during each valve-conducting period. In general, highest efficiency inversion is obtained with "switching" type circuits. In these circuits the valves are either switched fully "on" or "off" during each cycle of operation.

Negative resistance oscillators usually operate in what approximates a switching mode. For example, in the tunnel diode or unijunction transistor oscillator, the device effective resistance quickly changes from a low value to a high value in a cyclic fashion. The time of conduction of greatest current through the device is determined by the time constant of the circuit and the device negative resistance characteristic.

In mechanical-switch or transistor-switch inverters, devices are used which have the ability to interrupt the flow of current. The current is controlled in an on-off manner by the signal supplied to the control element of the switching device. With these switching type circuits, it is necessary to provide alternate paths for inductive load current flow. These are required to minimize voltage transients resulting from attempts to force abrupt changes in the current through inductive portions of the circuit. Thus, in transistor inverters, although commutation circuits are not required to transfer current from one transistor to the next, feedback rectifiers or other additional circuit elements must be provided when these inverters supply inductive loads.

36

2.1 RECTIFICATION

One of the most basic rectifier circuits is shown in Figure 2.1(*a*). This is most properly referred to as a full-wave single-way circuit. Although this is a lengthy name, it rather precisely defines the circuit operation. (In rectifiers, the expression "single way" means that current flows in only one direction through the transformer secondary windings.) The waveshapes for this circuit are indicated in Figure 2.1(*b*). These waveforms assume the following characteristics.

(1) *Ideal rectifiers.* Zero forward drop and infinite back resistance.
(2) *Ideal transformer.* Zero leakage reactance, zero winding resistance, and zero magnetizing current.
(3) Pure resistance load.

With modern semiconductor rectifiers and transformers, the waveforms in actual circuits are negligibly different from the waveforms considering ideal circuit components.

The operation of this basic rectifier circuit is quite straightforward. The ideal rectifier elements deliver power to the load during the entire interval when the voltage across them is in the forward direction. The load current is transferred from one electric valve or rectifier to another, automatically, when the forward voltage on one rectifier drops to zero and a forward voltage begins to be impressed upon the second rectifier. The operation of the rectifier devices is analogous to the action of check valves in fluid flow systems. It is very important to note that the simple rectifier does not have the ability to hold off forward voltage and, thus, conduction must occur whenever a forward voltage is applied. For this reason the voltage, e_{REC}, is either zero or negative. As stated previously, the rectifier forward voltage drop is negligible for the ideal element. In this center tap-type circuit, the peak inverse voltage on the valves is $2E_{1-n,m}$.

The average load voltage is given by the following.

$$E_d = \frac{1}{2\pi} \int_0^{2\pi} e_d(\omega t)\, d(\omega t)$$

$$= \frac{1}{\pi} \int_0^{\pi} E_{1-n,m} \sin \omega t\, d(\omega t)$$

$$= \frac{1}{\pi} \left[-E_{1-n,m} \cos \omega t \right]_0^{\pi}$$

$$= \frac{2E_{1-n,m}}{\pi} \tag{2.1}$$

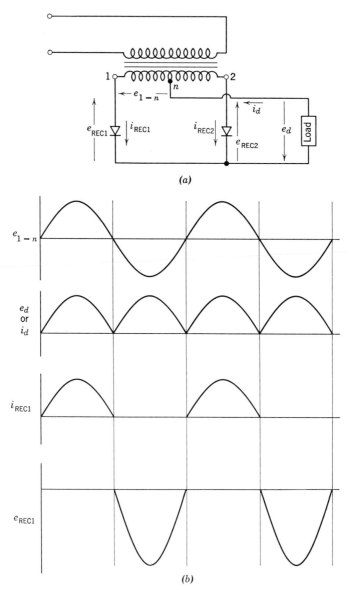

Figure 2.1   (*a*) Full-wave single-way rectifier. (*b*) Waveforms.

The average load current is

$$I_d = \frac{2E_{1-n,m}}{\pi R} \tag{2.2}$$

It is interesting to note that the effective current through each half of the transformer winding is

$$
\begin{aligned}
I_{\text{REC},e} &= \sqrt{\frac{1}{2\pi} \int_0^{2\pi} i_{\text{REC1}}^2(\omega t)\, d(\omega t)} \\
&= \sqrt{\frac{1}{2\pi} \int_0^{\pi} I_{\text{REC1},m}^2 \sin^2 \omega t\, d(\omega t)} \\
&= \sqrt{\frac{I_{\text{REC1},m}^2}{2\pi} \left[\frac{\omega t}{2} - \frac{\sin 2\omega t}{4} \right]_0^{\pi}} \\
&= \sqrt{\frac{I_{\text{REC1},m}^2}{2\pi} \left(\frac{\pi}{2} \right)} \\
&= \frac{I_{\text{REC1},m}}{2}
\end{aligned}
\tag{2.3}
$$

Therefore, the total transformer secondary effective volt-ampere rating is $E_{1-n,e} I_{\text{REC1},m}$; the rating for a single secondary winding for the same voltage, but carrying sinusoidal current as in a double-way bridge circuit, is $E_{1-n,e} I_{\text{REC1},m}/\sqrt{2}$. Thus, this center-tap circuit of Figure 2.1 does not make most effective use of the transformer secondary. However, it involves only two rectifying elements rather than four as in bridge circuits, and there is only one rectifier forward voltage drop in series with the load at any given instance of time.

There are many types of single-phase and polyphase rectifier circuits, each of which have advantages for certain applications. The center-tap circuit of Figure 2.1 is possibly the simplest circuit which illustrates the important principles involved in the rectification process that distinguish this process from inversion.

2.2 INVERSION

The conversion of d-c to a-c power may be accomplished with electric valves through which the conduction of current can be controlled. In this case, the valves must have the ability to hold off applied forward voltage, and the instant of time when conduction takes place must be controllable. Thus, more complicated devices are required than simple two-terminal rectifiers.

While the transfer of current from one valve to another occurs naturally and automatically in a rectifier circuit, the reliable accomplishment of this current transfer is one of the basic problems of the inversion process. The term commutation is used as the name for this current transfer from one electric valve to another in rectifier and inverter circuits. In general, complete commutation may involve a number of events. The most important of these events are (*1*) the reduction of forward current to zero in one valve, (*2*) the delay of the reapplication of forward voltage to this valve until it has regained its forward-blocking capability, and (*3*) the buildup of forward current in a second valve.* These events may occur concurrently or in sequence. In practical low-frequency inverters, circuit constants generally determine the time interval required for complete commutation. For high-frequency inverters, the characteristics of the controlled rectifying elements establish the time required for the complete commutation process.

The means for producing reliable commutation is generally one of the most difficult problems in inverters, and it is therefore the subject of much of the discussion throughout this book. The particular technique which is used to reduce the current to zero in a conducting valve and delay the reapplication of forward voltage to this valve until it has regained its forward-blocking capability is the principal difference between many types of inverter circuits.

The simpler inverters use valves which can interrupt the current flow in response to a signal input, such as the mechanical switch, vacuum tube, or transistor. These inverters are of two types: those in which the electric valves are gradually changed from on to off, and those in which the valves abruptly interrupt the current. Inverters of the latter type, having a switching mode of operation, generally require circuit elements which help to reduce the current to zero when a valve is turned off; or it is necessary to provide low-impedance circuits to which the current is diverted, so as to prevent large transients when inductance is present in the circuits.

Inverter circuits can be self-excited where the circuit is self-oscillatory or separately excited where a signal oscillator is used as a driver for the power inverter. The self-excited type inverter will often require some separately applied transient to start. In separately driven inverters,

---

* The time interval required for the controlled rectifier to regain its forward-blocking capability after forward conduction is referred to as the turn-off time. The turn-on time is the interval between the initiation of a gate turn on signal and the reduction of the controlled rectifier forward voltage to 10 per cent of its blocking value.

starting transients will occur, depending upon the position in the driving oscillator cycle when the d-c power is applied to the inverter. The separately excited inverter must be designed to accommodate the starting transient currents which may flow due to inductive loads and transformer magnetizing current.

2.3 ELECTRONIC AMPLIFIERS AND OSCILLATORS (WITH GRADUALLY VARYING VALVE CURRENT)[1,2]

Possibly the simplest electronic-type inverter is a Class-A amplifier using either vacuum tubes or transistors. This kind of circuit is not generally referred to as an inverter, but it does convert d-c power from a plate power supply to a-c power in a load. The simple Class-A amplifier, shown in Figure 2.2(a), can be considered as a driven or separately excited inverter. In separately excited schemes, the basic frequency is established by a signal generator output—in this case, the a-c voltage applied to the grid-cathode circuit. As shown in Figure 2.2(b), assuming linear tube characteristics and a pure resistance load, the load voltage is sinusoidally modulated by the grid-signal voltage. There is, therefore, an a-c component of voltage at the load but, in addition, there is a rather large d-c component. In this kind of inverter circuit, the electric valve is forced to conduct in accordance with the applied grid voltage. This system has 25 per cent maximum theoretical efficiency, and the standby losses with zero signal input are very high. This maximum theoretical efficiency is determined as follows.

$$E_{l,m}\bigg|_{max} = \frac{E_{bb}}{2} \qquad (2.4)$$

$$I_{b,m}\bigg|_{max} = I_{b0} \qquad (2.5)$$

where I_{b0} is the quiescent plate current with zero signal output.

$$\eta = \frac{E_{l,e}I_{b,e}}{E_{bb}I_{b0}} \qquad (2.6)$$

$$\eta\bigg|_{max} = \frac{\dfrac{E_{bb}}{2\sqrt{2}}\dfrac{I_{b0}}{\sqrt{2}}}{E_{bb}I_{b0}} = 25 \text{ per cent} \qquad (2.7)$$

There is considerable voltage drop across the valve for a large portion of the cycle and considerable d-c power loss in the load, which account for the poor operating efficiency. The d-c component in the load can be

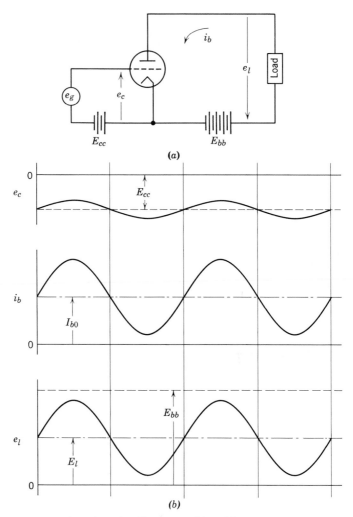

Figure 2.2   (a) Class-A amplifier. (b) Waveforms.

eliminated and the efficiency increased by the use of capacitance or transformer coupling to the load.

A modification of the Class-A amplifier which has more practical value for power amplifiers and oscillators is the Class C tuned amplifier shown in Figure 2.3(a). The approximate waveshapes for this amplifier are shown in Figure 2.3(b), assuming a high Q tuned circuit. For Class-B or -C operation, the negative grid bias is increased over that for Class-A operation. Class-C operation permits high efficiency, 80 per cent in

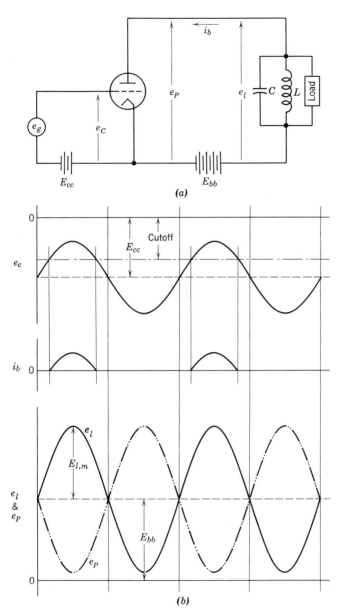

Figure 2.3 (a) Class C tuned amplifier. (b) Waveforms.

practical circuits, since plate current flows for less than 180° of the cycle during the time when the plate voltage is low. Physically, the Class C tuned amplifier may be considered a high Q resonant circuit sustained oscillation where the losses in the resonant circuit are made up by the pulses of tube current which flow during a small portion of each cycle. This, of course, would imply a nonlinear oscillation, but if the stored energy in the tuned circuit is high relative to the energy dissipated each cycle, the output voltage approaches a sine wave as shown in Figure 2.3(b).[1]

This Class C tuned amplifier is again a separately excited type inverter, and the electric valves conduct whenever the applied grid voltage exceeds the cutoff value.

These are but two of many electronic oscillator circuits which are widely used and extensively covered in previous technical literature. The self-excited inverters employing similar principles, are the Hartley, Colpitts, tuned-plate, tuned-grid, etc., oscillators. In general, the self-excited oscillators are amplifiers with feedback, designed in such manner that there is 180° phase shift and unity gain around the closed loop at the resonant frequency.

2.4 NEGATIVE RESISTANCE OSCILLATORS[3-7]

This class of inverter circuits converts d-c power to a-c power generally using a negative resistance circuit element or elements. Examples of this approach are unijunction transistor, glow tube, or tunnel diode oscillators. In most cases, this kind of inverter generates a nonsinusoidal waveshape. Figure 2.4(a) shows the basic unijunction transistor oscillator circuit. This is a type of relaxation oscillator which generates a saw-tooth wave at the unijunction transistor emitter. The circuit functions as an oscillator because the emitter input characteristic has a negative resistance region as indicated in Figure 2.4(b). The waveshapes for this circuit are shown in Figure 2.4(c). When voltage E is applied, the capacitor C_T starts to charge exponentially toward the supply voltage E through resistor R_T. This charging interval corresponds to operation along path $A–B$ of the emitter characteristic shown in Figure 2.4(b). The emitter is reverse biased during the charging interval so that the input current I_E is very low. When the emitter voltage reaches the peak-point voltage, the emitter becomes forward biased, and the dynamic resistance between the emitter and base one drops to a low value. The input current I_E abruptly rises along path $B–C$ to a value equal to the peak-point voltage appearing on capacitor C_T divided by the emitter-to-base-one saturation resistance plus resistance R_1. The capacitor now begins to discharge until the valley-point voltage at D is reached. At this point, the emitter ceases to conduct

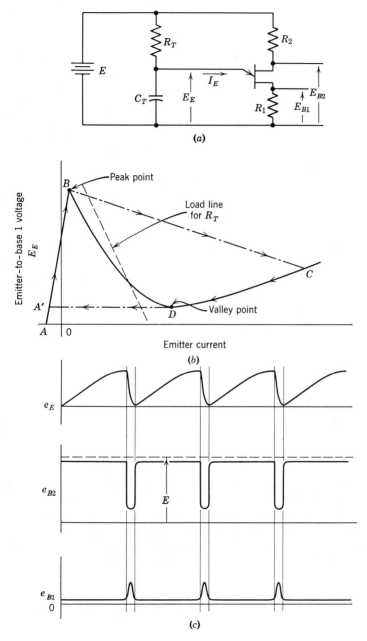

Figure 2.4   (*a*) Unijunction transistor oscillator.   (*b*) Unijunction transistor characteristic.   (*c*) Circuit waveforms.

so that the emitter current drops to that at point A', which is the starting point for the next cycle of operation.

The basic requirement for oscillation of the circuit in Figure 2.4(a) is that the load line for R_T intersect the negative resistance region of the emitter characteristic. The operating frequency is established by the peak-point voltage and the capacitor-charging time. This kind of oscillator is a self-excited arrangement and, generally, is limited to low-power applications because of the present availability of devices with negative resistance characteristics. In addition, these oscillators generally have quite low efficiency but, due to their simplicity, they are frequently used to drive other inverter circuits.

The unijunction transistor circuit, described, operates in a similar fashion to oscillators, using two terminal negative resistance devices like the tunnel diode. In the unijunction transistor, the emitter-to-base-one input has the negative resistance characteristic. Since this device has three terminals, it does have additional control possibilities over simple two-terminal devices. The emitter-to-base-one characteristic is modified as the magnitude of the base-to-base voltage is adjusted, because the emitter peak-point voltage is a fixed fraction of the base-to-base voltage. This property is often used in other unijunction transistor circuits particularly for synchronizing circuits or inhibiting operation until a specific time.

2.5   MECHANICAL-SWITCH INVERTERS

Considerably higher-efficiency inverters are practical with switching-type circuits than with even the Class C-oscillator schemes. This is true because there is negligible power loss in an ideal switching element. When it is open, there is zero current through it and, when it is closed, there is negligible voltage drop across a perfect switch. The mechanical vibrator is a simple illustration of an inverter where the switching elements are abruptly switched from "off" to "on" during each cycle. Since the mechanical switch has extremely low leakage current when open and very low contact drop when closed, it closely approximates an ideal switch in these respects. A circuit diagram of one mechanical-switch approach is shown in Figure 2.5(a). It is assumed that separate circuits or devices cause the mechanical switch to continuously vibrate. The R-C components are to suppress voltage transients produced by transformer or load inductance. Figure 2.5(b) shows the waveshapes for the case where there is appreciable time during which neither contact is closed. Some off-time is generally provided to prevent the possibility of having both contacts closed. The waveshapes of Figure 2.5(b) show the damped oscillations

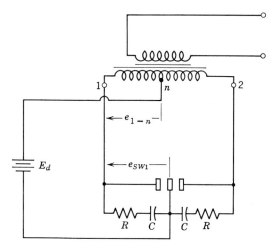

Figure 2.5 (*a*) Mechanical-switch inverter.

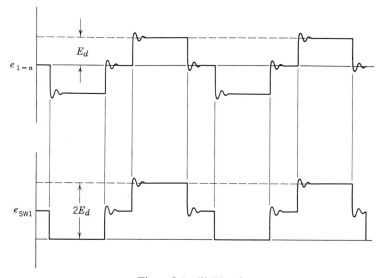

Figure 2.5 (*b*) Waveforms.

produced by circuit inductance ringing with the surge-absorbing capacitance across the contacts.

In this inverter, the conducting period of the electric valves is controlled by the opening and closing of the switch contacts. A particular switch holds off a forward voltage of $2E_d$ during the majority of the half-cycle in which it is turned off.

2.6   TRANSISTOR-SWITCH INVERTERS[8, 9]

These inverters are similar to the mechanical-switch type. However, the transistor is a more quiet and reliable switch than most mechanical devices. It also has very low forward drop and negligible leakage current such that it can approach the efficiencies possible with a perfect switch.

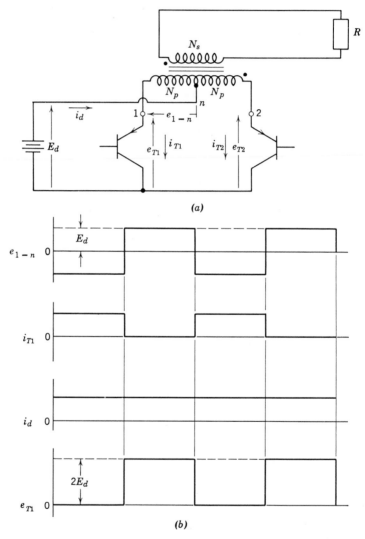

Figure 2.6   (a) Transistor-switch inverter. (b) Waveforms for transistor-switch inverter.

A separately excited transistor inverter is discussed in some detail to show the important principles of operation of transistor-switch inverters, and to indicate practical methods for aiding commutation with inductive loads.

Single-Phase Circuit with Ideal Transformer and Pure Resistance Load

Figure 2.6(a) shows a simple transistor inverter circuit which is assumed to be separately excited from a signal generator. This means that a square wave signal would be supplied to the base of each transistor. The flow of excessive current from the d-c source is prevented by making sure that both transistors are not on at the same time. It is assumed here that one transistor is turned off at the same instant that the next one is turned on. There is not a need for added commutation components in this inverter as the current is forced to zero in T_1 and transferred to T_2 when the transistors are switched by their base-driving signals. This circuit operates in a manner that is analogous to a mechanical vibrator-type system, delivering square waves of a-c power output. Some of the voltage and current waveshapes are shown in Fig. 2.6(b).

With Inductive Load

As most practical loads contain inductance, it is more realistic to examine the operation with inductive loads. However, these loads create rather severe problems. Since the current through an inductive load cannot be instantaneously reversed, it is necessary to add some path for load-current flow during the transistor-switching period. Figures 2.7(a) and 2.7(b) show a circuit and its operation under steady-state conditions with a resistance connected in parallel with the load inductance to prevent high voltage from appearing across the transistors when they are switched off. (A series resistor will not provide a circuit which will limit the voltage during transistor switching.) The parallel resistor is one solution to the problem but, as will be indicated, it is not a practical solution for most power inverter systems.

The minimum value of resistance is used to produce the waveshapes in Figure 2.7(b). For this case, at the end of each half-cycle the inductance current is just equal to the resistance current. This inductive current must be equal to or less than this, or the waveshapes will depart from those in Figure 2.7(b). The half-cycle average current through the resistance is

$$I_R = \frac{E_d}{R} \frac{N_s}{N_p} \tag{2.8}$$

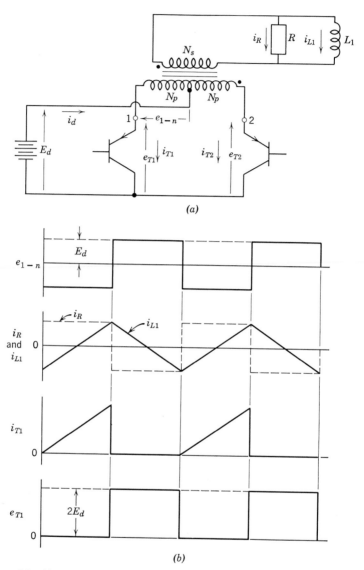

(a)

(b)

Figure 2.7   (a) Transistor-switch inverter with inductive load.   (b) Waveforms.   (c) Waveforms, assuming R twice that for Figure 2.7(b).

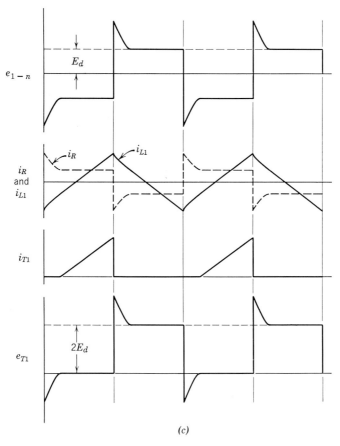

(c)

The half-cycle average current through the load inductance is determined as follows.

$$i_{L1} = \frac{1}{L_1} \int e_{1-n} \frac{N_s}{N_p} dt \qquad (2.9)$$

$$\Delta i_{L1} = \frac{E_d}{L_1} \frac{N_s}{N_p} \Delta t \qquad (2.10)$$

$$I_{L1} = \frac{1}{2} \frac{E_d}{L_1} \frac{N_s}{N_p} \frac{T}{4} = \frac{E_d}{8fL_1} \frac{N_s}{N_p} \qquad (2.11)$$

and for the waveshapes shown

$$\frac{E_d}{L_1} \frac{N_s}{N_p} \frac{T}{4} = \frac{E_d}{R} \frac{N_s}{N_p} \qquad (2.12)$$

$$\frac{E_d}{4fL_1} \frac{N_s}{N_p} = \frac{E_d}{R} \frac{N_s}{N_p} \qquad (2.13)$$

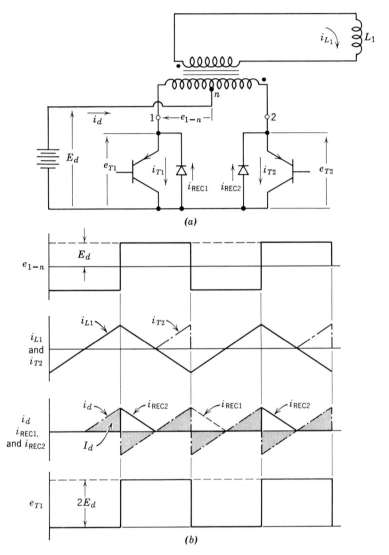

Figure 2.8 (*a*) Transistor-switch inverter with feedback rectifiers. (*b*) Waveforms.

Therefore

$$I_R = \frac{E_d}{R}\frac{N_s}{N_p} = \frac{2E_d}{8fL_1}\frac{N_s}{N_p} = 2I_{L1} \qquad (2.14)$$

This is, therefore, not a very practical arrangement, since twice as many average amperes of resistive load are required as are delivered to the inductive load to maintain the waveshapes in Figure 2.7(b).

Figure 2.7(c) shows the waveshapes which result for steady-state conditions when the resistance across the inductive load is doubled. The important point for this case is that at the end of a given half-cycle, the inductive current must momentarily circulate through the resistance after the particular transistor is switched off. During the first instant after switching, i_{L1} must remain the same. Since the primary and secondary ampere turns must be equal in the perfect transformer and, since the transistors are assumed open in the inverse direction, all of the inductive current must circulate through resistance R producing the voltage shown across the transformer primary winding. During this period of time there is no current flowing in the d-c circuit. The inductive current decays exponentially with a time constant of L/R until the voltage across R reaches the value $E_d(N_s/N_p)$. After this time is reached, the circuit operates as it did in Figure 2.7(a).

Although the power dissipated in the resistor is now reduced, the maximum voltage across each transistor is increased so that higher-voltage transistors are required.

With Feedback Rectifiers

Figure 2.8(a) shows a more desirable method of handling inductive loads. Diode rectifiers are connected across each transistor. They now provide a path for the peak inductive load current when a transistor is switched off. The waveshapes are as shown in Figure 2.8(b). It is interesting to see that the average current drawn from the battery is zero, as it should be, since there is no power consumed by the purely inductive load. This, therefore, results in limiting the voltage across each transistor to that of the circuit with resistance load, as shown in Figure 2.6.

REFERENCES

1. M.I.T. Electrical Engineering Staff, *Applied Electronics*, John Wiley & Sons, New York, 1943.
2. R. R. Benedict, *Introduction to Industrial Electronics*, Prentice–Hall, New York, 1951.
3. I. A. Lesk and V. P. Mathis, "The Double Base Diode—A New Semiconductor Device," *IRE Convention Record*, Part 6, March 1953, pp. 2–8.

4. T. P. Sylvan, "Design Fundamentals of Unijunction Transistor Relaxation Oscillators," *Electronic Equipment Engineering*, December 1957, pp. 20–23.

5. I. A. Lesk, "Non-Linear Resistance Device," U.S. Patent 2,769,926, November 6, 1956.

6. V. P. Mathis, "Sawtooth Wave Generator," U.S. Patent 2,792,499, May 14, 1957.

7. R. W. Aldrich, "Semiconductor Network," U.S. Patent 2,780,752, February 5, 1957.

8. R. L. Bright, G. F. Pittman, Jr., and G. H. Royer, "Transistors as On-Off Switches in Saturable Core Circuits," *Electrical Manufacturing*, December 1954, pp. 79–82.

9. G. H. Royer, "A Switching Transistor DC to AC Converter Having an Output Frequency Proportional to the DC Input Voltage," *AIEE Transactions*, July 1955.

Chapter Three

Phase-Controlled Rectifiers and A-C Line Voltage Commutated Inverters

The simplest inverters include switching devices which have the ability to interrupt the flow of current. However, higher power rating controlled electric valves generally do not have this interrupting ability. Power transistors are not presently available with current and voltage ratings as high as for silicon-controlled rectifier devices. In addition, considerable "drive" power is required to provide fast and reliable switching action in present-day transistors. These factors have led to renewed interest in controlled rectifier-type inverters with particular emphasis on circuits using silicon-controlled rectifiers.

In general, all rectifiers can be used as inverters so that there are numerous types of single-phase and polyphase circuit arrangements. To change a given rectifier to an inverter, two of the more basic modifications that are required are: (1) the two terminal rectifying elements must be replaced with controlled-rectifier devices, and (2) a reliable means of commutation must be incorporated into the circuit. As discussed in Chapter 2, an inverter requires electric valves which are capable of holding off appreciable forward voltage and through which the conduction of current can be controlled. In the plain rectifier, the process of transferring current from one conducting valve to the next occurs naturally and automatically. This commutation process generally is more difficult to reliably accomplish in an inverter. The a-c line voltage commutated inverters discussed in this chapter most nearly approach the operation of rectifiers. In fact, when controlled-rectifying elements are used, a given circuit can operate either as a phase-controlled rectifier or an inverter. With a gradual variation in the firing angle of the controlled-rectifier devices, a circuit will change from rectifier to inverter operation with a corresponding reversal of power flow. The commutation process when operating as an

55

inverter is accomplished automatically by the instantaneous relationships existing between the applied a-c line voltages. Thus, no added circuit elements are required to provide reliable commutation. However, as commutation is provided by the a-c line voltage, this type inverter can only operate into an a-c system where the voltage waveshape is maintained relatively independent of the circuit operation.

In this chapter, a full-wave single-way rectifier with d-c circuit inductance is discussed, leading to the discussions of a similar circuit as a phase-controlled rectifier or an a-c line commutated inverter. The six-phase double-way, star, and double-wye circuits are then considered to convey a basic understanding of the principles of operation of a-c line voltage commutated inverters.

3.1 FULL-WAVE SINGLE-WAY RECTIFIER[1-4]

Ideal Rectifier Circuit with Large Inductive Load

The rectifier circuit in Figure 3.1(a) is the same as that discussed in Chapter 2 except load inductance is added. Figure 3.1(b) shows the circuit waveforms, assuming ideal circuit components and a d-c inductance large enough to obtain negligible ripple in the d-c current. The d-c current is constant in magnitude, and each rectifier conducts a square block of current for one half-cycle. Transfer of current from one conducting rectifying element to the next occurs abruptly at the end of each half-cycle when the a-c voltage applied to one rectifying device becomes negative and the voltage applied to the other device becomes positive.

For a steady-state condition the average inductive voltage across the inductance must be zero as shown by the shaded areas. This may be seen from the following.

$$e_{Ld} = N \frac{d\phi}{dt} \times 10^{-8} \tag{3.1}$$

where N = number of turns
ϕ = core flux

$$\phi = \frac{10^8}{N} \int e_{Ld} \, dt \tag{3.2}$$

or

$$\Delta\phi \Big|_0^{2\pi} = \frac{10^8}{N} \int_0^{2\pi} e_{Ld} \, dt \tag{3.3}$$

In a steady-state condition, there must be no net flux change in the inductor over a one-cycle interval. That is, the positive $\Delta\phi$ must equal

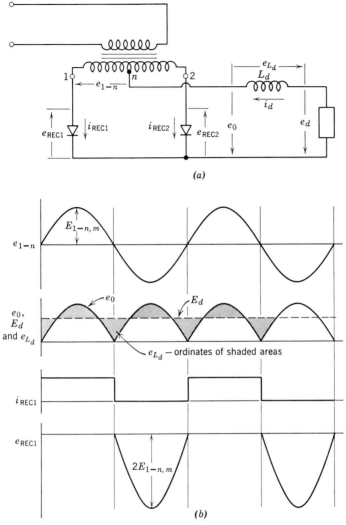

Figure 3.1 (a) Full-wave single-way rectifier. (b) Waveforms, assuming negligible ripple in the d-c current.

the negative $\Delta\phi$ to remain at a steady-state average flux level. Thus, as shown from equation (3.3), if $\Delta\phi\Big|_0^{2\pi}$ is equal to zero, the average voltage across the inductor must be zero.

The average d-c load voltage and current are the same as without the d-c inductance, or as given in equations (1.1) and (1.2) of Chapter 1.

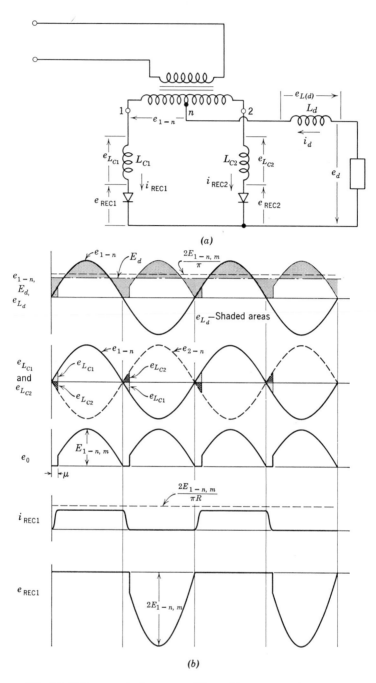

Figure 3.2 (a) Full-wave single-way rectifier with commutating reactance. (b) Waveforms.

58

Including Transformer Secondary Leakage Reactances

Figure 3.2(a) shows the circuit of Figure 3.1(a) with leakage reactances L_{c1} and L_{c2} added and, again, assuming ideal rectifying elements and a large enough inductance to produce negligible ripple in the d-c current. It is important to note that the reactances shown in Figure 3.2(a), and similar reactances throughout this chapter, are generally the transformer and a-c supply circuit reactances. They are indicated as separate circuit elements directly in series with the rectifying elements for clarity in understanding the effect of a-c circuit inductance on the operation of the rectifier.

The reactances in series with the rectifying elements make it impossible for the d-c load current to instantaneously transfer from one conducting rectifying device to the other. Immediately after the a-c voltage reaches zero, the current starts to gradually transfer from one valve to the next as shown in Figure 3.2(b). During the period when both rectifying elements are conducting, the transformer secondary winding is shorted through the leakage reactances. Since these reactances are normally equal, the potential of point n in Figure 3.2(a) will be equal to the potential of the common cathode connecting point for the rectifying devices, assuming that the d-c reactor is sufficiently large to maintain the d-c load current constant. The interval μ shown in Figure 3.2(b) dependent upon the magnitude of the leakage reactances, applied a-c voltage, and d-c load current is necessary for complete commutation or current transfer. Assuming the load current is constant during the interval μ, the current will increase through one rectifying element and decrease through the other element at the same rate, since the sum of the two currents at all instants must equal the constant load current. The primary effect of the a-c circuit reactances is to require a finite time interval μ for current transfer, during which the load voltage is reduced. In general, during the commutating interval, the rectifier output voltage is the average of the a-c voltages applied to the two conducting valves. In this case, the a-c voltages supplied to the two valves are equal, but of opposite sign, so that the average voltage E_0 is equal to zero during the commutation interval.

The current in the two rectifiers will change in accordance with the following expressions during the commutation interval μ shown in Figure 3.2(b).

$$i_{\text{REC1}} = \frac{1}{L_{c1}} \int e_{1-n} \, dt \tag{3.4}$$

$$i_{\text{REC2}} = I_d + \frac{1}{L_{c2}} \int e_{2-n} \, dt \tag{3.5}$$

(*Note.* e_{2-n} is negative during interval μ in Figure 3.2(*b*).) The average output voltage is

$$E_0 = \frac{1}{\pi} \int_{\mu}^{\pi} E_{1-n,m} \sin \omega t \, d(\omega t)$$

$$E_0 = \frac{E_{1-n,m}}{\pi} \left[-\cos \omega t \right]_{\mu}^{\pi}$$

$$E_0 = \frac{E_{1-n,m}}{\pi} (1 + \cos \mu) \tag{3.6}$$

and the output current is

$$I_d = \frac{E_{1-n,m}}{\pi R} (1 + \cos \mu) \tag{3.7}$$

The output voltage as a function of the leakage reactance is a useful expression, as this shows the effective regulation produced by reactance in the circuit. From equation (3.4),

$$I_d = \frac{1}{L_{c1}} \int_0^{\mu/\omega} e_{1-n} \, dt$$

or

$$I_d = \frac{1}{L_{c1}} \int_0^{\mu/\omega} E_{1-n,m} \sin \omega t \, dt$$

$$I_d = \frac{E_{1-n,m}}{L_{c1}} \left[-\frac{\cos \omega t}{\omega} \right]_0^{\mu/\omega}$$

$$I_d = \frac{E_{1-n,m}}{\omega L_{c1}} [1 - \cos \mu]$$

$$\cos \mu = 1 - \frac{I_d \omega L_{c1}}{E_{1-n,m}} \tag{3.8}$$

and, substituting into (3.6),

$$E_0 = \frac{E_{1-n,m}}{\pi} \left[1 + 1 - \frac{I_d \omega L_{c1}}{E_{1-n,m}} \right]$$

$$E_0 = \frac{2E_{1-n,m}}{\pi} - \frac{I_d \omega L_{c1}}{\pi} \tag{3.9}$$

The second term of equation (3.9), which represents the regulation due to a-c reactance or the average voltage across the a-c reactance, can be developed in another manner as follows.

$$e_{L_{c1}} = L_{c1} \frac{di_{\text{REC1}}}{dt}$$

$$e_{L_{c1}} \, dt = L_{c1} \, di_{\text{REC1}} \tag{3.10}$$

and

$$\frac{1}{T/2} \int_0^{T/2} e_{L_{c1}} \, dt = \frac{2}{T} \int_0^{I_d} L_{c1} \, di_{\text{REC1}}$$

$$E_{L_{c1}} = \frac{2}{T} L_{c1} I_d$$

$$E_{L_{c1}} = 2f L_{c1} I_d$$

$$E_{L_{c1}} = \frac{I_d \omega L_{c1}}{\pi} \tag{3.11}$$

The expression for the rectifier regulation, equation (3.9), may also be written in terms of the per-cent or per-unit commutating reactance as follows. Let the rated effective value of the a-c voltage applied to the transformer equal one per unit voltage and then, assuming $2N$ total turns on the center-tapped transformer winding and N turns on the other winding,

$$E_{1-n,e,\text{pu}} = 1.0 \tag{3.12}$$

$$E_{1-n,m,\text{pu}} = \sqrt{2} \tag{3.13}$$

Also, let the rated effective value of the a-c current in the transformer winding connected to the a-c source of voltage equal one per unit current: then, assuming negligible ripple in the d-c current, very short commutating intervals, and the transformer winding turns ratio above,

$$I_{d,\text{pu}} = 1.0 \tag{3.14}$$

As one per unit impedance is the per unit effective value of the transformer primary voltage divided by the per unit effective value of transformer primary current, the reactances associated with L_{c1} and L_{c2}, Figure 3.2(a), must be converted to an equivalent primary reactance to determine the per unit commutating reactance. Therefore, assuming $X_{c1} = X_{c2}$,

$$X_{c1,\text{pu}} = \frac{2X_{c1}}{4} \tag{3.15}$$

By substituting the per unit quantities in equation (3.9),

$$E_{0,pu} = \frac{2\sqrt{2}}{\pi} - \frac{2X_{c1,pu}}{\pi}$$

$$E_{0,pu} = \frac{2\sqrt{2}}{\pi}\left[1 - \frac{X_{c1,pu}}{\sqrt{2}}\right]$$

3.2 FULL-WAVE, SINGLE-WAY, PHASE-CONTROLLED RECTIFIER, AND A-C LINE VOLTAGE COMMUTATED INVERTER[5-7]

Phase-controlled rectifier operation and a-c line commutated inverter operation are very similar, and either type of operation may be obtained with the same circuit. When operated as a rectifier, the circuit supplies power to a d-c load. When operated as an inverter, a source of d-c voltage and power is required to force current through the circuit to deliver power to the a-c end of the inverter.

The valves in a simple rectifier circuit, as shown in Figure 3.1(a), conduct in sequence. Without phase control, each valve conducts during the portion of the cycle in which its anode voltage is the most positive. When a rectifier is phase-controlled, the turn-on of each valve is delayed, forcing the transfer of current from one valve to the next to be delayed. This delay forces the current to flow in a valve which has lower positive average voltage during its conducting period. The delay may be enough to make the average voltage negative during the valve's conducting period. The delay of current transfer can be any amount up to nearly 180°. As the delay is increased from 0 to 90°, the average d-c voltage is reduced to zero. When the delay of current transfer is more than 90°, the average d-c voltage becomes negative; and, thus, a source of d-c power is required to force current through the circuit. This mode of operation is called inverting, as the flow of power is from the d-c to the a-c circuit. In some respects this change in name is unnecessary, as this type of inversion is only rectification with so much phase control that the d-c voltage is reduced below zero or is negative.

The circuit of Figure 3.3(a) is similar to the rectifier circuit previously discussed, except controlled rectifying elements are used and a counter emf or battery-type load is assumed. The waveforms in Figures 3.3(b), 3.3(c), and 3.3(d) show the operation of this circuit for three different amounts of phase control. The important assumptions which result in these waveshapes are listed below.

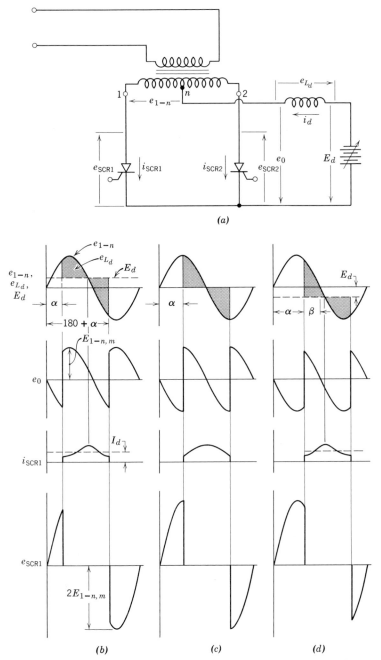

Figure 3.3 (a) Full-wave, single-way, phase-controlled rectifier and a-c line commutated inverter. (b) Waveforms with $\alpha = 60°$. (c) Waveforms with $\alpha = 90°$. (d) Waveforms with $\alpha = 120°$.

63

(1) Ideal controlled rectifying elements—zero forward drop when turned on, infinite forward resistance when turned off, and infinite back resistance.

(2) Ideal transformer—zero leakage reactance, zero winding resistance, and zero magnetizing current.

(3) Sufficient d-c circuit inductance to maintain continuous d-c current.

(4) A battery load which is adjusted to a d-c voltage equal to the average rectifier or inverter d-c voltage at a given firing angle for the rectifying elements.

(5) For inverter operation, a stiff a-c voltage bus is assumed. That is, the a-c voltage is maintained independent of the inverter circuit.

With silicon-controlled rectifiers and well-designed transformers, the waveforms for actual circuits are negligibly different from those assuming ideal components.

The important waveforms with the circuit operating as a phase-controlled rectifier to obtain approximately one-half maximum d-c voltage, are shown in Figure 3.3(b). The rectifying elements are gated on α degrees after the anode voltage becomes positive.* The average voltage across the d-c inductance must again be zero, as indicated by the shaded areas on the figure. The battery voltage must be as indicated for this particular firing angle to maintain the average voltage across the inductance equal to zero. An arbitrary magnitude of d-c current is assumed to illustrate the current waveshape. The average output voltage is given as follows.

$$E_0 = \frac{1}{\pi} \int_\alpha^{\pi+\alpha} E_{1-n,m} \sin \omega t \, d(\omega t)$$

$$E_0 = \frac{E_{1-n,m}}{\pi} \left[-\cos \omega t \right]_\alpha^{\pi+\alpha}$$

$$E_0 = \frac{2E_{1-n,m}}{\pi} \cos \alpha \tag{3.17}$$

Figure 3.3(c) shows the waveforms when α is 90°, in which case the d-c output voltage E_0 is zero. This is the changeover point from operation as a phase-controlled rectifier to operation as an a-c line voltage commutated inverter. Figure 3.3(d) shows the operation with the rectifying elements fired at the phase angle to obtain approximately one-half the maximum inverting voltage. When the phase-control angle α is increased,

* The angle α is used to measure the phase-control angle of retard from the firing angle which produces maximum rectifier d-c voltage. The angle β, which is equal to 180°−α, is used to measure the phase-control angle of advance from the firing angle which produces maximum d-c voltage for inverter operation.

as illustrated by the waveforms in Figures 3.3(b) to 3.3(d), the firing angle of the controlled rectifying elements is further retarded from the firing point to provide maximum d-c voltage for rectifier operation. Forward voltage is held off by the controlled rectifying elements for a greater portion of each cycle. Conversely, the time interval is reduced during which negative voltage appears on a given rectifying element immediately after its conducting period. For the maximum phase-control angle of retard for inverter operation, the angle β must be at least great enough to

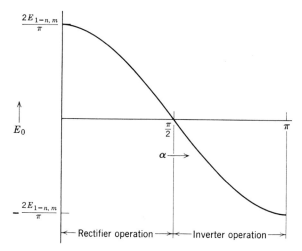

Figure 3.4 Average d-c voltage E_0 vs. firing angle α.

produce complete commutation, as described in Chapter 2. Thus the time interval of negative voltage shown in the e_{SCR1} waveform must be at least sufficient to accomplish the principal events for complete commutation including (*1*) the reduction of forward current to zero in SCR1, (*2*) the delay of the reapplication of forward voltage to this valve until it has regained its forward blocking capability, and (*3*) the buildup of forward current in SCR2.

As mentioned previously, the a-c supply is considered to be a stiff a-c voltage bus. This forces the a-c voltage to remain sinusoidal, as indicated by the waveforms in Figure 3.3. The battery voltage must be set at the value shown for each particular firing angle to result in zero average voltage on the d-c inductance. The expression for the d-c output voltage is again equation (3.17) for the inverter operation with α between 90 and 180°. Figure 3.4 is a plot of the relationship in equation (3.17), and the regions of rectifier and inverter operation are indicated.

As stated previously, the waveshapes in Figure 3.3 are for a circuit that

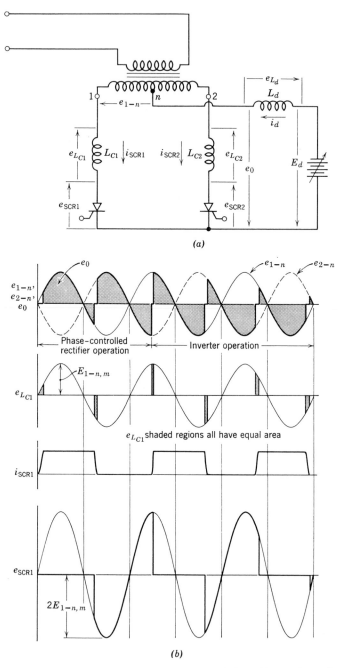

Figure 3.5 (a) Full-wave, single-way, phase-controlled rectifier and a-c line commutated inverter with commutating reactance. (b) Waveforms with varying firing angle.

66

has negligible a-c circuit reactance and resistance. Reactance in the a-c circuit produces a finite commutation time and substantial circuit regulation. This voltage regulation in volts at a given d-c current is the same as for the simple rectifier. The commutation time is especially important in inverter operation. The firing angle α must be less than 180° by enough to permit complete commutation or transfer of current from one valve to the next while the a-c voltage has the correct polarity for commutation. If commutation is not complete in time, the circuit operates as a rectifier, usually circulating excessive current in both the d-c and the a-c circuits. Figure 3.5(a) is similar to Figure 3.3(a) except for the addition of a-c reactance. The waveforms for Figure 3.5(a) are shown in Figure 3.5(b). The firing angle is delayed further from one half-cycle to the next to show the operation both as a phase-controlled rectifier and an a-c line voltage commutated inverter. Except for the addition of the commutating reactance, the same assumptions were used for the waveforms shown in Figures 3.3 and 3.5. It is important to note that the average voltage across the leakage reactance or the voltage area lost from the a-c supply wave is a constant value for a given d-c current. Therefore, the commutating angle varies, depending upon the instantaneous voltages at a particular firing angle. As indicated in Figure 3.5(b), commutating reactance reduces the time interval during which negative voltage appears on the rectifying elements, since some time is required to change the current in the a-c reactances. Thus, there is less delay time before the reapplication of forward voltage to a controlled rectifying element.

3.3 SIX-PHASE DOUBLE-WAY RECTIFIER[1-4]

With Large Inductive Load

The six-phase double-way circuit of Figure 3.6(a), often called a three-phase bridge rectifier, is widely used. This particular circuit provides double-way conduction in the transformer windings, relatively low d-c ripple, and 120° conduction in the rectifying elements. In Figure 3.6(a), and in subsequent figures, the rectifying devices are numbered in their normal conduction sequence. The waveforms shown in Figure 3.6(b) assume negligible ripple in the d-c current, zero a-c circuit reactance, and ideal rectifier elements.

The average output voltage is

$$E_0 = \frac{1}{\pi/6} \int_{\pi/6}^{\pi/3} [E_{1-n,m} \sin \omega t - E_{2-n,m} \sin (\omega t - 120°)] \, d(\omega t) \quad (3.18)$$

and if

$$E_{1-n,m} = E_{2-n,m}$$

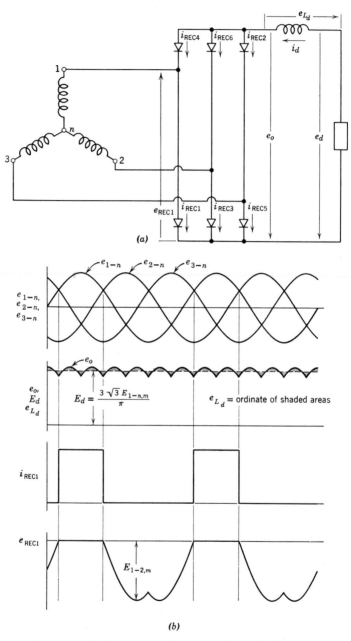

Figure 3.6　(*a*) Six-phase double-way rectifier. (*b*) Waveforms.

then

$$E_0 = \frac{6}{\pi} E_{1-n,m} \left[-\cos \omega t + \cos (\omega t - 120°) \right]_{\pi/6}^{\pi/3}$$

$$E_0 = \frac{6E_{1-n,m}}{\pi} \left[-\frac{1}{2} + \frac{1}{2} + \frac{\sqrt{3}}{2} - 0 \right]$$

$$E_0 = \frac{3\sqrt{3}E_{1-n,m}}{\pi} \tag{3.19}$$

As noted, the peak-to-peak ripple in the d-c output is rather low. When expressed as a per cent of the average output voltage E_0, the ripple is as follows.

$$\text{Per cent peak-to-peak ripple} = \frac{\left(2 \times \frac{\sqrt{3}}{2} - 1.5\right) E_{1-n,m}}{3 \frac{\sqrt{3}}{\pi} E_{1-n,m}} \times 100 \text{ per cent}$$

$$= \frac{(1.732 - 1.5)\pi}{3\sqrt{3}} \times 100 \text{ per cent}$$

$$= 14 \text{ per cent} \tag{3.20}$$

The important differences between the operation of this six-phase double-way circuit and the single-phase arrangement in Figure 3.1 are: (1) the conduction angle is 120° instead of 180°: (2) there is lower ripple in the d-c output voltage: and (3) the average d-c voltage is greater for the same a-c voltage and peak-inverse voltage on the valves. Again, the shaded areas above and below the E_d line must be equal in Figure 3.6(b) in a steady-state condition. It should be particularly noted here, as for all other plain rectifiers, that the voltage is always negative across a given rectifying device unless it is conducting. This, of course, is consistent with the principle that no forward voltage can be held off by simple rectifying elements.

Including Transformer Secondary Leakage Reactance

Figure 3.7 shows the effect of transformer leakage reactance on the six-phase double-way rectifier. Commutation takes place in sequence between the rectifying elements connected to the same d-c output terminal of the rectifier. Immediately prior to the commutating interval μ in Figure 3.7(b), rectifier 5 is the only rectifier conducting current to the positive d-c output bus. When e_{1-n}, the a-c voltage supplied to rectifier

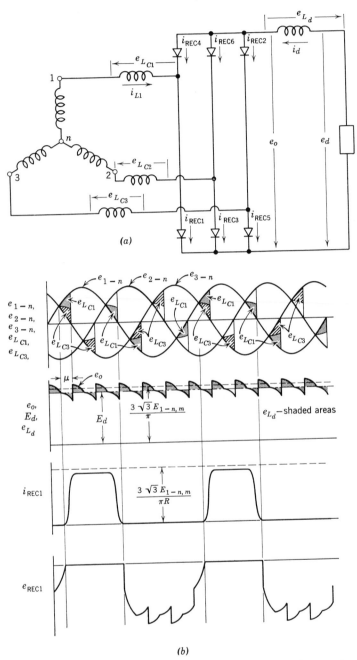

Figure 3.7 (a) Six-phase double-way rectifier with commutating reactance. (b) Waveforms.

1, begins to exceed the voltage applied to rectifier 5, e_{3-n}, the current begins to transfer to rectifier 1 because of its greater applied a-c voltage. The rate of current transfer is dependent upon the difference between the a-c voltages, e_{1-n} and e_{3-n}, applied to the two conducting valves and upon the a-c circuit reactances. During the commutating interval, the voltage of the positive output bus with respect to the transformer secondary neutral is the average of a-c voltages e_{1-n} and e_{3-n} applied to the two conducting valves, assuming equal leakage reactances. In this instance, the output voltage during commutation is not generally zero, since both a-c line voltages are positive during reasonable commutating intervals. The effect of the leakage reactance is to reduce the output voltage during the commutating interval similar to the effect of leakage reactance on the rectifier in Figure 3.2(a).

The current in the two conducting rectifiers during the commutation interval, μ in Figure 3.7(b) assuming equal leakage reactances, will change in accordance with the following expressions.

$$i_{REC1} = \frac{1}{2L_{c1}} \int (e_{1-n} - e_{3-n}) \, dt \tag{3.21}$$

$$i_{REC3} = \frac{1}{2L_{c3}} \int (e_{3-n} - e_{1-n}) \, dt \tag{3.22}$$

The average output voltage is

$$E_0 = \frac{1}{\frac{\pi}{3}} \int_{\pi/6}^{\pi/2} [E_{1-n,m} \sin \omega t - E_{2-n,m} \sin (\omega t - 120°)] \, d(\omega t)$$

$$- \frac{1}{\frac{\pi}{3} \times 2} \int_{\pi/6}^{\pi/6+\mu} [E_{1-n,m} \sin \omega t - E_{3-n,m} \sin (\omega t - 240°)] \, d(\omega t)$$

$$E_0 = \frac{3}{\pi} E_{1-n,m} \left[-\cos \omega t + \cos (\omega t - 120°) \right]_{\pi/6}^{\pi/2}$$

$$- \frac{3}{2\pi} E_{1-n,m} \left[-\cos \omega t + \cos (\omega t - 240°) \right]_{\pi/6}^{\pi/6+\mu}$$

$$E_0 = \frac{3}{\pi} E_{1-n,m} \left[\frac{\sqrt{3}}{2} + \frac{\sqrt{3}}{2} \right]$$

$$- \frac{3}{2\pi} E_{1-n,m} \left[-\cos \left(\frac{\pi}{6} + \mu \right) + \cos (\mu - 210°) + \frac{\sqrt{3}}{2} + \frac{\sqrt{3}}{2} \right]$$

$$E_0 = \frac{3\sqrt{3}}{\pi} E_{1-n,m} - \frac{3\sqrt{3}}{2\pi} E_{1-n,m}$$

$$- \frac{3}{2\pi} E_{1-n,m}[\cos(\mu - 210°) - \cos(\mu + 30°)]$$

$$E_0 = \frac{3\sqrt{3}}{2\pi} E_{1-n,m}\left[1 - \frac{\cos(\mu - 210°) - \cos(\mu + 30°)}{\sqrt{3}}\right]$$

$$E_0 = \frac{3\sqrt{3}}{2\pi} E_{1-n,m}\left[1 - \frac{\cos\mu\cos 210° + \sin\mu\sin 210°}{\sqrt{3}}\right.$$

$$\left. - \frac{-\cos\mu\cos 30° + \sin\mu\sin 30°}{\sqrt{3}}\right]$$

$$E_0 = \frac{3\sqrt{3}}{2\pi} E_{1-n,m}(1 + \cos\mu) \tag{3.23}$$

In this analysis, it is assumed that there is no commutation overlap. When there is such overlap, the voltage regulation is increased. The commutating angle μ for a given reactance and at a given d-c current may be determined as follows. During the commutation interval μ

$$\Delta i_{\text{REC1}} = \frac{1}{2L_{c1}} \int_{\pi/6\omega}^{\pi/6\omega+\mu/\omega} [E_{1-n,m}\sin\omega t - E_{3-n,m}\sin(\omega t - 240°)]\, dt$$

$$\Delta i_{\text{REC1}} = \frac{E_{1-n,m}}{2L_{c1}}\left[-\frac{\cos\omega t}{\omega} + \frac{\cos(\omega t - 240°)}{\omega}\right]_{\pi/6\omega}^{\pi/6\omega+\mu/\omega}$$

$$\Delta i_{\text{REC1}} = \frac{E_{1-n,m}}{2L_{c1}\omega}\left[-\cos(\mu + 30°) + \cos(\mu - 210°)\right.$$

$$\left. + \cos 30° - \cos(-210°)\right]$$

$$\Delta i_{\text{REC1}} = \frac{E_{1-n,m}}{2\omega L_{c1}}\left[-\cos\mu\cos 30° + \sin\mu\sin 30° + \cos\mu\cos 210°\right.$$

$$\left. + \sin\mu\sin 210° + \cos 30° - \cos(-210°)\right] \tag{3.24}$$

and, since $\Delta i_{\text{REC1}} = I_d$, then

$$I_d = \frac{\sqrt{3}E_{1-n,m}}{2\omega L_{c1}}[1 - \cos\mu] \tag{3.25}$$

$$\cos\mu = 1 - \frac{2\omega L_{c1}I_d}{\sqrt{3}E_{1-n,m}} \tag{3.26}$$

Substituting equation (3.26) into equation (3.23) gives

$$E_0 = \frac{3\sqrt{3}}{2\pi} E_{1-n,m} \left[1 + 1 - \frac{2\omega L_{c1} I_d}{\sqrt{3} E_{1-n,m}} \right]$$

$$E_0 = \frac{3\sqrt{3}}{\pi} E_{1-n,m} - \frac{3\omega L_{c1} I_d}{\pi} \qquad (3.27)$$

The expression for rectifier regulation in terms of the per unit a-c circuit reactance may be developed in a similar manner to that used in Section 3.2. Let the rated effective value of the normal rated a-c line current equal one per unit current. Then

$$E_{1-n,e,\mathrm{pu}} = 1.0 \qquad (3.28)$$

$$E_{1-n,m,\mathrm{pu}} = \sqrt{2} \qquad (3.29)$$

$$I_{L1,e,\mathrm{pu}} = 1.0 \qquad (3.30)$$

For the circuit of Figure 3.7(a), the a-c line current is related to the d-c current as follows.

$$I_{L1,e} = \sqrt{\frac{1}{T/2} \int_0^{T/2} i_{L1}^2 \, dt}$$

$$I_{L1,e} \approx \sqrt{\frac{2}{T} \int_{T/12}^{5T/12} I_d^2 \, dt}$$

$$I_{L1,e} \approx \sqrt{\frac{2}{T} I_d^2 \frac{4T}{12}}$$

$$I_{L1,e} \approx \sqrt{\tfrac{2}{3}} I_d \qquad (3.31)$$

and

$$I_{d,\mathrm{pu}} = \frac{1}{\sqrt{\tfrac{2}{3}}} \qquad (3.32)$$

Then, substituting into equation (3.27),

$$E_{0,\mathrm{pu}} = \frac{3\sqrt{3}}{\pi} \sqrt{2} - \frac{3X_{c1,\mathrm{pu}}}{\pi} \frac{1}{\sqrt{\tfrac{2}{3}}}$$

$$E_{0,\mathrm{pu}} = \frac{3\sqrt{3}\sqrt{2}}{\pi} \left[1 - \frac{X_{c1,\mathrm{pu}}}{2} \right] \qquad (3.33)$$

3.4 SIX-PHASE DOUBLE-WAY CONTROLLED RECTIFIER AND A-C LINE VOLTAGE COMMUTATED INVERTER

Figure 3.8(a) shows a six-phase circuit arrangement which operates in a similar manner to the controlled rectifier discussed in Section 3.2.

Figures 3.8(*b*) through 3.8(*f*) illustrate the waveforms for different firing angles and assuming an ideal transformer, perfect rectifiers, and negligible ripple in the d-c current. In successive figures, the firing angle is increased. The principal differences between this six-phase circuit and the circuit in Section 3.2 are again that there is less ripple in the d-c output voltage, 120° conduction through the rectifying elements, and more d-c voltage for a given a-c voltage input and peak inverse voltage on the valves. The basic operation is similar to that of Figure 3.3(*a*). An increasing amount of forward voltage must be held off by the controlled-rectifier devices as

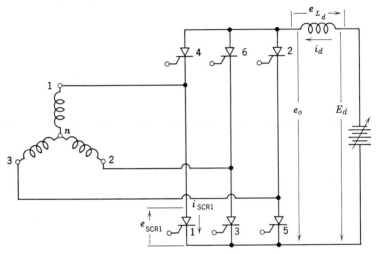

Figure 3.8(*a*) Six-phase double-way controlled rectifier and a-c line voltage commutated inverter.

the firing angle α is increased. Also, the anode-to-cathode voltage of the controlled-rectifying elements is again negative for a smaller portion of each cycle as the phase-control angle is increased. For reliable inverter operation, the minimum angle β, as illustrated in Figure 3.8(*f*), must be such as to provide the negative voltage shown on SCR1 at least for sufficient time to accomplish complete commutation for all possible a-c line voltage disturbances and the maximum load current.

The average output voltage is

$$E_0 = \frac{1}{\frac{\pi}{3}} \int_{\pi/6+\alpha}^{\pi/2+\alpha} [E_{1-n,m} \sin \omega t - E_{2-n,m} \sin (\omega t - 120°)] \, d(\omega t) \quad (3.34)$$

If

$$E_{1-n,m} = E_{2-n,m}$$

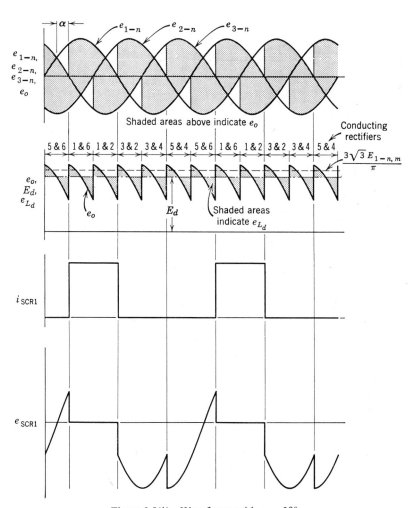

Figure 3.8(b)   Waveforms with $\alpha = 30°$.

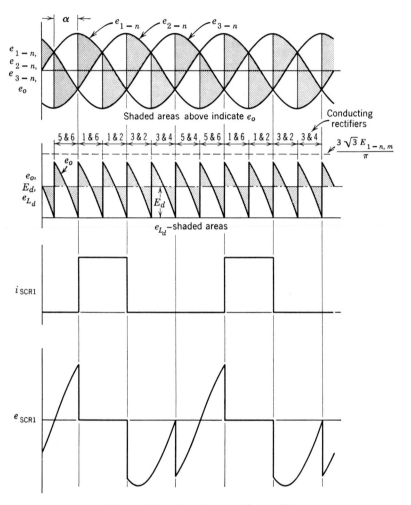

Figure 3.8(c)   Waveforms with $\alpha = 60°$.

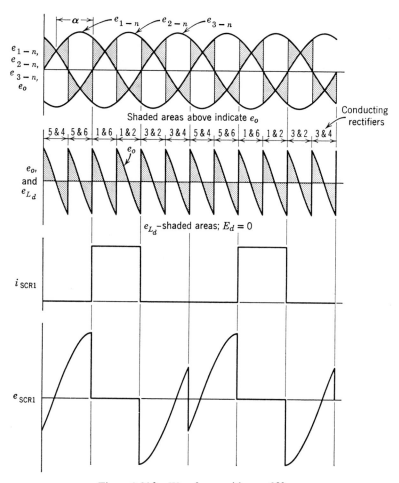

Figure 3.8(d)   Waveforms with $\alpha = 90°$.

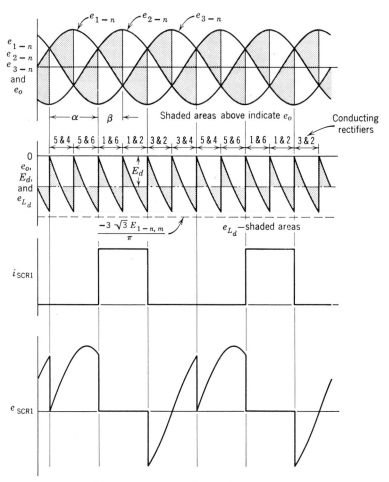

Figure 3.8(e)   Waveforms with $\alpha = 120°$.

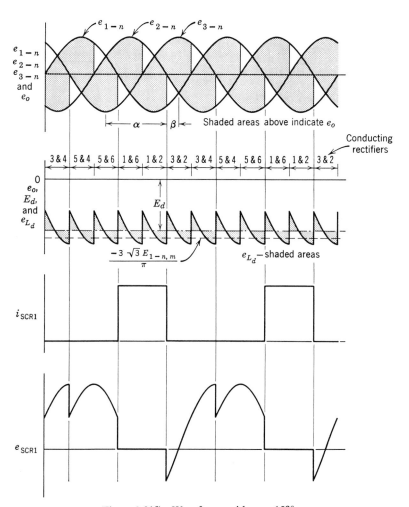

Figure 3.8(f)   Waveforms with $\alpha = 150°$.

then

$$E_0 = \frac{3E_{1-n,m}}{\pi} \left[-\cos \omega t + \cos (\omega t - 120°) \right]_{\pi/6+\alpha}^{\pi/2+\alpha}$$

$$E_0 = \frac{3E_{1-n,m}}{\pi} [-\cos (\alpha + 90°) + \cos (\alpha - 30°)$$

$$+ \cos (\alpha + 30°) - \cos (\alpha - 90°)]$$

$$E_0 = \frac{3E_{1-n,m}}{\pi} [-2 \cos \alpha \cos 90° + 2 \cos \alpha \cos 30°]$$

$$E_0 = \frac{3\sqrt{3}E_{1-n,m}}{\pi} \cos \alpha \tag{3.35}$$

Thus, the average output voltage as a function of the firing angle varies in the same manner as for the single-phase circuit.

The average output voltage E_0 for several firing angles is listed in Table 3.1.

Table 3.1

α (degrees)	E_0
0	$1.65\ E_{1-n,m}$
30	$1.43\ E_{1-n,m}$
60	$0.83\ E_{1-n,m}$
90	0
120	$-0.83\ E_{1-n,m}$
150	$-1.43\ E_{1-n,m}$
180	$-1.65\ E_{1-n,m}$

Figures 3.9(a) and 3.9(b) show the circuit and waveforms including the effect of a-c circuit reactance. Again the firing angle is delayed further from one half-cycle to the next to show the operation changing from a phase-controlled rectifier to an a-c line voltage commutated inverter.

3.5   SIX-PHASE STAR-CONTROLLED RECTIFIER AND A-C LINE VOLTAGE COMMUTATED INVERTER

This circuit is shown in Figure 3.10(a). It is a single way-type circuit similar to the full-wave single-way arrangement of Section 3.2: that is, current flows in only one direction through the transformer windings connected to the rectifying elements. The waveforms are shown in Figure

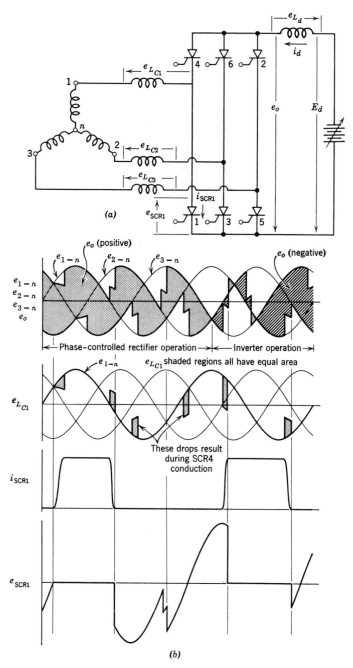

Figure 3.9 (a) Six-phase double-way controlled rectifier and a-c line voltage commutated inverter with commutating reactance. (b) Waveforms for varying firing angle.

81

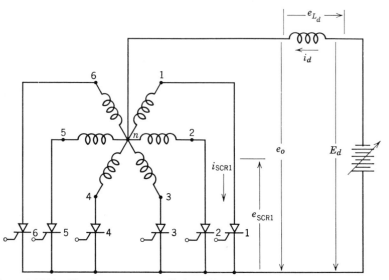

Figure 3.10(*a*)   Six-phase star-controlled rectifier and a-c line voltage commutated inverter.

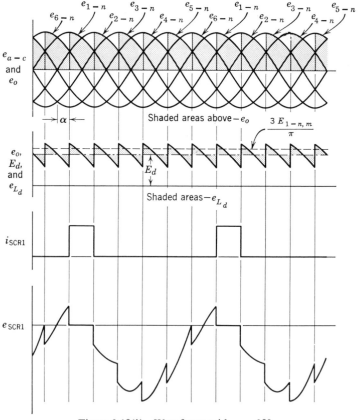

Figure 3.10(*b*)   Waveforms with $\alpha = 30°$.

3.10(b) for operation as a phase-controlled rectifier and in Figure 3.10(c) for inverter operation. The six-phase star circuit has lower ripple, higher d-c voltage for a given a-c voltage and peak inverse voltage on the controlled rectifying elements, but poorer utility of the transformer and

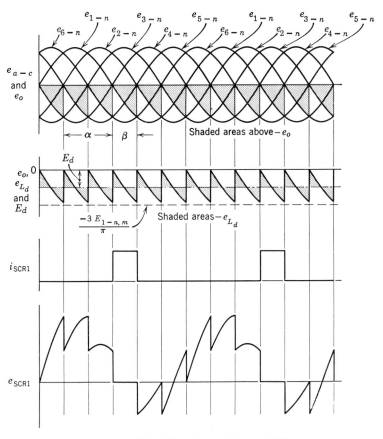

Figure 3.10(c)   Waveforms with $\alpha = 120°$.

valves than the full-wave single-way circuit of Figure 3.3. The poorer utility results from the shorter conducting interval of 60°, which requires more winding and rectifying element rating than 180° conduction.

The single-way circuit in Figure 3.10 also has poorer transformer utility than the double-way circuit in Figure 3.8. In general, single-way circuits have poorer transformer utility than double-way arrangements. The single-way circuits have the advantage that there is only one rectifying element in series with the d-c current at any instant of time. Therefore, the six-phase star arrangement in Figure 3.10 is used in preference to a circuit similar to Figure 3.8 where the d-c voltage is low and only one valve drop is desired; or in situations where it is advantageous to have only one cathode potential for all valves.

It is interesting to note that again the same relationship can be developed for the average output voltage as a function of the firing angle. This is shown as follows.

$$E_0 = \frac{1}{\pi/3} \int_{\pi/3+\alpha}^{2\pi/3+\alpha} E_{1-n,m} \sin \omega t \, d(\omega t) \tag{3.36}$$

$$E_0 = \frac{3E_{1-n,m}}{\pi} \left[-\cos \omega t \right]_{\pi/3+\alpha}^{2\pi/3+\alpha}$$

$$E_0 = \frac{3E_{1-n,m}}{\pi} \left[\cos (\alpha + 60°) - \cos (\alpha + 120°) \right]$$

$$E_0 = \frac{3E_{1-n,m}}{\pi} \left[\cos \alpha \cos 60° - \sin \alpha \sin 60° \right.$$

$$\left. - \cos \alpha \cos 120° + \sin \alpha \sin 120° \right]$$

$$E_0 = \frac{3E_{1-n,m}}{\pi} \cos \alpha \tag{3.37}$$

3.6   SIX-PHASE DOUBLE-WYE CONTROLLED RECTIFIER AND A-C LINE VOLTAGE COMMUTATED INVERTER

The circuit of Figure 3.11(a) is another single way-type circuit that is useful when rectifier forward voltage drop must be minimized. Conduction occurs for 120° as in the three-phase bridge, thus better utilizing the rectifiers and transformers than the six-phase star circuit. In addition, each rectifier carries only one-half the peak load current, as two rectifiers are always conducting. The circuit is often discussed as two three-phase single-way circuits in parallel. Although the interphase transformer is an added circuit element, it is not large, as only the difference voltage from points n_1 to n_2 is across this transformer and this voltage is a third harmonic of the supply frequency. This circuit has been quite widely used for large mercury arc rectifiers primarily because there is only one arc drop in series with the d-c circuit, all valves have a common cathode potential, and it has good transformer and valve utility.

The waveforms of the circuit for rectifier and inverter operation are shown in Figures 3.11(b) and 3.11(c).

The average output voltage is determined by the following.

$$E_0 = \frac{1}{2} \frac{1}{\pi/3} \int_{\pi/6+\alpha}^{\pi/2+\alpha} [E_{1-n1,m} \sin \omega t + E_{6-n2,m} \sin (\omega t + 60°)] \, d(\omega t) \quad (3.38)$$

If

$$E_{1-n1} = E_{6-n2}$$

then

$$E_0 = \frac{3}{2\pi} E_{1-n1,m} \left[-\cos \omega t - \cos (\omega t + 60°) \right]_{\pi/6+\alpha}^{\pi/2+\alpha}$$

$$E_0 = \frac{3E_{1-n1,m}}{2\pi} [-\cos (\alpha + 90°) - \cos (\alpha + 150°)$$

$$+ \cos (\alpha + 30°) + \cos (\alpha + 90°)]$$

$$E_0 = \frac{3E_{1-n1,m}}{2\pi} [-\cos \alpha \cos 150° + \sin \alpha \sin 150°$$

$$+ \cos \alpha \cos 30° - \sin \alpha \sin 30°]$$

$$E_0 = \frac{3E_{1-n1,m}}{2\pi} [\sqrt{3} \cos \alpha]$$

$$E_0 = \frac{3\sqrt{3}E_{1-n1,m}}{2\pi} \cos \alpha \quad (3.39)$$

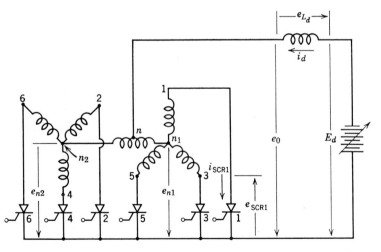

Figure 3.11(a) Six-phase double-wye controlled rectifier and a-c line voltage commutated inverter.

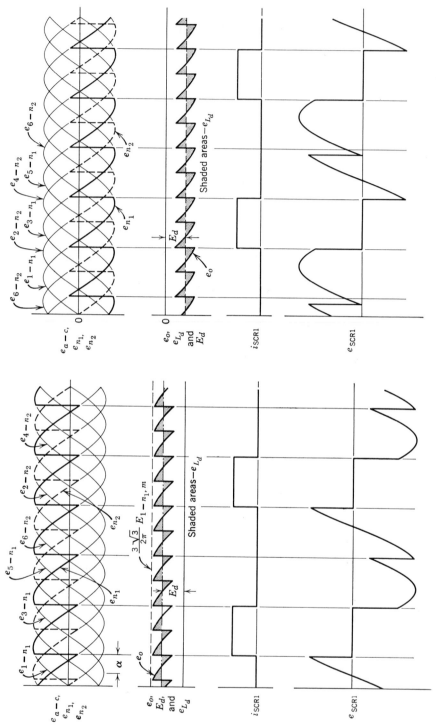

Figure 3.11(b) Waveforms with $\alpha = 45°$.

Figure 3.11(c) Waveforms with $\alpha = 135°$.

3.7   FREQUENCY DIVIDER[8, 9]

An important application of a-c line voltage commutation is the static frequency changer or frequency divider. Figure 3.12 shows an elementary version of a single-phase circuit. When the proper gating signals are provided, the output frequency may be lower than the input. If SCR1

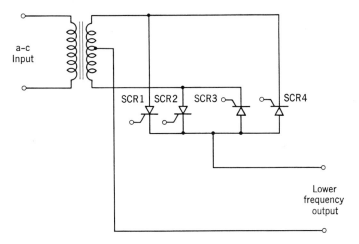

Figure 3.12   Single-phase frequency divider.

and SCR2 are gated "on" for a number of successive half-cycles of the a-c supply, the average output voltage is positive at the upper terminal during this interval. SCR3 and SCR4 are gated "on" for a similar number of successive half-cycles of the input to reverse the output polarity. Thus, an alternating output is provided which is at a lower frequency than the a-c input. When the frequency ratio is several to one or more, the output voltage can be phase controlled on successive half-cycles of the input frequency. With the proper phase control the minimum filtering is required to produce a reasonably sinusoidal output voltage. There are again many single- and polyphase arrangements which may be used to provide a frequency divider using a-c line voltage commutation.

REFERENCES

1. H. Rissik, *Mercury-Arc Current Converters*, Pitman and Sons, London, 1935.
2. O. K. Martl and H. Winograd, *Mercury Arc Power Rectifiers—Theory and Practice*, McGraw–Hill, New York, 1930.

3. D. C. Prince and F. B. Vodges, *Principles of Mercury Arc Rectifiers and their Circuits*, McGraw–Hill, New York, 1947.
4. M.I.T. Electrical Engineering Staff, *Applied Electronics*, John Wiley & Sons, New York. 1943.
5. F. W. Gutzwiller, "An All-Solid State Phase Controlled Rectifier System," AIEE Paper 59–217, 1959.
6. F. W. Gutzwiller, "Phase-Controlling Kilowatts with Semiconductors," *Control Engineering*, May, 1959.
7. F. W. Gutzwiller et al., *Silicon Controlled Rectifier Manual*, Second Edition, General Electric Company, Auburn, N.Y., 1961.
8. E. F. W. Alexanderson and A. H. Mittag, "The 'Thyratron' Motor," *AIEE Transactions*, Vol. 53, 1934, pp. 1517–23.
9. S. C. Caldwell, L. R. Peaslee, and D. L. Plette, "The Frequency Converter Approach to a Variable Speed Constant Frequency System," AIEE CP–60–1076, San Diego, Calif., August 8–12, 1960.

Chapter Four

Parallel Capacitor-Commutated Inverters

The parallel capacitor-commutated inverter is one of the most well-known inverters requiring additional circuit elements to accomplish the commutation process. The basic single-phase circuit is described and analyzed in this chapter. A rigorous mathematical analysis is extremely complex. However, an approximate analysis, based on the fundamental operating principles of the inverter, gives results closely approximating the actual situation for most practical circuits.

An analog computer simulation of the basic parallel inverter permits a study of the effect of changing important circuit parameters. Waveforms for the simulated circuit over a wide range of circuit parameters are presented and discussed.

The parallel inverter requires a rather large commutating capacitor to reliably handle inductive loads; the circuit output voltage changes considerably with load impedance, and care must be exercised to insure reliable starting. Load voltage feedback and related techniques may be used in modifications of the basic parallel inverter to permit reliable operation over a wider inductive load range.

4.1 COMMUTATING PRINCIPLE

The commutating principle of the parallel inverter is illustrated in Figure 4.1. The term "parallel capacitor-commutated inverter" is used to indicate an inverter which is commutated by a capacitor connected in parallel with the load. In Figure 4.1 the capacitor is not directly in parallel with the load, but this simplified circuit illustrates the commutation action obtained in more efficient parallel capacitor inverters.

When the SCR in Figure 4.1 is gated on, the capacitor will charge with

89

the polarity shown exponentially approaching the d-c source voltage. When switch SW is closed, the capacitor is connected across the SCR in a direction to provide a negative anode-cathode voltage, thereby diverting the load current through the capacitor, turning off the SCR. The capacitor

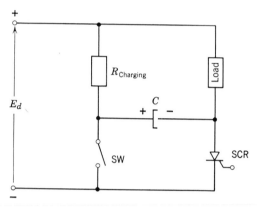

Figure 4.1   Circuit to illustrate commutating principle of parallel inverter.

size and the voltage to which it is charged must be sufficient to divert the maximum load current from the SCR for the time interval required for the SCR to regain its ability to hold off forward voltage.

4.2   SINGLE-PHASE INVERTER[1-6]

A well-known single-phase circuit using parallel capacitor commutation is shown in Figure 4.2. This circuit is similar to the mechanical-switch and transistor-switch circuits discussed in Chapter 2. It differs from these only by the addition of the commutating capacitor and the d-c reactor. In the circuits referred to in Chapter 2, the current was forced to zero in a particular switching element when the switch was opened. Thus, the initiation of current transfer from one valve to the next was forced to occur because of the ability of the mechanical or transistor switch to interrupt forward current. The parallel capacitor-commutated inverter of Figure 4.2 can use valves which do not have the ability to interrupt forward current; that is, controlled-rectifier devices. The current-switching action (that is, the transfer of current from one valve to the next) is obtained by the action of the commutating capacitor. This commutation is initiated when the next SCR to conduct is gated on. The d-c reactor prevents excessive current from flowing into the commutating capacitor during the switching intervals.

One method of looking at the circuit in Figure 4.2 is to consider that the d-c current is alternately switched from one half to the other half of the transformer primary winding. The d-c source alternately supplies the d-c current to the two halves of the transformer primary producing magnetomotive forces of opposite polarity. This is equivalent to an alternating current in a single primary winding and, thus, the transformer secondary winding delivers an alternating current to the load. In general,

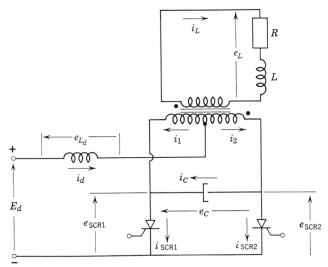

Figure 4.2   Parallel capacitor-commutated inverter.

for the circuit in Figure 4.2, the d-c reactor is large enough to prevent cyclic variations in the d-c input current. When this input current is commutated from one SCR to the other, the switched d-c current is equivalent to a square wave current supplied to the a-c portion of the circuit—the transformer, capacitor, and load. During the conducting period for each SCR, the average voltage across one half of the transformer center tapped winding must equal the d-c source voltage for steady-state conditions.

A second method of describing the operation of Figure 4.2 is to consider the SCR'S as switches which alternately connect the d-c source voltage to one half and then the other half of the transformer primary winding. This is equivalent to having an alternating voltage impressed on a single primary winding so that an alternating voltage is delivered to the load.

In general, parallel inverters can be analyzed as having the d-c current switched in polarity with respect to the a-c portion of the circuit—the transformer, capacitor, and load. Certain extreme values of circuit constants will make the circuit most straightforward to analyze by either the voltage-switched or current-switched method. For example, with a large d-c reactor and relatively large commutating angles, the a-c portion of the circuit of Figure 4.2 can be analyzed as being driven by a square wave-shape of current. When the commutating angle is small and the load is practically all resistive, the circuit may be analyzed by considering a square waveshape of voltage applied to the a-c portion of the circuit.

In the circuit of Figure 4.2 with SCR1 conducting, the capacitor is charged positively on the right-hand end. When SCR2 is turned on, the capacitor voltage provides a reverse anode-cathode voltage for turning off SCR1. For operation with a very large d-c inductance, essentially a constant current flows from the d-c source for a given load. Then with pure resistance load and a small capacitor, approximately a square wave current flows into the load. If the inductive load is added in parallel with the resistive load, the current in the inductance reaches a maximum value in the latter portion of each half-cycle. The input square wave of current cannot supply this maximum inductive load current. It is supplied by the capacitor, thereby producing considerable loss in voltage on the capacitor during the latter portion of each half-cycle. This lower capacitor voltage produces a smaller commutating angle. In most practical circuits, commutation is obtained when the capacitor kva exceeds the inductive load kva. Thus, in general, the commutating capacitor must not only be large enough to provide commutation with pure resistance load but, in addition, it must supply the inductive load kva. This gives rise to one of the basic disadvantages of this inverter, that is, rather large capacitors are required to handle inductive load. Also, the waveshape and magnitude of the output voltage are changed quite radically if the inductive load is removed while the same capacitance value is retained.

Circuit Operating Principles

Several basic principles relating to the operation of the circuit in Figure 4.2 are discussed in this section. The principles discussed are important to develop a sound understanding of the circuit. They are also very useful in any mathematical analysis of the circuit operation.

As mentioned in the previous section, with a large d-c inductance, the d-c source forces an equivalent square wave current into the a-c portion of the circuit in Figure 4.2. This square wave current produces an a-c load voltage out of phase with the current because of the combined reactance of the commutating capacitor and the load. The fundamental component of the square wave input current can be considered to have both a power and a reactive component. The power component is in phase with the fundamental a-c load voltage, and the reactive component is 90° out of phase with the fundamental load voltage. The fundamental power is consumed by the load resistance while the reactive input supplies the reactive kva of the load and the commutating capacitor. A similar situation is true for each harmonic contained in the square wave input current. These facts lead to the following specific operating principles.

(1) *The power supplied from the d-c source must equal the power delivered to the load.* This principle is very well known, and is true for any rectifier or inverter assuming negligible losses. In addition, this principle is true for the fundamental component, each harmonic, and the total power assuming constant *L-R-C* parameters. In equation form, the principle is

$$E_d I_d = \frac{E_{L,e}^2}{R'} \tag{4.1}$$

and

$$E_{L,e(n)} I_{d,e(n)} \cos \theta_n = \frac{E_{L,e(n)}^2}{R'} \tag{4.2}$$

where $E_{L,e(n)}$ = effective value of fundamental ($n = 1$) or nth harmonic component of a-c load voltage.

$I_{d,e(n)}$ = effective value of nth harmonic component of d-c source current

θ_n = power factor angle between particular component of a-c load voltage and d-c source current

R' = resistance of equivalent parallel R'-L' to series R-L shown in Figure 4.2

(2) *The fundamental and each harmonic component of reactive kva supplied by the source must equal the corresponding components of reactive kva delivered to the total a-c circuit.* This principle is very similar to the first one. It is most clearly understood by considering the circuit of Figure 4.2 with a very large d-c reactor. The equivalent circuit is then a square wave current generator supplying the complete a-c circuit as shown in Figure 4.3. The commutating capacitor has been reflected to the transformer secondary winding, assuming the total primary turns are twice the secondary turns. In addition, the series *R-L* load has been changed to the equivalent shunt R'-L' load. This principle, in equation form, is

$$E_{L,e1} I_{d,e1} \sin \theta_1 = \frac{E_{L,e1}^2}{X_{C1}'} \tag{4.3}$$

$$E_{L,e(n)} I_{d,e(n)} \sin \theta_n = \frac{E_{L,e(n)}^2}{X_{C(n)}'} \tag{4.4}$$

where the quantities are for the fundamental ($n = 1$) or particular harmonic component and where X_C' is the net capacitive reactance of the combination of $4C$ and L' for the particular component. A net capacitive reactance is generally required for proper operation of the circuit.

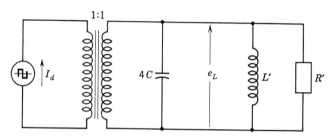

Figure 4.3   Equivalent circuit of Figure 4.2 with a large d-c reactor.

(3) *For steady-state conditions during the conducting period of each SCR, the average voltage on one half of the transformer primary must equal the d-c source voltage.* The difference between the d-c source voltage and the voltage across one half the transformer primary is the voltage across the

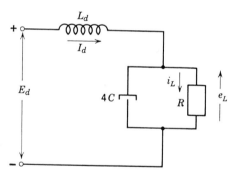

Figure 4.4   Equivalent circuit of Figure 4.2 during each half-cycle.

d-c reactor. This principle is true, since the circuit is completely symmetrical on each half-cycle as shown in Figure 4.4, and there can be no average voltage across the d-c reactor for steady-state operation.

These principles give rise to the following additional facts. For operation with sinusoidal output voltage, $E_d/E_{L,e}$ plotted as a function of the commutating angle is the same cosine curve as shown in Figure 3.4

(Chapter 3). The commutating angle of the circuit, that is, the time interval immediately after conduction when negative anode-cathode voltage appears across a valve, is also equal to the circuit power factor angle. In fact, this approximation is quite good for output voltage waveshapes, departing considerably from a sine wave as will be shown in a subsequent section.

Circuit Analysis[3],[4]

A detailed analysis of the circuit in Figure 4.2 may be carried out by solving the differential equations describing the circuit operation. However, this analysis is quite complex even for the case where there is negligible ripple in the d-c current and negligible load inductance, as shown in the appendix to this chapter.

The equations resulting from the assumption of a large d-c reactor and negligible load inductance analysis are as follows

$$e_L = \frac{E_d(1 + \epsilon^{-T/2\tau} - 2\epsilon^{-t/\tau})}{1 + \epsilon^{-T/2\tau} - \dfrac{4\tau}{T}(1 - \epsilon^{-T/2\tau})} \tag{4.27}*$$

$$i_2 = \frac{E_d}{R} \frac{\epsilon^{-t/\tau}}{1 + \epsilon^{-T/2\tau} - \dfrac{4\tau}{T}(1 - \epsilon^{-T/2\tau})} \tag{4.35}$$

$$i_L = \frac{E_d}{R} \frac{1 + \epsilon^{-T/2\tau} - 2\epsilon^{-t/\tau}}{1 + \epsilon^{-T/2\tau} - \dfrac{4\tau}{T}(1 - \epsilon^{-T/2\tau})} \tag{4.36}$$

$$i_1 = \frac{E_d}{R} \frac{1 + \epsilon^{-T/2\tau} - \epsilon^{-t/\tau}}{1 + \epsilon^{-T/2\tau} - \dfrac{4\tau}{T}(1 - \epsilon^{-T/2\tau})} \tag{4.37}$$

where $\tau = 4RC$.

The effect on the load voltage of changing load resistance or commutating capacitance can be determined by substituting different values of τ

---

* The equation numbers that do not appear in this chapter are used in the appendix for the equations involved in the detailed analysis of Figure 4.2.

into equation (4.27) as follows.

$$\tau = 4RC = \frac{T/2}{3}$$

$$e_L(t) = \frac{E_d(1 + \epsilon^{-3} - 2\epsilon^{-t/T/6})}{1 + \epsilon^{-3} - \frac{2}{3}(1 - \epsilon^{-3})}$$

$$e_L(t) = \frac{E_d(1 + 0.05 - 2\epsilon^{-t/T/6})}{\frac{1}{3} + \frac{5}{3}\epsilon^{-3}}$$

$$= \frac{3E_d(1.05 - 2\epsilon^{-t/T/6})}{1.25}$$

$$e_L(t) = E_d(2.52 - 4.80\epsilon^{-t/T/6})$$

$$e_L(t)\Big|_{t=0} = -2.28E_d$$

$$e_L(t)\Big|_{t=T/2} = E_d(2.52 - 4.80\epsilon^{-3}) = 2.28E_d$$

$$\tau = 4RC = \frac{T/2}{6}$$

$$e_L(t) = \frac{E_d(1 + \epsilon^{-6} - 2\epsilon^{-t/T/12})}{1 + \epsilon^{-6} - \frac{1}{3}(1 - \epsilon^{-6})}$$

$$= \frac{E_d(1 + 0.0025 - 2\epsilon^{-t/T/12})}{\frac{2}{3} + \frac{4}{3}\epsilon^{-6}}$$

$$= \frac{3E_d(1.0025 - 2\epsilon^{-t/T/12})}{2.01}$$

$$e_L(t) = E_d(1.5 - 2.99\epsilon^{-t/T/12})$$

$$e_L(t)\Big|_{t=0} = -1.49E_d$$

Waveforms of the load voltage and the controlled-rectifier voltage for

$$\tau \ll T/2, \quad \tau = T/12, \quad \text{and} \quad \tau = T/6$$

are shown in Figure 4.5. As indicated in this figure, the increase of the load resistance or commutating capacitance appreciably changes the wave-shape and peak magnitude of the load voltage. The magnitude of the load voltage must increase as $4RC$ is increased in order to maintain the areas

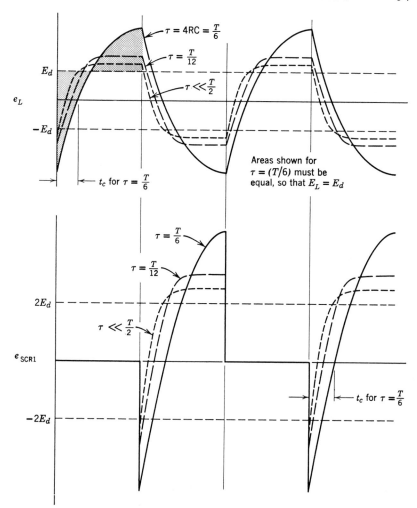

Figure 4.5 Waveforms of Figure 4.2, neglecting load inductance and assuming a large d-c reactor.

equal as shown in Figure 4.5. With these areas equal, the average load voltage during the conducting interval for each valve is equal to E_d so that there is no average voltage across L_d.

The time interval for an important part of commutation is the interval during which the anode-cathode voltage is negative immediately after conduction of a given SCR. In the parallel inverter, when there is negligible reactance in the commutating circuit, the negative anode-cathode voltage interval is essentially the complete commutating period. This time interval must be sufficient to allow the SCR to regain its forward blocking capability. As shown in Figure 4.5, this time interval increases with increasing load resistance or commutating capacitance. This means that for a certain load, the commutating capacitance must be at least the amount required to provide the necessary commutating time.

The commutating time may be calculated for the case of constant d-c current and pure resistance load by setting equation (4.27) equal to zero as follows.

$$e_L \Big|_{t=t_c} = 0 = \frac{E_d(1 + \epsilon^{-T/2\tau} - 2\epsilon^{-t_c/\tau})}{1 + \epsilon^{-T/2\tau} - \frac{4\tau}{T}(1 - \epsilon^{-T/2\tau})} \tag{4.38}$$

$$2\epsilon^{-t_c/\tau} = 1 + \epsilon^{-T/2\tau}$$

$$-\frac{t_c}{\tau} = \ln \frac{\epsilon^{-T/2\tau} + 1}{2}$$

$$t_c = \tau \ln \frac{2}{\epsilon^{-T/2\tau} + 1}$$

The commutating time, expressed in radians, or the commutating angle is

$$\mu = \frac{2\pi}{T} t_c = \frac{2\pi\tau}{T} \ln \frac{2}{\epsilon^{-T/2\tau} + 1} \tag{4.39}$$

As indicated in the previous section, the commutating angle can be quite closely approximated by the circuit power factor angle for the fundamental components even when the circuit output voltage departs considerably from a sine wave. This is shown as follows.

$$\text{Fundamental reactive a-c current} = \frac{E_{L,e1}}{X_{C1}'} \tag{4.45}$$

$$\text{Fundamental resistive a-c current} = \frac{E_{L,e1}}{R} \tag{4.46}$$

$$\tan \theta_1 = \frac{\dfrac{E_{L,e1}}{X_{C1}'}}{\dfrac{E_{L,e1}}{R}} = \frac{R}{X_{C1}'} \tag{4.47}$$

with negligible load inductance

$$\tan \theta_1 = 4\omega RC \tag{4.48}$$

Using equations (4.39) and (4.48), actual values of μ and θ are tabulated in Table 4.1.

As shown in the appendix to this chapter the expression for the load voltage is quite complex in its general form when smaller values

Table 4.1   Comparison of exact and approximate methods of calculating commutating angle

$\mu = \dfrac{2\pi}{T}(4RC)\ln\dfrac{2}{\epsilon^{-T/8RC}+1}$				$\tan\theta_1 = 4\omega RC = \dfrac{2\pi}{T}(4RC)$	
4RC	$\epsilon^{-T/8RC}$	$\ln\left(\dfrac{2}{\epsilon^{-T/8RC}+1}\right)$	μ		θ_1
			radians	degrees	degrees
$T/20$	0.000045	0.693	0.0693π	12.5	17.5
$T/10$	0.0067	0.687	0.138π	25	32
$T/8$	0.0183	0.676	0.169π	30.5	38
$T/4$	0.135	0.565	0.283π	51	57.5
$T/2$	0.368	0.380	0.38π	68	72
T	0.606	0.22	0.44π	79	81

of L_d are used and when load inductance is involved. The form of the solution under these conditions may be obtained from the form of the inverse transform for equations (4.17) and (4.18) in the appendix. For the case with pure resistance load but finite L_d, from equation (4.18),

$$e_L(t) = K_A + K_B\epsilon^{-t/8RC}\sin(\omega_1 t + \Psi_1) \qquad (4.49)$$

When load inductance is included, from equation (4.17),

$$e_L(t) = K_C + K_D\epsilon^{-\gamma_1 t} + K_E\epsilon^{-\gamma_2 t}\sin(\omega_2 t + \Psi_2) \qquad (4.50)$$

The complete solutions of these equations are quite involved.

In most practical cases, it is much simpler to carry out an approximate analysis using the basic principles discussed in the previous section. When a relatively large inductive load is present, the output waveform is nearly sinusoidal for many practical circuits. For the case when the a-c voltage waveshape approaches a sine wave, the approximate analysis is

$$I_{d,e1}E_{L,e}\sin\theta = \frac{E_{L,e}^2}{X_C'} \qquad (4.3)$$

$$I_{d,e1}E_{L,e}\cos\theta = \frac{E_{L,e}^2}{R'} \qquad (4.51)$$

where R' and X_L' are equivalent parallel load resistance and inductance, as shown in Figure 4.3, and

$$\frac{1}{X_C'} = \frac{1}{X_C} - \frac{1}{X_L}$$

$$\frac{E_{L,e}^2}{R'} = E_d I_d \qquad (4.1)$$

For a square wave current,

$$I_{d,e1} = \frac{4}{\sqrt{2}T} \int_0^{T/2} I_d \sin \omega t \, dt$$

$$= \frac{2\sqrt{2} I_d}{T} \left[-\frac{\cos \omega t}{\omega} \right]_0^{T/2}$$

$$= \frac{2\sqrt{2}}{\pi} I_d \tag{4.52}$$

Combining (4.3) and (4.51),

$$\left(\frac{I_{d,e1}}{E_{L,e}}\right)^2 = \frac{1}{R'^2} + \frac{1}{X_C'^2}$$

or,

$$E_{L,e}^2 = \frac{I_{d,e1}^2 R'^2 X_C'^2}{R'^2 + X_C'^2} \tag{4.53}$$

Combining with (4.1) and (4.52),

$$E_d I_d R' = E_{L,e}^2 = \frac{8}{\pi^2} I_d^2 \frac{R'^2 X_C'^2}{R'^2 + X_C'^2}$$

$$I_d = \frac{\pi^2}{8} \frac{E_d}{R'} \frac{R'^2 + X_C'^2}{X_C'^2} \tag{4.54}$$

From (4.1)

$$E_{L,e} = \sqrt{E_d I_d R'} = \frac{\pi}{\sqrt{8}} E_d \frac{\sqrt{R'^2 + X_C'^2}}{X_C'} \tag{4.55}$$

Equation (4.55) may be developed in another manner as follows.

$$\frac{E_d}{E_{L,e}} = \frac{2\sqrt{2}}{\pi} \cos \theta \tag{4.56}$$

This is the relationship defined by the cosine curve in Figure 3.4 (Chapter 3), which is valid here for operation with sinusoidal a-c voltage. When $\theta = 0$, equation (4.56) gives the average value of a rectified sinusoidal a-c wave. From equations (4.45), (4.46), and (4.47),

$$\cos \theta = \frac{X_C'}{\sqrt{R'^2 + X_C'^2}} \tag{4.57}$$

Combining (4.56) and (4.57),

$$E_{L,e} = \frac{\pi}{\sqrt{8}} E_d \frac{\sqrt{R'^2 + X_C'^2}}{X_C'}$$

which is equation (4.55).

Analog Computer Simulation

An analog computer can be used very effectively to examine the effect of changing circuit parameters on the circuit waveshapes for Figure 4.2. The equations developed in the previous section can be solved with a fairly simple computer setup, and the actual waveshapes can be observed by recording the voltages at appropriate points. The computer simulation enables one to readily observe the waveshapes for many variations in circuit parameters for idealized circuit components.

A Goodyear L3 computer was used to set up a simulated version of the circuit of Figure 4.2. Simple relays were used to simulate the controlled rectifiers. This is an accurate representation as long as the switching times for the controlled rectifiers are very short relative to the inverter period, which is generally true with present SCR devices for inverters operating at approximately 400 cps or less. Assuming a 60-cps inverter, the time scale was reduced by 300:1 on the computer setup so that low-frequency response recorders could satisfactorily be used and so that the relay switching times were negligible with respect to the inverter period. Three computer amplifiers were used to provide a square wave oscillator to drive the relays analogous to square wave gating signals on the actual controlled rectifiers.

The computer diagram used to represent the circuit of Figure 4.2, is shown in Figure 4.6. In addition, the relationships between actual circuit parameters and computer parameters are noted on the diagram. The load voltage e_L is subtracted from E_d and the difference supplies the input to integrator amplifier 13. Amplifier 21 reverses the sign only so that its output is proportional to i_d. The scale factors are such that the actual output of amplifier 21 is plus 100 i_d. The voltage proportional to 100 i_d minus a voltage proportional to 100 i_L is then supplied to integrator amplifier 14 to produce positive e_L from the output of sign-reversing amplifier 15. The load voltage e_L is supplied to the amplifier 23 arrangement to generate 100 i_L. Amplifier 24 is used as a sign reverser only for convenience in recording. This computer simulator requires two single-pole double-throw relays to simulate the two controlled rectifiers. It is necessary to switch i_d from one half of the transformer primary to the next half of the primary on alternate half-cycles, and it is necessary to subtract minus e_L and then plus e_L from E_d on alternate half-cycles. The third relay is used to permit recording e_L for one half-cycle only to simulate the voltage across a controlled rectifier. Computer amplifiers 1, 2, and 3 are connected to form the square wave oscillator to drive the relays.

Figures 4.7 through 4.16 show waveshapes for circuit parameters,

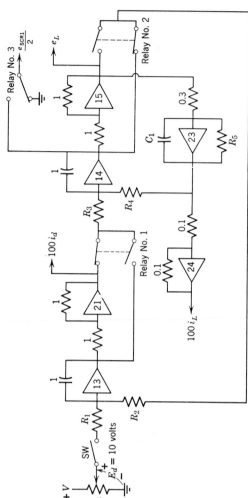

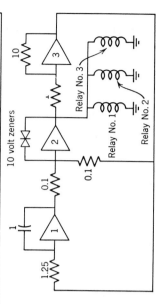

Figure 4.6 Analog computer diagram for Figure 4.2. (1) All resistors are in megohms and capacitors are in μfarads. (2) Relays are shown in positions for half-cycle when SCR1 is "on." (3) Assuming 300:1 reduced time base from 60 cps.

$$L_d = \tfrac{1}{3}R_1 \text{ (henries)}$$

where R_1 is in megohms and $R_1 = R_2$.

$$C = \frac{R_3}{0.12} \text{ (μfarads)}$$

where R_3 is in megohms and $R_3 = R_4$.

$$R = \frac{30}{R_5} \text{ (ohms)}$$

where R_5 is in megohms.

$$L = 0.1C_1 \text{ (henries)}$$

where C_1 is in μfarads.

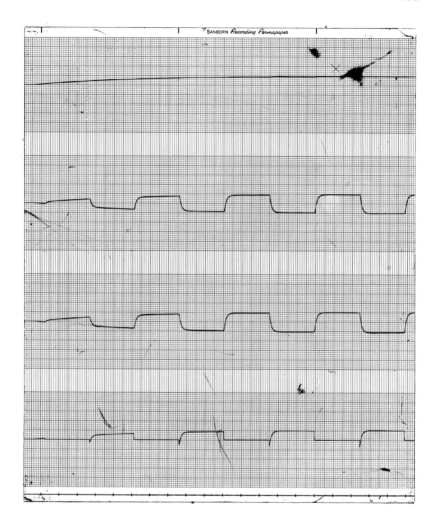

Figure 4.7 $L_d = 1.67$ henries; $C = 0.8$ μfarads; $R = 100\ \Omega$ and $L = 0$.

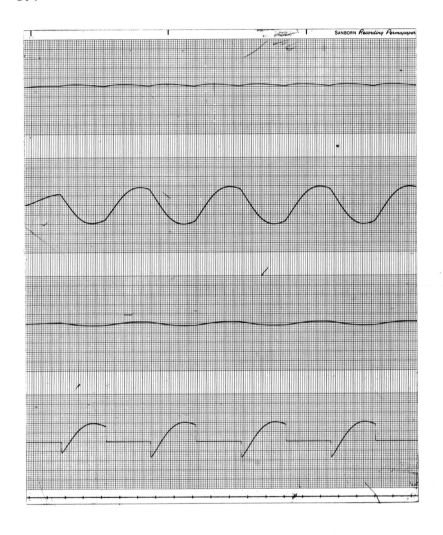

Figure 4.8　$L_d = 1.67$ henries; $C = 0.8\ \mu$farads; $R = 1000\ \Omega$; and $L = 0$.

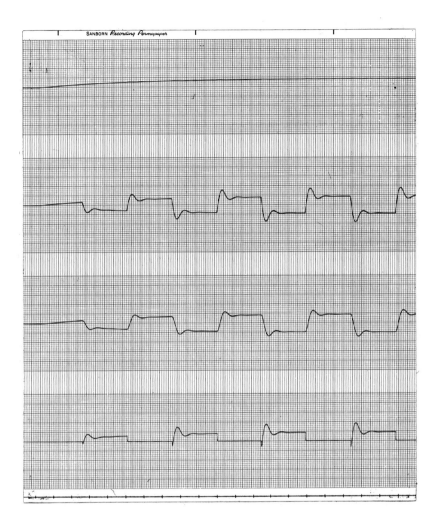

Figure 4.9 $L_d = 1.67$ henries; $C = 0.8$ μfarads; $R = 100$ Ω; and $L = 0.05$ henries.

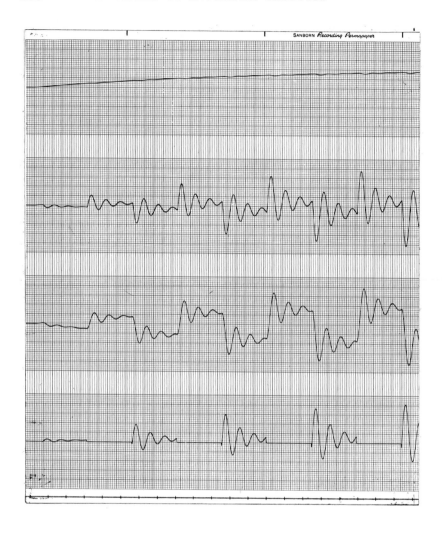

Figure 4.10 $L_d = 1.67$ henries; $C = 0.8\,\mu$farads; $R = 30\,\Omega$; $L = 0.05$ henries.

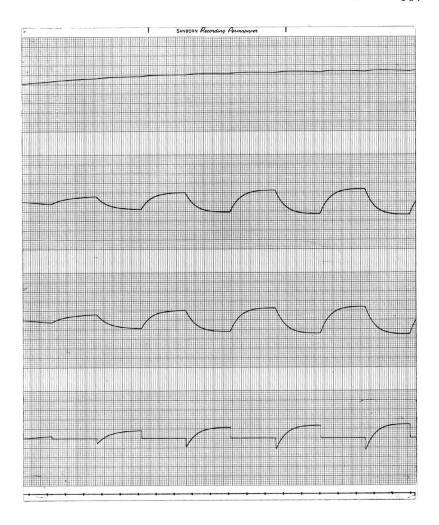

Figure 4.11 $L_d = 1.67$ henries; $C = 4\,\mu$farads; $R = 100\,\Omega$; and $L = 0$.

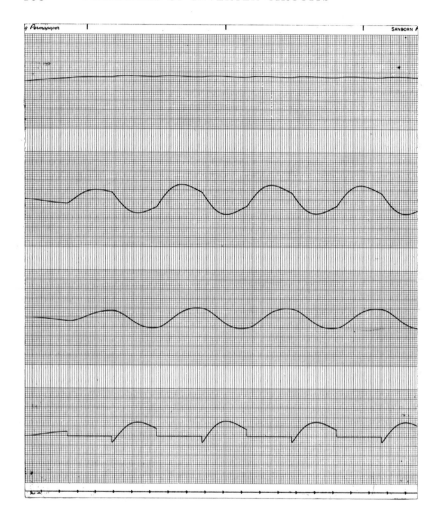

Figure 4.12 $L_d = 1.67$ henries; $C = 4\,\mu$farads; $R = 100\,\Omega$; and $L = 0.3$ henries.

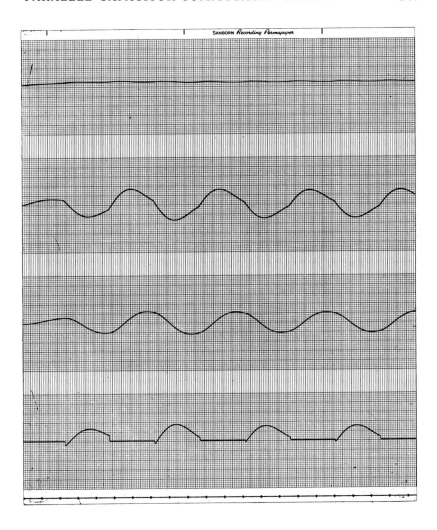

Figure 4.13 $L_d = 1.67$ henries; $C = 4\,\mu$farads; $R = 75\,\Omega$; and $L = 0.3$ henries.

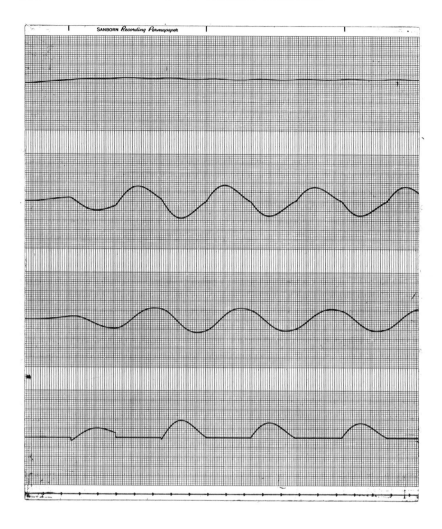

Figure 4.14 $L_d = 1.67$ henries; $C = 4\,\mu$farads; $R = 60\,\Omega$;   and $L = 0.3$ henries.

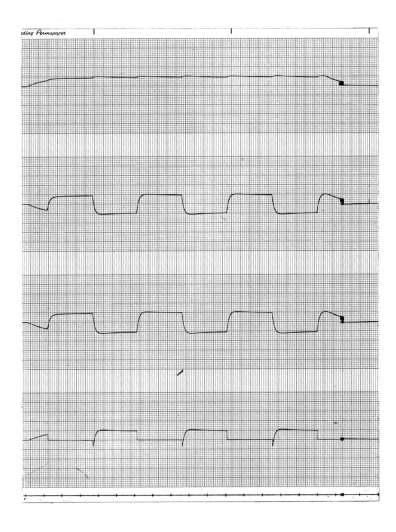

Figure 4.15 $L_d = 0.33$ henries; $C = 0.8\,\mu$farads; $R = 100\,\Omega$;   and $L = 0$.

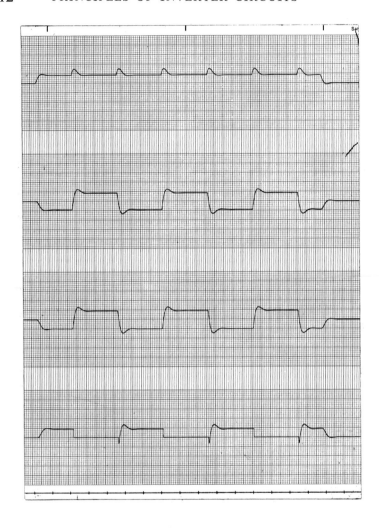

Figure 4.16 $L_d = 0.05$ henries; $C = 0.8\,\mu$farads; $R = 100\,\Omega$; and $L = 0$.

illustrating the most significant features of the inverter in Figure 4.2. On all recordings, the following calibrations were used.

Upper trace, i_d: 0.02 amp/small division
Second trace, e_L: 2 v/small division
Third trace, i_L: 0.02 amp/small division
Lower trace, e_{SCR}: 4 v/small division
E_d set at 10 v
Chart speed: 10 small divisions/second

Much can be learned about the operation of the parallel inverter circuit in Figure 4.2 by studying the waveshapes indicated on the recordings. Some of the most important facts indicated on these recordings are as follows.

(1) For pure resistance load, when the load resistance is increased, the commutating angle is increased and the peak amplitude of the load voltage is increased, thus resulting in greater voltage on the SCR's.

(2) The commutating angle is increased by using a larger commutating capacitor with all other circuit parameters constant.

(3) When inductive load is added with other parameters constant, the commutating angle is reduced and the load voltage can become quite oscillatory. For an actual circuit the parameters used for Figure 4.14 would result in a commutating failure since the commutating angle is zero and there is no negative voltage to turn off the controlled rectifiers.

(4) When L_d is reduced, the load voltage becomes quite oscillatory for high resistance loads, and this again may result in high peak voltages on the controlled rectifiers.

Figures 4.17 and 4.18 show the starting transients for two sets of circuit parameters. In both cases, the relays were first energized by applying power to all computer amplifiers simulating the application of gating signals to the controlled rectifiers and then the d-c power was applied by closing switch SW in Figure 4.6. For different transients the d-c power switch was closed at a different instant relative to the relay switching. As shown on these recordings, for some starting transients, the commutating angle may approach zero, which would result in failure to start. In particular, in the actual circuit if the d-c power is supplied just an instant prior to switching, the inverter would generally fail, as there would be no voltage on the commutating capacitor to turn off the appropriate SCR.

Figures 4.17 and 4.18 therefore indicate another potential difficulty with the parallel inverter—reliable starting. Saturation of the transformer may also occur at starting, depending on the residual flux in the core and the particular instant when the d-c power is supplied. For this reason, an air gap is often used in the transformer, or it is operated at low flux

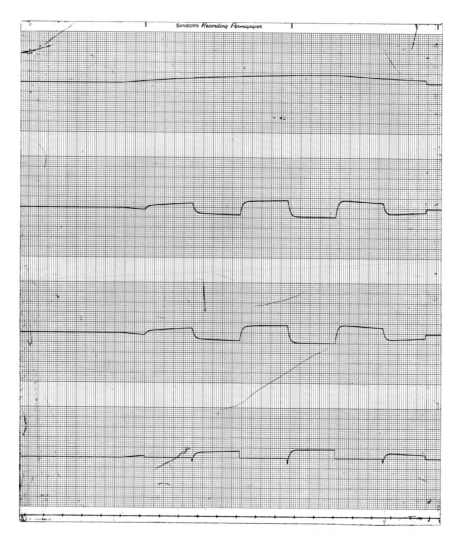

Figure 4.17 Starting transients. $L_d = 1.6$ henries; $C = 0.8\ \mu$farads; $R = 100\ \Omega$; and $L = 0$.

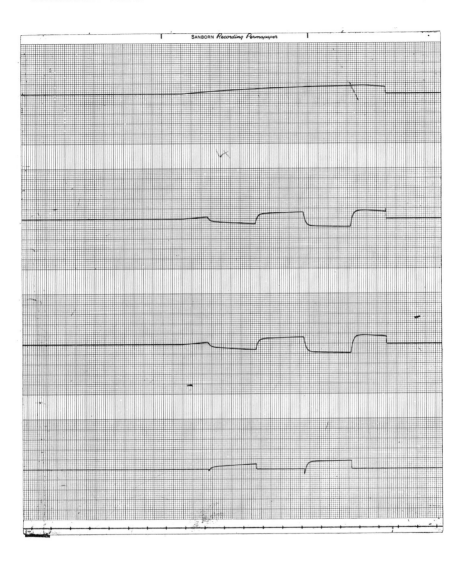

Figure 4.17 (*continued*)

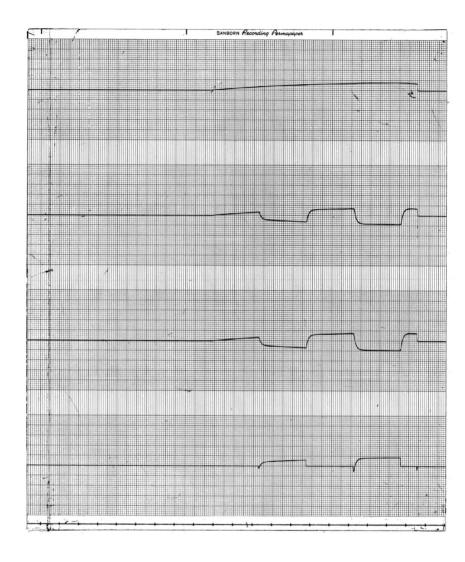

Figure 4.17   (*continued*)

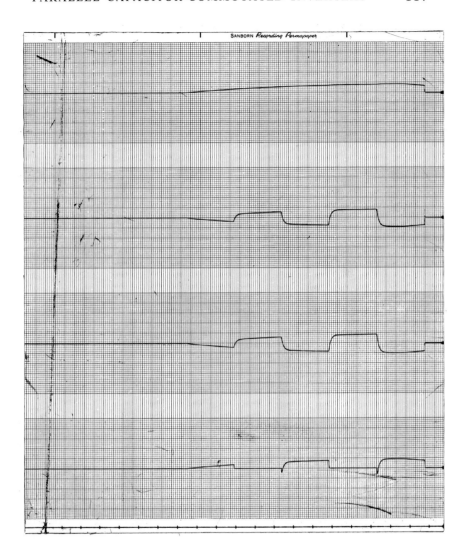

Figure 4.17   (*concluded*)

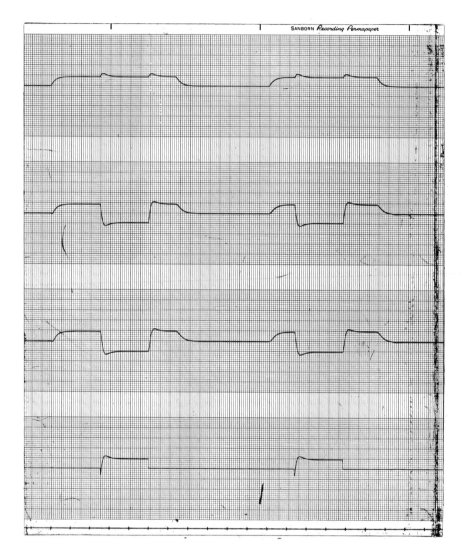

Figure 4.18 Starting transients. $L_d = 0.1$ henries; $C = 0.8$ μfarads; $R = 100\ \Omega$; and $L = 0.1$ henries.

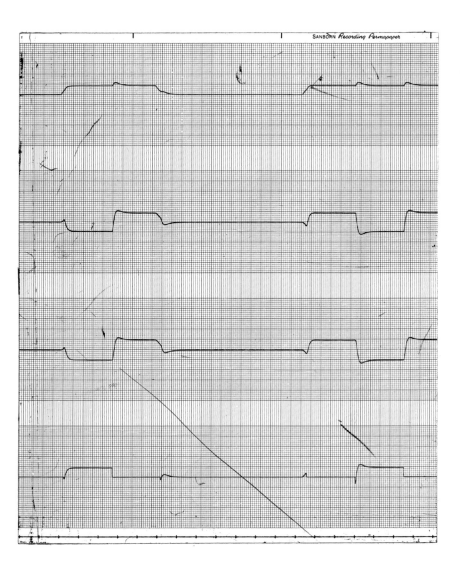

density to prevent it from saturating during start-up. Circuits that generate pulses for SCR gating can be used to provide more reliable starting, as then one SCR cannot be turned on immediately following another. It is also important to note that the circuit of Figure 4.2 must be turned off by disconnecting the d-c power rather than by turning off the gating power. If the gating power is removed first, current will continue to flow through the SCR remaining on, resulting in saturation of the transformer and excessive d-c current.

REFERENCES

1. H. Rissik, *Mercury-Arc Current Converters*, Pitman and Sons, London, 1935.
2. R. R. Benedict, *Introduction to Industrial Electronics*, Prentice-Hall, New York, 1951.
3. C. F. Wagner, "Parallel Inverter with Resistance Load," *AIEE Transactions*, November 1935, pp. 1227–1235.
4. C. F. Wagner, "Parallel Inverter with Inductive Load," *AIEE Transactions*, September 1936, pp. 970–980.
5. C. C. Herskind, "Grid Controlled Rectifiers and Inverters," *AIEE Transactions*, June 1934, pp. 926–935.
6. F. W. Gutzwiller, "Silicon Controlled Rectifier Circuit Including a Variable Frequency Oscillator," U.S. Patent 3,040,270, June 19, 1962.

APPENDIX.   CIRCUIT ANALYSIS OF FIGURE 4.2, CHAPTER 4

For this analysis, the following assumptions are made.

(1) The transformer is an ideal transformer with negligible magnetizing current and negligible winding resistance and reactance. Each half of the primary winding has a number of turns equal to the secondary turns.

(2) The inductance L_d has negligible resistance.

(3) SCR1 and SCR2 are ideal controlled rectifiers with zero forward resistance when on, infinite forward resistance when off, and infinite reverse resistance.

(4) E_d is a pure d-c voltage or ideal battery-type source.

(5) The load contains resistance R and inductance L.

These assumptions simplify the analysis, and the results obtained are very similar to those obtained with practical circuits.

For Figure 4.2, during the time in which SCR1 is conducting and SCR2 is turned off, the following equations may be written.

$$i_1 - i_2 = i_L \tag{4.5}$$

$$i_1 + i_2 = i_d \tag{4.6}$$

$$i_C = i_2 = C \frac{de_C}{dt} \tag{4.7}$$

$$e_C = 2e_L \tag{4.8}$$

$$e_L = Ri_L + L\frac{di_L}{dt} \tag{4.9}$$

$$i_d = \frac{1}{L_d} \int \left(E_d - \frac{e_C}{2}\right) dt \tag{4.10}$$

or

$$i_d = \frac{1}{L_d} \int (E_d - e_L)\, dt$$

Subtracting (4.5) from (4.6),

$$2i_2 = i_d - i_L \tag{4.11}$$

Combining (4.7) and (4.8),

$$i_2 = 2C\frac{de_L}{dt} \tag{4.12}$$

Combining (4.11) and (4.12),

$$i_d - i_L = 4C\frac{de_L}{dt} \tag{4.13}$$

Combining (4.10) and (4.13),

$$\frac{1}{L_d} \int (E_d - e_L)\, dt - i_L = 4C\frac{de_L}{dt} \tag{4.14}$$

Now, by Laplace transforms (4.9) and (4.14) become

$$e_L(s) = Ri_L(s) + sLi_L(s) - i_L(0+) \tag{4.15}$$

$$\frac{E_d}{s^2 L_d} - \frac{e_L(s)}{sL_d} + \frac{i_d(0+)}{s} - i_L(s) = 4sCe_L(s) - 2i_2(0+) \tag{4.16}$$

Combining (4.15) and (4.16),

$$\frac{E_d}{s^2 L_d} - \frac{e_L(s)}{sL_d} + \frac{i_d(0+)}{s} - \frac{e_L(s) + i_L(0+)}{R + sL} = 4sCe_L(s) - 2i_2(0+)$$

$$e_L(s)\left[4sC + \frac{1}{R + sL} + \frac{1}{sL_d}\right] = \frac{E_d}{s^2 L_d} + \frac{i_d(0+)}{s} + 2i_2(0+) - \frac{i_L(0+)}{R + sL}$$

$$e_L(s) = \frac{\dfrac{E_d(R + sL)}{s} + \left[\dfrac{i_d(0+)}{s} + 2i_2(0+) - \dfrac{i_L(0+)}{R + sL}\right](R + sL)sL_d}{4s_L^3 L_d C + 4s^2 RL_d C + s(L_d + L) + R}$$

$$e_L(s) = \frac{\dfrac{E_d\left(s + \dfrac{R}{L}\right)}{4sL_d C} + \left[2i_2(0+) + \dfrac{i_d(0+)}{s} - \dfrac{i_L(0+)}{R + sL}\right]\dfrac{\left(s + \dfrac{R}{L}\right)s}{4C}}{s^3 + \dfrac{R}{L}s^2 + \dfrac{L_d + L}{4LL_d C}s + \dfrac{R}{4LL_d C}} \tag{4.17}$$

The inverse transform of this equation is quite complex so that a special case will be analyzed.

Equation (4.17), *assuming a pure resistance load*, is

$$e_L(s) = \frac{\dfrac{E_d}{4sL_dC} + \left[2i_2(0+) + \dfrac{i_d(0+)}{s}\right]\dfrac{s}{4C}}{s^2 + \dfrac{s}{4RC} + \dfrac{1}{4L_dC}} \tag{4.18}$$

$$e_L(s) = \frac{\dfrac{E_d}{4L_dC} + [2si_2(0+) + i_d(0+)]\dfrac{s}{4C}}{s\left[s^2 + \dfrac{s}{4RC} + \dfrac{1}{4L_dC}\right]}$$

Then $e_L(s)$, *assuming a pure resistance load and a very large* L_d, is

$$e_L(s) = \frac{[2si_2(0+) + I_d]\dfrac{s}{4C}}{s\left[s^2 + \dfrac{s}{4RC}\right]} \tag{4.19}$$

$$e_L(s) = \frac{2si_2(0+) + I_d}{4sC\left(s + \dfrac{1}{4RC}\right)}$$

Equation (4.19) can be rewritten to find the inverse Laplace transform rather simply as follows.

$$e_L(s) = \frac{K_1}{s} + \frac{K_2}{s + \dfrac{1}{4RC}} \tag{4.20}$$

$$K_1 = se_L(s)\Big|_{s=0} = I_dR$$

$$K_2 = \text{unknown as } i_2(0+) \text{ is unknown}$$

The inverse transform is

$$e_L(t) = I_dR + K_2\epsilon^{-t/4RC} \tag{4.21}$$

but it is known that

$$e_L(t)\Big|_{t=0} = -e_L(t)\Big|_{t=T/2} \tag{4.22}$$

$$I_dR + K_2 = -I_dR - K_2\exp\left(-\frac{T/2}{4RC}\right)$$

and,

$$K_2 = \frac{-2I_dR}{1 + \exp\left(-\dfrac{T/2}{4RC}\right)}$$

Then

$$e_L(t) = I_dR\left[1 - \frac{2\exp\left(-\dfrac{t}{4RC}\right)}{1 + \exp\left(-\dfrac{T/2}{4RC}\right)}\right] \qquad (4.23)$$

$$= I_dR\left(1 - \frac{2\epsilon^{-t/\tau}}{1 + \epsilon^{-\frac{T}{2\tau}}}\right) \quad \text{where} \quad \tau = 4RC$$

The actual value of I_d is not known but may be determined as follows. The average voltage across L_d must be zero in steady state so that

$$\frac{1}{T/2}\int_0^{T/2}(E_d - e_L)\,dt = 0 \qquad (4.24)$$

$$\int_0^{T/2}\left[E_d - I_dR\left(1 - \frac{2\epsilon^{-t/\tau}}{1 + \epsilon^{-T/2\tau}}\right)\right]dt = 0$$

$$\left[E_dt - I_dRt + \frac{\dfrac{2I_dR}{-1/\tau}\epsilon^{-t/\tau}}{1 + \epsilon^{-T/2\tau}}\right]_0^{T/2} = 0$$

$$E_dT/2 - I_dRT/2 - \frac{2\tau I_dR\epsilon^{-T/2\tau}}{1 + \epsilon^{-T/2\tau}} + \frac{2\tau I_dR}{1 + \epsilon^{-T/2\tau}} = 0$$

$$I_dR = \frac{E_dT/2}{\dfrac{T}{2} + \dfrac{2\tau\epsilon^{-T/2\tau}}{1 + \epsilon^{-T/2\tau}} - \dfrac{2\tau}{1 + \epsilon^{-T/2\tau}}}$$

$$I_d = \frac{E_d}{R}\frac{T/2(1 + \epsilon^{-T/2\tau})}{T/2(1 + \epsilon^{-T/2\tau}) - 2\tau(1 - \epsilon^{-T/2\tau})}$$

$$I_d = \frac{E_d}{R}\frac{T/2\tau(1 + \epsilon^{-T/2\tau})}{T/2\tau(1 + \epsilon^{-T/2\tau}) - 2(1 - \epsilon^{-T/2\tau})} \qquad (4.25)$$

or

$$I_d = \frac{E_d/R}{1 + \dfrac{16RC}{T}\left(\dfrac{\epsilon^{-T/8RC} - 1}{\epsilon^{-T/8RC} + 1}\right)} \qquad (4.26)$$

Then, substituting (4.25) in (4.23),

$$e_L(t) = \frac{E_d T/2\tau(1 + \epsilon^{-T/2\tau})}{T/2\tau(1 + \epsilon^{-T/2\tau}) - 2(1 - \epsilon^{-T/2\tau})}\left(1 - \frac{2\epsilon^{-t/\tau}}{1 + \epsilon^{-T/2\tau}}\right) \quad (4.27)$$

$$e_L(t) = \frac{E_d T/2\tau(1 + \epsilon^{-T/2\tau} - 2\epsilon^{-t/\tau})}{T/2\tau(1 + \epsilon^{-T/2\tau}) - 2(1 - \epsilon^{-T/2\tau})}$$

$$e_L(t) = \frac{E_d(1 + \epsilon^{-T/2\tau} - 2\epsilon^{-t/\tau})}{1 + \epsilon^{-T/2\tau} - \dfrac{4\tau}{T}(1 - \epsilon^{-T/2\tau})}$$

Equation (4.27) may be developed in another fashion by analyzing the equivalent circuit of Figure 4.4. This is the equivalent circuit during the interval when one SCR is conducting. Assuming a very large d-c reactor, pure resistance load, and neglecting the initial conditions,

$$I_d = \frac{e_L}{R} + 4C\frac{de_L}{dt} \quad (4.28)$$

$$e_L(s) = \frac{I_d/s}{\dfrac{1}{R} + 4sC} = \frac{I_d/4C}{s\left(s + \dfrac{1}{4RC}\right)} \quad (4.29)$$

$$e_L(t) = K_1 + K_2\epsilon^{-t/4RC} \quad (4.30)$$

The conditions defined by equations (4.22) and (4.24) permit solving for K_1 and K_2

$$e_L(t)\Big|_{t=0} = e_L(t)\Big|_{t=T/2} \quad (4.22)$$

$$\frac{1}{T/2}\int_0^{T/2}[E_d - e_L(t)]\,dt = 0 \quad (4.24)$$

Then

$$K_1 + K_2 = -K_1 - K_2\exp\left(\frac{-T/2}{4RC}\right)$$

or

$$K_1 = -\frac{K_2}{2}\left[1 + \exp\left(-\frac{T/2}{4RC}\right)\right] \quad (4.31)$$

and

$$\left[E_d t - K_1 t - K_2(-4RC)\epsilon^{-t/4RC}\right]_0^{T/2} = 0$$

$$(E_d - K_1)T/2 + 4K_2RC\exp\left(-\frac{T/2}{4RC}\right) - 4K_2RC = 0 \quad (4.32)$$

Combining (4.31) and (4.32),

$$E_d T/2 + \frac{K_2 T}{4}\left[1 + \exp\left(-\frac{T/2}{4RC}\right)\right] - 4K_2 RC\left[1 - \exp\left(-\frac{T/2}{4RC}\right)\right] = 0$$

$$K_2 = \frac{-E_d T/2}{T/4[1 + \exp(-T/2\tau)] - \tau[1 - \exp(-T/2\tau)]} \tag{4.33}$$

where $\tau = 4RC$, and

$$K_1 = \frac{E_d T/4(1 + \epsilon^{-T/2\tau})}{T/4(1 + \epsilon^{-T/2\tau}) - \tau(1 - \epsilon^{-T/2\tau})} \tag{4.34}$$

and, again,

$$e_L(t) = \frac{E_d[1 + \epsilon^{-T/2\tau} - 2\epsilon^{-t/\tau}]}{1 + \epsilon^{-T/2\tau} - \frac{4\tau}{T}(1 - \epsilon^{-T/2\tau})} \tag{4.27}$$

From (4.12),

$$i_2 = 2C\frac{de_L}{dt} = \frac{E_d}{R}\frac{\epsilon^{-t/\tau}}{1 + \epsilon^{-T/2\tau} - \frac{4\tau}{T}(1 - \epsilon^{-T/2\tau})} \tag{4.35}$$

From (4.11),

$$i_L = I_d - 2i_2 = \frac{E_d}{R}\frac{(1 + \epsilon^{-T/2\tau})}{1 + \epsilon^{-T/2\tau} - \frac{4\tau}{T}(1 - \epsilon^{-T/2\tau})}$$

$$- 2\frac{E_d}{R}\frac{\epsilon^{-t/\tau}}{1 + \epsilon^{-T/2\tau} - \frac{4\tau}{T}(1 - \epsilon^{-T/2\tau})}$$

$$i_L = \frac{E_d}{R}\frac{1 + \epsilon^{-T/2\tau} - 2\epsilon^{-t/\tau}}{1 + \epsilon^{-T/2\tau} - \frac{4\tau}{T}(1 - \epsilon^{-T/2\tau})} \tag{4.36}$$

From (4.5),

$$i_1 = i_2 + i_L = \frac{E_d}{R}\frac{1 + \epsilon^{-T/2\tau} - \epsilon^{-t/\tau}}{1 + \epsilon^{-T/2\tau} - \frac{4\tau}{T}(1 - \epsilon^{-T/2\tau})} \tag{4.37}$$

Equation (4.27) may be checked in another way by determining the circuit fundamental component power factor as follows. From equation (4.27),

$$e_L(t) = E_d(K_{10} - 2K_{11}\epsilon^{-t/4RC}) \tag{4.40}$$

where

$$K_{10} = \frac{1 + \epsilon^{-T/8RC}}{1 + \epsilon^{-T/8RC} - \frac{16RC}{T}(1 - \epsilon^{-T/8RC})}$$

$$K_{11} = \frac{K_{10}}{1 + \epsilon^{-T/8RC}}$$

The power factor of the fundamental component of $e_L(t)$ is determined as follows.

$$e_L(t)\bigg|_{\substack{\text{fund.}\\\text{comp.}}} = A_1 \sin \omega t + B_1 \cos \omega t \qquad (4.41)$$

$$= \sqrt{A_1{}^2 + B_1{}^2} \sin(\omega t - \theta)$$

where $\tan \theta = -B_1/A_1$.

From Fourier analysis, and since the wave is symmetrical on each half-cycle,

$$A_1 = \frac{4}{T}\int_0^{T/2} E_d(K_{10} - 2K_{11}\epsilon^{-t/4RC}) \sin \omega t \, dt$$

$$= \frac{4E_d}{T}\int_0^{T/2} (K_{10} \sin \omega t - 2K_{11}\epsilon^{-t/4RC} \sin \omega t) \, dt$$

$$= \frac{4E_d}{T}\left[-\frac{K_{10} \cos \omega t}{\omega} - 2K_{11}\epsilon^{-t/4RC} \frac{\left(-\dfrac{1}{4RC}\sin \omega t - \omega \cos \omega t\right)}{\left(\dfrac{1}{4RC}\right)^2 + \omega^2} \right]_0^{T/2}$$

$$= \frac{4E_d}{T}\left[\frac{2K_{10}}{\omega} - \frac{2K_{11}\epsilon^{-T/8RC}(\omega)}{\left(\dfrac{1}{4RC}\right)^2 + \omega^2} + \frac{2K_{11}(-\omega)}{\left(\dfrac{1}{4RC}\right)^2 + \omega^2} \right]$$

$$A_1 = \frac{4E_d}{T}\left[\frac{2K_{10}}{\omega} - \frac{2K_{11}\omega}{\omega^2 + \left(\dfrac{1}{4RC}\right)^2}(\epsilon^{-T/8RC} + 1) \right] \qquad (4.42)$$

$$B_1 = \frac{4}{T}\int_0^{T/2} E_d(K_{10} - 2K_{11}\epsilon^{-t/4RC}) \cos \omega t \, dt$$

$$= \frac{4E_d}{T}\left[\frac{K_{10} \sin \omega t}{\omega} - 2K_{11}\epsilon^{-t/4RC} \frac{\left(-\dfrac{1}{4RC}\cos \omega t + \omega \sin \omega t\right)}{\omega^2 + \left(\dfrac{1}{4RC}\right)^2} \right]_0^{T/2}$$

$$= \frac{4E_d}{T}\left[-\frac{2K_{11}\epsilon^{-T/8RC}\left(\dfrac{1}{4RC}\right)}{\omega^2 + \left(\dfrac{1}{4RC}\right)^2} + \frac{2K_{11}\left(-\dfrac{1}{4RC}\right)}{\omega^2 + \left(\dfrac{1}{4RC}\right)^2} \right]$$

$$= -\frac{8E_d K_{11}}{4RCT\left[\omega^2 + \left(\dfrac{1}{4RC}\right)^2\right]}(\epsilon^{-T/8RC} + 1) \qquad (4.43)$$

From (4.42) and (4.43),

$$\tan \theta = \frac{-B_1}{A_1} = \frac{\dfrac{2K_{11}}{4RC\left[\omega^2 + \left(\dfrac{1}{4RC}\right)^2\right]}(\epsilon^{-T/8RC} + 1)}{\dfrac{2K_{10}}{\omega} - \dfrac{2K_{11}\omega}{\omega^2 + \left(\dfrac{1}{4RC}\right)^2}(\epsilon^{-T/8RC} + 1)}$$

$$\tan \theta = \frac{\dfrac{1}{4RC}(\epsilon^{-T/8RC} + 1)}{\dfrac{K_{10}}{K_{11}\omega}\left[\omega^2 + \left(\dfrac{1}{4RC}\right)^2\right] - \omega(\epsilon^{-T/8RC} + 1)}$$

From (4.40),

$$\frac{K_{10}}{K_{11}} = \epsilon^{-T/8RC} + 1$$

Thus,

$$\tan \theta = \frac{\dfrac{\omega}{4RC}}{\omega^2 + \left(\dfrac{1}{4RC}\right)^2 - \omega^2} = 4\omega RC \qquad (4.44)$$

This checks with equation (4.48) in this chapter.

Chapter Five

Series Capacitor-Commutated Inverters

The series capacitor-commutated inverter generally involves a series L-C resonant circuit to provide commutation. In the most basic inverter, when a controlled rectifier device is turned "on," an oscillatory pulse of current flows. The resonant frequency of the circuit determines the duration of the damped sinusoidal pulse of current through the controlled rectifying element in series with the resonant circuit and the load.

The operation of the basic series capacitor-commutated inverter is described in this chapter, including a mathematical analysis to determine the steady-state circuit voltages and currents as a function of time. Next, several modified single-phase circuits are discussed. These circuits extend the operating range of the basic series inverter. Additional single-phase and polyphase circuits are then presented to illustrate alternate versions of the series capacitor-commutated inverter.

The simple series inverter can produce a very nearly sinusoidal output voltage waveform when supplying a relatively fixed load. When the load current is increased, the amplitude of the voltage across the capacitor becomes larger. Feedback circuits can be added to the simple series inverter to alleviate this problem and increase the practical operating load range.

In general, the series inverter can be turned off more easily than the parallel inverter, as it is not necessary to turn one SCR "on" to turn the previously conducting SCR "off." The SCR gating excitation may be interrupted to stop the series inverter. This eliminates the need for power switching in certain applications since the inverter may be turned on and off by properly switching the gate circuits, although care must be taken to avoid undesirable starting transients.

128

5.1 BASIC SERIES CAPACITOR-COMMUTATED INVERTER[1-3]

The basic series capacitor-commutated inverter arrangement is shown in Figure 5.1. The term "series capacitor inverter" is used to indicate an inverter that is commutated by a capacitor in series with the load.

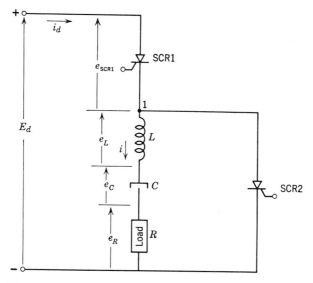

Figure 5.1 Basic series capacitor-commutated inverter.

In Figure 5.1, when SCR1 is gated on, with SCR2 off, essentially a series resonant circuit is connected to the d-c supply. If there is no loss in the system, the condition for $R = 0$, the capacitor voltage sinusoidally builds up to $2E_d$ during the first half-cycle of oscillation and then conduction stops, since the current drops below the SCR holding current. The current is a sinusoidal pulse reaching a maximum when the capacitor is charged to E_d and going to zero when the final voltage of $2E_d$ appears across the capacitor.

On the second half-cycle of inverter operation, when SCR2 is closed, the equivalent circuit is again the same series resonant circuit, now with no d-c supply voltage, but with an initial voltage on the capacitor of $2E_d$. The current is again a sinusoidal pulse, and the capacitor charges to a final voltage of negative $2E_d$, assuming $R = 0$. When SCR1 is turned on for the second time, the conditions are the same as for the first half-cycle except that the initial capacitor voltage is negative $2E_d$ rather than zero. At the end of the third half-cycle, the capacitor voltage is positive $4E_d$, assuming negligible loss. This operation continues on subsequent half-cycles and thus, for the condition of $R = 0$, the capacitor voltage would continually increase on successive cycles of inverter operation. In the practical case, the resistance in the circuit and/or the load resistance provide loss so as to prevent continual buildup of voltage across the capacitor. The ratio of the inductive reactance X_L, or the capacitive reactance X_C, to the resistance R is the Q of the circuit. The fundamental component of the capacitor voltage builds up exponentially to a steady-state value with a time constant of Q/π cycles when the circuit is operated at its resonant frequency.

An important feature of the circuit in Figure 5.1 is that it will only operate properly when the current is zero for at least the recovery time of the SCR's at the end of each half-cycle of operation. This condition is required to enable the SCR's to regain their forward blocking capability at the end of each conducting half-cycle.

5.2   CIRCUIT ANALYSIS

Figure 5.1 may be analyzed by solving the differential equations describing the circuit operation. This analysis is shown in the appendix to this chapter. For steady-state operation, assuming the current i reaches zero exactly at the end of each half-cycle of the inverter operating frequency, the equations are as follows.

During the interval when SCR1 is conducting and SCR2 is off ($0 \leq t \leq T/2$)

$$i(t) = \frac{\dfrac{E_d}{L}}{\left[1 - \exp\left(-\dfrac{R}{2L}\dfrac{T}{2}\right)\right]\left(\dfrac{1}{LC} - \dfrac{R^2}{4L^2}\right)^{1/2}} \epsilon^{-(R/2L)t} \times \sin\sqrt{\frac{1}{LC} - \frac{R^2}{4L^2}}\,t \quad (5.18)^*$$

$$e_L(t) = \frac{E_d}{\left[1 - \exp\left(-\dfrac{R}{2L}\dfrac{T}{2}\right)\right]} \epsilon^{-(R/2L)t} \times \left[\cos\sqrt{\frac{1}{LC} - \frac{R^2}{4L^2}}\,t - \frac{\sin\sqrt{1/LC - R^2/4L^2}\,t}{\dfrac{2L}{R}\sqrt{1/LC - R^2/4L^2}}\right] \quad (5.19)$$

---

* The equation numbers that do not appear in this chapter are used in the appendix for the equations involved in the analysis of Figure 5.1.

$$e_C(t) = E_d - \frac{E_d}{\left[1 - \exp\left(-\frac{R}{2L}\frac{T}{2}\right)\right]}\epsilon^{-(R/2L)t}$$

$$\times \left[\cos\sqrt{\frac{1}{LC} - \frac{R^2}{4L^2}}\,t + \frac{\sin\sqrt{1/LC - R^2/4L^2}\,t}{\frac{2L}{R}\sqrt{1/LC - R^2/4L^2}}\right] \quad (5.20)$$

During the interval when SCR2 is conducting and SCR1 is off $(T/2 \leq t \leq T)$,

$$i(t) = \frac{-\dfrac{E_d}{L}}{\left[1 - \exp\left(-\dfrac{R}{2L}\dfrac{T}{2}\right)\right]\left(\dfrac{1}{LC} - \dfrac{R^2}{4L^2}\right)^{1/2}}$$

$$\times \left\{\exp\left[-\frac{R}{2L}\left(t - \frac{T}{2}\right)\right]\right\}\sin\left(\frac{1}{LC} - \frac{R^2}{4L^2}\right)^{1/2}\left(t - \frac{T}{2}\right) \quad (5.21)$$

$$e_L(t) = \frac{-E_d}{\left[1 - \exp\left(-\dfrac{R}{2L}\dfrac{T}{2}\right)\right]}\exp\left[-\frac{R}{2L}\left(t - \frac{T}{2}\right)\right]$$

$$\times \left[\cos\sqrt{\frac{1}{LC} - \frac{R^2}{4L^2}}\left(t - \frac{T}{2}\right) - \frac{\sin\sqrt{1/LC - R^2/4L^2}(t - T/2)}{(2L/R)\sqrt{1/LC - R^2/4L^2}}\right] \quad (5.22)$$

$$e_C(t) = \frac{E_d}{\left[1 - \exp\left(-\dfrac{R}{2L}\dfrac{T}{2}\right)\right]}\exp\left[-\frac{R}{2L}\left(t - \frac{T}{2}\right)\right]$$

$$\times \left[\cos\sqrt{\frac{1}{LC} - \frac{R^2}{4L^2}}\left(t - \frac{T}{2}\right) + \frac{\sin\sqrt{1/LC - R^2/4L^2}(t - T/2)}{(2L/R)\sqrt{1/LC - R^2/4L^2}}\right] \quad (5.23)$$

Figure 5.2 shows the form of the waveshape of voltage e_C for various values of $2L/R$, assuming that $\sqrt{1/LC - R^2/4L^2}$ is held constant and equal to the inverter operating frequency. When the circuit resistance R is small, the resonant frequency is not appreciably affected when this resistance is changed. For these conditions $R^2/4L^2 \ll 1/LC$ so that the circuit resonant frequency $\sqrt{1/LC - R^2/4L^2}$ is approximately $\sqrt{1/LC}$. Thus, in Figure 5.2, for the greater amplitude waveforms which result when R is small, essentially all other circuit parameters remain constant as resistance R is varied to change from one waveform to the next.

When resistance R is large or critical damping is approached, as R is varied with a fixed L, it is necessary to change C considerably to maintain the circuit resonant frequency constant. This is the case for the lower amplitude curves in Figure 5.2.

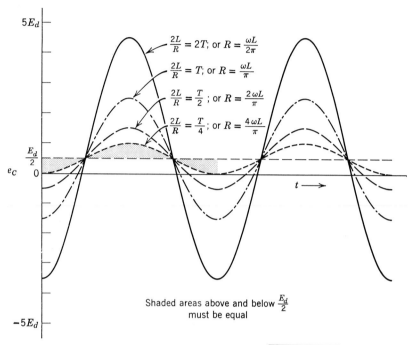

$$\frac{2L}{R} = 2T; \text{ or } R = \frac{\omega L}{2\pi}$$

$$\frac{2L}{R} = T; \text{ or } R = \frac{\omega L}{\pi}$$

$$\frac{2L}{R} = \frac{T}{2}; \text{ or } R = \frac{2\omega L}{\pi}$$

$$\frac{2L}{R} = \frac{T}{4}; \text{ or } R = \frac{4\omega L}{\pi}$$

Shaded areas above and below $\frac{E_d}{2}$ must be equal

Figure 5.2 e_C Waveforms for Figure 5.1, assuming $\sqrt{1/LC - R^2/4L^2}$ is a constant and equal to the inverter frequency.

For the case of critical damping, R is equal to $2X_L$ or $2X_C$, which may be shown as follows. With critical damping,

$$\frac{R^2}{4L^2} = \frac{1}{LC} \tag{5.31}$$

or

$$R = 2L\sqrt{\frac{1}{LC}} = 2\omega_0 L = 2X_L \tag{5.32}$$

where $\omega_0 = 1/\sqrt{LC}$ = circuit resonant frequency with $R^2/4L^2 \ll 1/LC$. Also, since

$$L = \frac{1}{\omega_0^2 C} \tag{5.33}$$

$$R = (2\omega_0)\left(\frac{1}{\omega_0^2 C}\right) = \frac{2}{\omega_0 C} = 2X_C \tag{5.34}$$

Figure 5.3 shows the important waveforms for the case in Figure 5.2 where $2L/R = T/2$ or $R = 2\omega L/\pi$.

In general, for a practical case, it is quite simple to carry out an approximate analysis of the circuit in Figure 5.2 without solving the circuit differential equations. This analysis is based on the fundamental principles relating to the operation of the circuit. SCR1 and SCR2 are assumed to

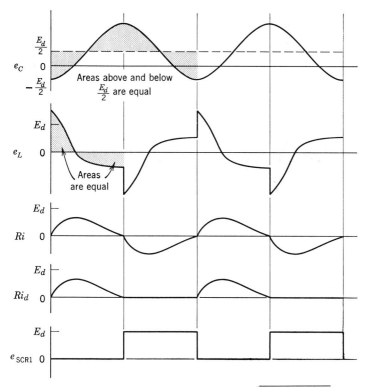

Figure 5.3   Waveforms for Figure 5.1, assuming $\sqrt{1/LC - R^2/4L^2}$ is a constant equal to the inverter frequency, and assuming $2L/R = T/2$ or $R = 2\omega L/\pi$.

alternately conduct for equal intervals of time. The voltage from point 1 to the negative d-c in Figure 5.2 has a magnitude E_d when SCR1 is conducting and zero during the conducting interval of SCR2. Thus, the voltage applied to the series L-R-C combination has a d-c component of $E_d/2$ and the a-c components of a square wave voltage with amplitude $E_d/2$. For steady-state operation, the average value of the applied voltage must appear across the capacitor, since no average current can flow through the capacitor (requiring that no average voltage appear across R) and no average voltage can appear across the inductor. In addition, for the case where the current i reaches zero precisely at the end of each half-cycle, there can be no average voltage across the inductor during *each* half-cycle. The a-c voltage across the load R may be determined by calculating the fraction of the fundamental and each harmonic in the applied square wave that appears across R.

The maximum value of the fundamental component of the applied square wave voltage with amplitude $E_d/2$ is

$$A_1 = \frac{4}{T} \int_0^{T/2} \frac{E_d}{2} \sin \omega t\, dt = \frac{2E_d}{\pi} \tag{5.35}$$

The effective value of the fundamental component of the voltage across R is

$$E_{R,e1} = \frac{\sqrt{2}}{\pi} E_d \frac{R}{\sqrt{R^2 + (\omega L - 1/\omega C)^2}}$$

$$= \frac{\sqrt{2} E_d}{\pi} \frac{1}{\sqrt{1 + \left(\dfrac{\omega L - 1/\omega C}{R}\right)^2}} \tag{5.36}$$

A similar analysis may be carried out for each harmonic, giving

$$E_{R,e(n)} = \frac{\sqrt{2} E_d}{n\pi} \frac{1}{\sqrt{1 + \left(\dfrac{n\omega L - 1/n\omega C}{R}\right)^2}} \tag{5.37}$$

where $n = 1, 3, 5, \ldots$ and $E_{R,e(n)}$ is the effective value of the particular harmonic voltage across the load R.

In a practical case, the circuit in Figure 5.2 is generally used to produce reasonably sinusoidal output voltage. For this mode of operation, X_L and X_C are essentially equal and, normally, several times R. Thus, the operating frequency is approximately $1/\sqrt{LC}$ at this resonant condition, and the current waveshape is also fairly close to a sine wave. The combined impedance of X_L and X_C is approximately zero. Therefore, all of the fundamental component in the applied square wave voltage, with magnitude $E_d/2$, is impressed on the load resistance. The load current is the fundamental component of the applied voltage divided by resistance R. The a-c voltage on X_C or X_L can be determined by multiplying this current times the reactance of X_C or X_L. This approximate analysis is a very simple and accurate way of getting results for many practical circuits. In general, this method of analysis will produce very little error for circuits with high Q.

The circuit in Figure 5.1 may also operate in the mode where the circuit resonant frequency $\sqrt{1/LC - R^2/4L^2}$ is greater than the inverter operating

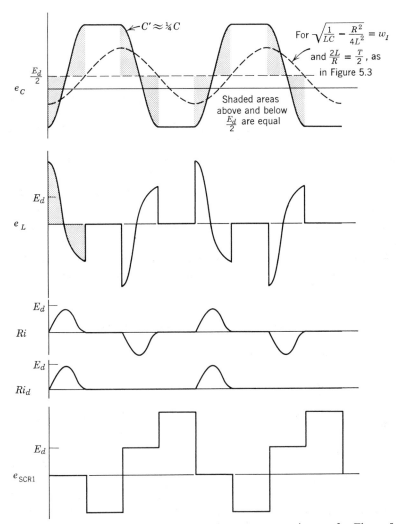

Figure 5.4   Waveforms for Figure 5.1 with the same assumptions as for Figure 5.3, except one-fourth the capacitance.

frequency. For this mode of operation, the current i will reach zero and remain at zero for some time before the next SCR to conduct is turned on. As indicated in the appendix, equations of the same form as (5.18) through (5.23) describe the circuit operation while current is flowing. After the current reaches zero, the capacitor voltage will remain at a fixed value until the beginning of the next SCR-conducting interval. Figure 5.4 shows the approximate waveforms with one-fourth the capacitance but all other constants the same as in Figure 5.3. The circuit resonant frequency has been increased to approximately twice the inverter operating frequency. This operation is very similar to that shown in Figures 5.2 and 5.3 except there is an off-time or holding interval during each half-cycle when neither SCR is conducting.

Circuit operation similar to that shown in Figure 5.4 would also result if the inverter frequency were reduced while retaining all other circuit constants which produced the waveforms in Figure 5.3. In fact, with the lower inverter frequency, the waveforms would be the same as in Figure 5.3 except with a holding interval on each half-cycle during which the capacitor voltage remained constant, similar to the condition in Figure 5.4.

The approximate analysis may also be used when the inverter frequency is less than $\sqrt{1/LC - R^2/4L^2}$ during the periods when either SCR is conducting. The actual rms currents and voltages must be calculated, including the effect of the holding intervals during each half-cycle.

As mentioned previously, the circuit in Figure 5.1. will operate properly only in the two modes described where the current i oscillates to zero before the next SCR to conduct is turned on.

5.3   MODIFICATIONS TO BASIC SERIES INVERTER

The performance of the circuit of Figure 5.1 can be improved by incorporating the modifications shown in Figure 5.5. In Figure 5.5(a), the inductor has been divided into two equal parts which are closely coupled on one core. The important change in the operation of this modified circuit is that now one SCR may be turned before the current in the other SCR oscillates to zero. When an SCR is turned on, a voltage is induced in the coupled winding of the reactor which reverses the voltage on the other SCR causing it to stop conducting. The operating range of the series inverter has, therefore, been extended to include operation when the inverter operating frequency is somewhat greater than the resonant frequency of the series L-R-C circuit. This increased operating range is obtained by an increase in the reactor size and complexity, as only one-half of the center-tapped reactor conducts current each half-cycle. The peak voltage on each SCR is also increased in Figure 5.5(a) over that in Figure 5.1.

In Figure 5.5(b) the capacitor is divided into two equal parts.[4] During steady-state operation, on the half-cycle when SCR1 is conducting, capacitor C_2 is charged by current from the d-c source and capacitor C_1 is discharged through the load. Both the charging and discharging currents flow through SCR1, L_1, and the load. During the next half-cycle when SCR2 is conducting, the roles of the two capacitors are reversed with both

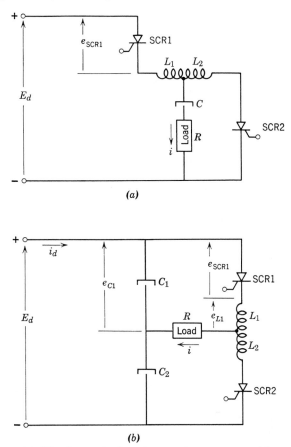

Figure 5.5   Modifications to basic series capacitor-commutated inverter.

currents flowing through the load in the opposite direction. On each half-cycle, one-half of the load current is obtained from the d-c source and one-half is obtained from a capacitor being discharged. If $C_1 = C_2 = C/2$, $L_1 = L_2 = L$, and E_d and R are equal for Figures 5.1 and 5.5, the average current drawn from the source on each half-cycle in Figure 5.5(b) is one-half the average d-c source current flowing during the half-cycles when SCR1 is "on" in Figures 5.1 or 5.5(a). The principal advantage of the circuit in Figure 5.5(b) over Figure 5.5(a) is that an equal amount of current is drawn from the source on both half-cycles resulting in a con-siderable reduction in the d-c source ripple current.

Figure 5.5(b) is similar to Figure 5.5(a) in that one SCR can be turned on before the current in the other SCR has oscillated to zero. Figure 5.6 shows typical waveforms to illustrate this operating mode for a practical

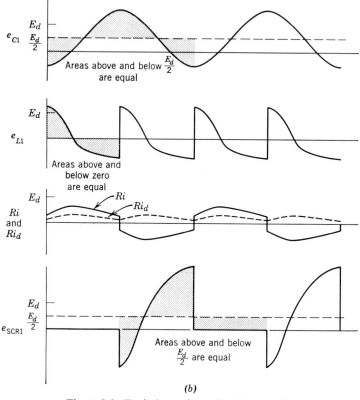

(b)

Figure 5.6   Typical waveforms for Figure 5.5(*b*).

version of the circuit in Figure 5.5(*b*). The large negative voltage on the SCR is obtained by the action of the center-tapped reactor. This provides improved commutation action, since there is a negative voltage across an SCR for an appreciable time interval at the beginning of each "off" half-cycle. Later, in the half-cycle, when the reactor voltage reverses, the center-tapped reactor voltage adds to the SCR positive voltage. The d-c input current flows during both half-cycles, and it is one-half the magnitude of the average load current.

5.4   OTHER SERIES CAPACITOR INVERTER CIRCUITS

Figure 5.7 is a circuit more suitable for low d-c source voltages than the circuits in Figures 5.1 or 5.5. The average d-c source current is equal to the sum of the average currents in the two SCR's. For the following discussion, it is assumed that the output transformer has a one-to-one-turns ratio from each primary to the load winding. The instantaneous load current is therefore equal to the d-c current. The center-tapped transformer connected to the capacitor causes one-half the input current to flow in the capacitor. Assuming ideal transformers, the complete circuit is equivalent to a series *L-R-C* circuit when one SCR is conducting. When the other SCR is conducting, the current is reversed in both the

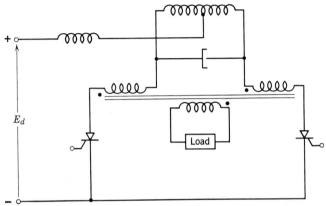

Figure 5.7   Series capacitor-commutated inverter for low d-c input voltage.

capacitor and the load. The circuit operates very much like Figure 5.5(*b*) in that the d-c source supplies power on both half-cycles. In addition, one SCR can be gated on before zero current is obtained in the other SCR. The principal differences between the circuits of Figures 5.7 and 5.5(*b*) are: (1) the average load voltage on a half-cycle, assuming the unity turns ratio output transformer, is nearly equal to the d-c source voltage for Figure 5.7; and (2) the average load current on each half-cycle is equal to the average d-c source current in Figure 5.7.

Figure 5.8 is a polyphase version of the series inverter. This circuit is most readily analyzed if the capacitors across the d-c are large so as to establish a constant voltage neutral. For this case, the capacitors in series with the load resonate with the center-tapped reactors to provide series capacitor-type commutation.

The circuits shown up to this point illustrate the basic types of series capacitor-commutated inverters. Many variations of each of these circuits are possible. Figure 5.9 illustrates one type of an important family of variations. This is a modification of Figure 5.5(*b*) to permit operation over a wider load range. The diode rectifiers provide clamping or limiting so that the junction point between the capacitors cannot swing more positive than the upper d-c bus or more negative than the lower d-c bus. This

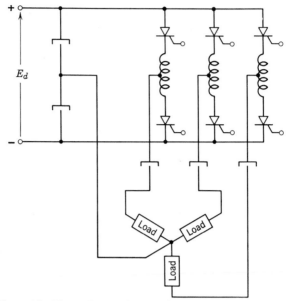

Figure 5.8   Three-phase series capacitor-commutated inverter.

permits operation over a wider load range without producing excessive voltage on the circuit components. In general, it may be more desirable not to limit the voltage swing as rigidly as in Figure 5.9 either by inserting resistance in series with the diode rectifiers or by inserting a small capacitor directly in series with the load. Several other similar clamping or limiting techniques can also be used to extend the operating load range of Figure 5.5(b); for example, the rectification of the load or reactor voltages and the feeding of the rectified power back to the d-c source.

An important point to re-emphasize is that essentially all rectifier circuits can operate as inverters with all types of commutation. This is

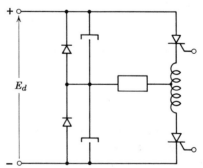

Figure 5.9   Series capacitor-commutated inverter with capacitor voltage limiting.

illustrated in the circuit of Figure 5.10. This is a similar power circuit to that used in the full-wave single-way rectifier or the parallel capacitor-commutated inverter shown in Figure 4.2 (Chapter 4). The circuit of Figure 5.10 operates as a series capacitor-commutated inverter. For low voltage d-c sources, it is particularly advantageous to have the capacitor in the transformer secondary, as in Figure 5.10, assuming that a step-up transformer is used. However, the transformer leakage reactance must be low to provide proper SCR commutation for this arrangement.

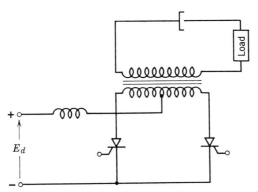

Figure 5.10   Full-wave single-way series capacitor-commutated inverter.

In any given inverter arrangement, it is also possible to use a combination of series and parallel capacitor commutation. Thus, there are not only many variations of each type inverter, but there are many possible "hybrid" circuits where commutation is accomplished by a combination of series and parallel capacitor-commutation techniques.

REFERENCES

1. C. A. Sabbah, "Series Parallel Static Converters," Part I, *General Electric Review*, May 1931, pp. 288–301.
2. C. A. Sabbah, "Series Parallel Static Converters," Part II, *General Electric Review*, October 1931, pp. 580–589.
3. C. A. Sabbah, "Series Parallel Static Converters," Part III, *General Electric Review*, December 1931, pp. 738–744.
4. H. R. Lowry, "Inverter Circuit," U.S. Patent 3,047,789, July 31, 1962.

APPENDIX.   CIRCUIT ANALYSIS OF FIGURE 5.1, CHAPTER 5

An analysis of the circuit in Figure 5.1 may be carried out as follows.

During the Interval When SCR1 Is On and SCR2 Is Off,
$(0 \le t \le T/2)$

$$E_d = Ri + L\frac{di}{dt} + \frac{1}{C}\int i\, dt \tag{5.1}$$

$$\frac{E_d}{s} = \left(R + sL + \frac{1}{sC}\right)i(s) - Li(0+) + \frac{e_C(0+)}{s}$$

$$i(s) = \frac{\dfrac{E_d}{s} + Li(0+) - \dfrac{e_C(0+)}{s}}{R + sL + \dfrac{1}{sC}}$$

$$i(s) = \frac{\dfrac{E_d - e_C(0+)}{L} + si(0+)}{s^2 + \dfrac{R}{L}s + \dfrac{1}{LC}} \tag{5.2}$$

For the circuit of Figure 5.1 to operate properly, the current must be zero prior to the end of each SCR conducting interval and therefore $i(0+)$ will be assumed zero. Then, from equation (5.2),

$$i(s) = \frac{\dfrac{E_d - e_C(0+)}{L}}{\left(s + \dfrac{R}{2L}\right)^2 + \left(\dfrac{1}{LC} - \dfrac{R^2}{4L^2}\right)} \tag{5.3}$$

$$i(t) = \frac{\dfrac{E_d - e_C(0+)}{L}}{\sqrt{1/LC - R^2/4L^2}}\,\epsilon^{-(R/2L)t}\sin\sqrt{\frac{1}{LC} - \frac{R^2}{4L^2}}\,t \tag{5.4}$$

when $R \to 0$,

$$i(t) = \frac{\dfrac{E_d - e_C(0+)}{L}}{\sqrt{1/LC}}\sin\sqrt{\frac{1}{LC}}\,t \tag{5.5}$$

which is the familiar expression for the sinusoidal oscillation of a series resonant circuit in response to an applied d-c voltage step. The voltage across the inductance L is

$$e_L(t) = L\frac{di}{dt} \tag{5.6}$$

From (5.4),

$$e_L(t) = \frac{E_d - e_C(0+)}{\sqrt{1/LC - R^2/4L^2}} \left[-\frac{R}{2L} \epsilon^{-(R/2L)t} \sin\sqrt{\frac{1}{LC} - \frac{R^2}{4L^2}}\, t \right.$$

$$\left. + \sqrt{\frac{1}{LC} - \frac{R^2}{4L^2}}\, \epsilon^{-(R/2L)t} \cos\sqrt{\frac{1}{LC} - \frac{R^2}{4L^2}}\, t \right]$$

$$e_L(t) = [E_d - e_C(0+)]\epsilon^{-(R/2L)t}$$

$$\times \left[\cos\sqrt{\frac{1}{LC} - \frac{R^2}{4L^2}}\, t - \frac{\sin\sqrt{1/LC - R^2/4L^2}\, t}{\sqrt{4L/R^2C - 1}} \right] \quad (5.7)$$

The voltage across the capacitor C is

$$e_C(t) = E_d - e_L - e_R \quad (5.8)$$

and, combining (5.4), (5.7), and (5.8),

$$e_C(t) = E_d - [E_d - e_C(0+)]\epsilon^{-(R/2L)t}$$

$$\times \left[\cos\sqrt{\frac{1}{LC} - \frac{R^2}{4L^2}}\, t - \frac{\sin\sqrt{1/LC - R^2/4L^2}\, t}{\sqrt{4L/R^2C - 1}} \right]$$

$$- \frac{2[E_d - e_C(0+)]}{\sqrt{4L/R^2C - 1}}\, \epsilon^{-(R/2L)t} \sin\sqrt{\frac{1}{LC} - \frac{R^2}{4L^2}}\, t$$

$$e_C(t) = E_d - [E_d - e_C(0+)]\epsilon^{-(R/2L)t}$$

$$\times \left[\cos\sqrt{\frac{1}{LC} - \frac{R^2}{4L^2}}\, t + \frac{\sin\sqrt{1/LC - R^2/4L^2}\, t}{\sqrt{4L/R^2C - 1}} \right] \quad (5.9)$$

During the Interval When SCR2 Is On and SCR1 Is Off,
$(T/2 \leq t \leq T)$.

From equations (5.4), (5.7), and (5.9),

$$i(t) = \frac{\dfrac{-e_C(T/2+)}{L}}{\sqrt{1/LC - R^2/4L^2}}\, \epsilon^{-(R/2L)(t-T/2)} \sin\sqrt{\frac{1}{LC} - \frac{R^2}{4L^2}}\left(t - \frac{T}{2} \right) \quad (5.10)$$

$$e_L(t) = -e_C\left(\frac{T}{2} + \right) \exp\left[-\frac{R}{2L}\left(t - \frac{T}{2} \right) \right]$$

$$\times \left[\cos\sqrt{\frac{1}{LC} - \frac{R^2}{4L^2}}\left(t - \frac{T}{2} \right) - \frac{\sin\sqrt{1/LC - R^2/4L^2}(t - T/2)}{\sqrt{4L/R^2C - 1}} \right] \quad (5.11)$$

$$e_C(t) = e_C\left(\frac{T}{2} +\right) \exp\left[-\frac{R}{2L}\left(t - \frac{T}{2}\right)\right]$$

$$\times \left[\cos\sqrt{\frac{1}{LC} - \frac{R^2}{4L^2}}\left(t - \frac{T}{2}\right) + \frac{\sin\sqrt{1/LC - R^2/4L^2}(t - T/2)}{\sqrt{4L/R^2C - 1}}\right] \quad (5.12)$$

Steady-State Operation Assuming i(t) Reaches Zero Exactly at the End of Each Half-Cycle of the Inverter Operating Frequency

When operating as in the above heading, the inverter operating frequency is

$$\omega = \sqrt{\frac{1}{LC} - \frac{R^2}{4L^2}} = \frac{2\pi}{T} \quad (5.13)$$

and from equation (5.9),

$$e_C(t)\Big|_{t=T/2} = E_d + [E_d - e_C(0+)] \exp\left(-\frac{R}{2L}\frac{T}{2}\right) = e_C\left(\frac{T}{2} +\right) \quad (5.14)$$

and from equation (5.12),

$$e_C(t)\Big|_{t=T} = e_C(0+) = -e_C\left(\frac{T}{2} +\right) \exp\left(-\frac{R}{2L}\frac{T}{2}\right) \quad (5.15)$$

Combining (5.14) and (5.15),

$$E_d + [E_d - e_C(0+)] \exp\left(-\frac{R}{2L}\frac{T}{2}\right) = \frac{-e_C(0+)}{\exp\left(-\frac{R}{2L}\frac{T}{2}\right)}$$

$$E_d\left[1 + \exp\left(-\frac{R}{2L}\frac{T}{2}\right)\right] = e_C(0+)\left[\exp\left(-\frac{R}{2L}\frac{T}{2}\right) - \exp\left(+\frac{R}{2L}\frac{T}{2}\right)\right]$$

$$e_C(0+) = \frac{E_d\left[1 + \exp\left(-\frac{R}{2L}\frac{T}{2}\right)\right]}{\exp\left(-\frac{R}{2L}\frac{T}{2}\right) - \exp\left(-\frac{R}{2L}\frac{T}{2}\right)}$$

$$e_C(0+) = \frac{E_d\left[1 + \exp\left(+\frac{R}{2L}\frac{T}{2}\right)\right]}{1 - \exp\left(+\frac{R}{L}\frac{T}{2}\right)} \quad (5.16)$$

and from (5.15)

$$e_C\left(\frac{T}{2}+\right) = -\frac{e_C(0+)}{\exp\left(-\frac{R}{2L}\frac{T}{2}\right)}$$

$$e_C\left(\frac{T}{2}+\right) = \frac{E_d\left[1+\exp\left(-\frac{R}{2L}\frac{T}{2}\right)\right]}{1-\exp\left(-\frac{R}{L}\frac{T}{2}\right)} \tag{5.17}$$

Equations (5.4), (5.7), (5.9), (5.10), (5.11), and (5.12), for the steady-state condition, become

$$0 \leq t \leq T/2$$

$$i(t) = \frac{E_d - \dfrac{E_d[1+\epsilon^{(R/2L)T/2}]}{1-\epsilon^{(R/L)T/2}}}{L\sqrt{1/LC - R^2/4L^2}} \epsilon^{-(R/2L)t}\sin\sqrt{\frac{1}{LC}-\frac{R^2}{4L^2}}\,t$$

$$i(t) = \frac{E_d \dfrac{1+\epsilon^{-(R/2L)T/2}}{1-\epsilon^{-(R/L)T/2}}}{L\sqrt{1/LC - R^2/4L^2}} \epsilon^{-(R/2L)t}\sin\sqrt{\frac{1}{LC}-\frac{R^2}{4L^2}}\,t \tag{5.18}$$

$$e_L(t) = \frac{E_d[1+\epsilon^{-(R/2L)T/2}]}{1-\epsilon^{-(R/L)T/2}}\epsilon^{-(R/2L)t}$$
$$\times\left[\cos\sqrt{\frac{1}{LC}-\frac{R^2}{4L^2}}\,t - \frac{\sin\sqrt{1/LC - R^2/4L^2}\,t}{\sqrt{4L/R^2C - 1}}\right] \tag{5.19}$$

$$e_C(t) = E_d - \frac{E_d[1+\epsilon^{-(R/2L)T/2}]}{1-\epsilon^{-(R/L)T/2}}\epsilon^{-(R/2L)t}$$
$$\times\left[\cos\sqrt{\frac{1}{LC}-\frac{R^2}{4L^2}}\,t + \frac{\sin\sqrt{1/LC - R^2/4L^2}\,t}{\sqrt{4L/R^2C - 1}}\right] \tag{5.20}$$

$$T/2 \leq t \leq T$$

$$i(t) = \frac{\dfrac{-E_d[1+\epsilon^{-(R/2L)T/2}]}{1-\epsilon^{-(R/L)T/2}}}{L\sqrt{1/LC - R^2/4L^2}}$$
$$\times\exp\left[-\frac{R}{2L}\left(t-\frac{T}{2}\right)\right]\sin\sqrt{\frac{1}{LC}-\frac{R^2}{4L^2}}\left(t-\frac{T}{2}\right) \tag{5.21}$$

$$e_L(t) = \frac{-E_d[1 + \epsilon^{-(R/2L)T/2}]}{1 - \epsilon^{-(R/L)T/2}} \exp\left[-\frac{R}{2L}\left(t - \frac{T}{2}\right)\right]$$

$$\times \left[\cos\sqrt{\frac{1}{LC} - \frac{R^2}{4L^2}}\left(t - \frac{T}{2}\right) - \frac{-\sin\sqrt{1/LC - R^2/4L^2}\,(t - T/2)}{\sqrt{4L/R^2C - 1}}\right]$$

$$(5.22)$$

$$e_C(t) = \frac{E_d[1 + \epsilon^{-(R/2L)T/2}]}{1 - \epsilon^{-(R/L)T/2}} \exp\left[-\frac{R}{2L}\left(t - \frac{T}{2}\right)\right]$$

$$\times \left[\cos\sqrt{\frac{1}{LC} - \frac{R^2}{4L^2}}\left(t - \frac{T}{2}\right) + \frac{\sin\sqrt{1/LC - R^2/4L^2}\,(t - T/2)}{\sqrt{4L/R^2C - 1}}\right]$$

$$(5.23)$$

The values of $e_C(0+)$ and $e_C(T/2+)$ from (5.16) and (5.17) for several values of $2L/R$ are listed below.

$2L/R$	$e_C(0^+)$	$e_C(T/2^+)$
$T/4$	$E_d\dfrac{(1 + \epsilon^2)}{1 - \epsilon^4} = -0.156E_d$	$E_d\dfrac{(1 + \epsilon^{-2})}{1 - \epsilon^{-4}} = 1.16E_d$
$T/2$	$-0.583E_d$	$1.58E_d$
T	$-1.54E_d$	$2.54E_d$
$2T$	$-3.52E_d$	$4.52E_d$

$e_C(t)$, from equation (5.20) for $t = T/4$, is also listed below.

$2L/R$	$e_C(t)\Big|_{t=T/4}$
$T/4$	$E_d\left[1 - 1.16\,\epsilon^{-1}\dfrac{4/T}{2\pi/T}\right] = 0.728E_d$
$T/2$	$0.695E_d$
T	$0.685E_d$
$2T$	$0.684E_d$

$Ri(t)$, using equations (5.17) and (5.18), is

$$Ri(t) = e(T/2+)\frac{R/L}{2\pi/T}\epsilon^{-(R/2L)t}\sin\frac{2\pi t}{T} \qquad (5.24)$$

$2L/R$	$Ri(t)\big|_{t=T/6}$	$Ri(t)\big|_{t=T/4}$
$T/4$	$0.66E_d$	$0.54E_d$
$T/2$	$0.62E_d$	$0.61E_d$
T	$0.58E_d$	$0.63E_d$
$2T$	$0.574E_d$	$0.635E_d$

Steady-State Operation of Figure 5.1 with $\sqrt{1/LC - R^2/4L^2} >$ Inverter Operating Frequency.

For this mode of operation, the current will reach zero, and remain at zero for some time before the next SCR is turned on. Equations (5.4), (5.7), (5.9), (5.10), (5.11), and (5.12) still describe the operation during the time when current is flowing. After the current has reached zero, the capacitor voltage will remain at its final value until the beginning of the next half-cycle of the inverter operating frequency.

Some understanding of the effects of changes in circuit parameters on this mode of operation can be obtained by examining the effect of changing the capacitor C.

Let the inverter operating frequency

$$\omega_I = \sqrt{\frac{1}{LC} - \frac{R^2}{4L^2}} \qquad (5.25)$$

for the case shown in Figure 5.2 (Chapter 5), where

$$\frac{2L}{R} = T/2 = \frac{\pi}{\omega_I} \qquad (5.26)$$

Then ω_R, the resonant frequency of the circuit, for decreasing capacitance, holding all other parameters constant, may be calculated as

$$\omega_R = \sqrt{\frac{1}{LaC} - \frac{R^2}{4L^2}} \qquad (5.27)$$

where a will be a fraction to indicate smaller capacitor values. Combining (5.26) and (5.27)

$$\omega_R{}^2 = \frac{1}{LaC} - \frac{\omega_I{}^2}{\pi^2} \qquad (5.28)$$

From (5.25) and (5.26),

$$\omega_I{}^2 = \frac{1}{LC} - \frac{R^2}{4L^2} = \frac{1}{LC} - \frac{\omega_I{}^2}{\pi^2}$$

or

$$\omega_I{}^2\left(1 + \frac{1}{\pi^2}\right) = \frac{1}{LC} \tag{5.29}$$

Combining (5.28) and (5.29),

$$\omega_R{}^2 = \frac{\omega_I{}^2\left(1 + \dfrac{1}{\pi^2}\right)}{a} - \frac{\omega_I{}^2}{\pi^2}$$

$$\omega_R = \omega_I\sqrt{\frac{1 + 1/\pi^2}{a} - \frac{1}{\pi^2}}$$

$$\omega_R = \omega_I\sqrt{\frac{1}{a} + \frac{1}{\pi^2}\left(\frac{1}{a} - 1\right)} \tag{5.30}$$

a	ω_R
1	ω_I
$\frac{1}{2}$	1.45 ω_I
$\frac{1}{4}$	2.07 ω_I
$\frac{1}{8}$ $(C' = \frac{1}{8}C)$	2.95 ω_I

Chapter Six

Harmonic-Commutated Inverters

The a-c line voltage commutated circuits discussed in Chapter 3 operate properly over a phase-control angle of from zero to 180°, assuming negligible commutating reactance and resistance. Phase-controlled rectifier operation is obtained over the range from zero to 90°, and phase-controlled inverter operation is obtained from 90 to 180°. Reliable commutation does not occur outside this 180° phase-control region. With a-c line voltage commutated circuits, some additional voltage is required to oppose the a-c system voltages to produce current transfer from one valve to the next beyond the normal 180° phase-control range. In this chapter, circuits are discussed which permit operation over the full 360° of phase control. These circuits include a harmonic of the a-c system voltage for providing commutation and are, therefore, referred to as harmonic-commutated inverters.[1-3]

With a-c line voltage commutation, a circuit places an inductive load on the a-c system. Such a circuit cannot supply inductive kva output. Harmonic commutation permits operation with phase control angles which place a leading or capacitive reactive load on the a-c lines. This means that a harmonic-commutated inverter, with phase-control angles between 180 and 270°, is capable of delivering inductive kva to an a-c system. With a harmonic-commutated inverter, it is not necessary to add capacitance until the total a-c circuit has a leading power factor, as is required in a parallel inverter.

When the harmonic voltage is supplied from a fixed or properly controlled voltage source, a harmonic-commutated inverter may supply static a-c loads with leading or lagging power factor. In this case it is not essential to connect the inverter to a stiff a-c bus where the a-c system voltage is nearly constant, contrary to the situation for line-commutated

149

inverters. However, the harmonic-commutating voltage must be quite large when the a-c system voltage is not reasonably constant or when the a-c system has high inductive reactance.

Harmonic commutation may be used to provide power factor improvement and reduced regulation for both rectifying and inverting. The principal disadvantage of this commutating technique is that it will not operate over a wide load range without a fixed harmonic voltage source or stabilization of the harmonic voltage by some means as, for example, with rectifiers to feed back power to the dc when the harmonic voltage exceeds a given value. In general, harmonic commutation is most useful for polyphase systems where the a-c system voltage is maintained relatively constant independent of the inverter, and then for extending the operating range beyond that possible with a-c line voltage commutation.

6.1   FOUR-PHASE HARMONIC-COMMUTATED INVERTER
WITH EXTERNAL HARMONIC SUPPLY

The circuit of Figure 6.1 is a four-phase harmonic commutated inverter. This is very similar to the corresponding rectifier with the addition of a d-c reactor, a harmonic voltage supply, and a two-winding transformer with a mid-tapped secondary to properly couple the harmonic voltage into the circuit.

The basic operation of this circuit can most easily be understood by assuming the following.

(1) Ideal transformers, reactors, and controlled rectifiers.
(2) Sufficient d-c circuit inductance to maintain continuous d-c current.
(3) A stiff a-c bus; that is, the a-c voltage is maintained independent of of the inverter circuit.
(4) A d-c supply voltage which is adjusted equal to the average value of e_0 for a given firing angle of the controlled rectifiers—with this assumption and assuming an ideal d-c reactor, the d-c current is indeterminate. In a practical case for a given current, the inverter d-c supply voltage is more than the e_0 for an ideal circuit by an amount equal to the resistance and reactance regulation of the circuit.
(5) A sinusoidal harmonic voltage source.

In the circuit of Figure 6.1, each of the four controlled-rectifier anodes has two a-c voltage components—a fundamental and a second harmonic. The fundamental voltage is transformed from the a-c output lines, and the second harmonic is transformer coupled from the externally generated

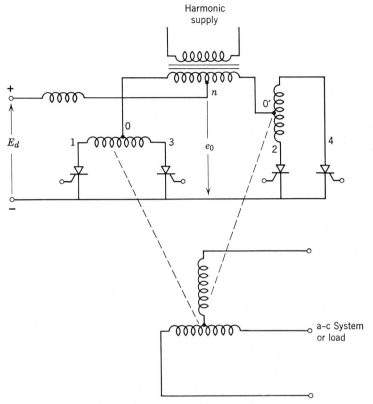

Figure 6.1   Four-phase harmonic-commutated inverter with external harmonic supply.

harmonic voltage supply. Figure 6.2 shows how the operation of the phase-controlled rectifier or inverter is extended by including the proper amount of second harmonic voltage at the appropriate phase angle. The waveforms in Figure 6.2 show operation with phase-control angles not normally obtainable in simple a-c line commutated circuits.

For the circuit of Figure 6.1 with zero harmonic voltage and a phase-control angle of 180°, SCR1 conducts during the interval from 225 to 315° in the e_{1-0} voltage wave. (The firing angle α is measured from the rectifier full on point, as in previous chapters.) After the end of the SCR1 conducting interval, $e_{2-0'}$ becomes less positive than e_{1-0} so that, normally, current could not be transferred from SCR1 to SCR2 for firing angles greater than 180°.

In Figure 6.2(a), the phase-control angle is approximately 195° or just slightly beyond the maximum phase-control angle of 180° for a-c line voltage commutation. For the conditions in Figure 6.2(a), a relatively

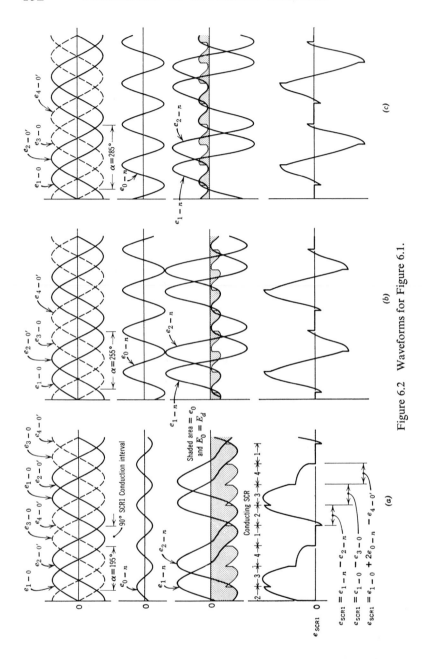

Figure 6.2 Waveforms for Figure 6.1.

small amplitude harmonic voltage produces commutation. At the end of the conducting period of SCR1, the harmonic voltage is added to e_{1-0} and $e_{2-0'}$ with the proper polarity to increase the magnitude of e_{1-n} and decrease the magnitude of e_{2-n}. As indicated by the e_{1-n} and e_{2-n} voltage waves, when SCR2 is gated on at the end of the SCR1 conducting interval, the current will be diverted from SCR1 to SCR2, since e_{2-n} is more positive than e_{1-n} at that instant. A rather small amplitude harmonic voltage produces commutation, since the e_{1-0} and $e_{2-0'}$ voltages are still quite close together at the end of the 90° conducting period for SCR1. With the conditions in Figure 6.2(a), the circuit operates as an inverter at just slightly less than maximum d-c voltage.

In Figure 6.2(b) the firing angle has been delayed further to approximately 255°. For this situation, somewhat greater amplitude harmonic voltage is required to provide commutation, since the e_{1-0} and $e_{2-0'}$ voltages are further apart at the end of the conducting period for SCR1. Thus, a larger e_{0-n} voltage is required to make e_{2-n} more positive than e_{1-n} at the point when current is transferred from SCR1 to SCR2. It is important to note that the harmonic voltage phase position has also been changed, going from Figure 6.2(a) to Figure 6.2(b). This is desirable so that commutation occurs when the harmonic voltage wave is at its peak value. For Figure 6.2(b), the circuit is still operating as an inverter but with a relatively low d-c voltage.

In Figure 6.2(c), the firing angle is approximately 285°. The harmonic voltage amplitude is the same as for Figure 6.2(b), but it has been shifted in phase so that the e_{0-n} negative maximum again occurs at the commutation point. It is not necessary to increase the harmonic amplitude, since the instantaneous difference between e_{1-0} and $e_{2-0'}$ at commutation was greater for the conditions in Figure 6.2(b). The conditions in Figure 6.2(c) result in rectifier operation with a relatively small d-c voltage. As indicated by the shaded area in Figure 6.2(c), the voltage E_0 has the opposite polarity from that for Figures 6.2(a) and 6.2(b).

A second harmonic of the a-c voltage is used to provide harmonic commutation for the circuit in Figure 6.1. The amplitude and phase of the harmonic voltage are adjusted when the firing angle is changed. With this arrangement, commutation can occur over the complete phase-control range from zero to 360°, rather than from zero to 180° only, as is the case with a-c line voltage commutation. When a harmonic-commutated circuit is operated as a rectifier with the phase-control angle $\alpha = 0$, the average rectifier output voltage can be reduced either by advancing or retarding the phase-control angle.

Figure 6.3 illustrates how the operation of the a-c line commutated inverter is extended by harmonic commutation. The reactive current in the a-c system varies in magnitude, as indicated by the appropriate curve in Figure 6.3. For the normal operating range of line commutation with α from zero to 180°, the a-c current lags the a-c voltage so that the circuit places an inductive load on the a-c system. For the additional phase-control range from $\alpha = 180°$ to $\alpha = 360°$, made possible with harmonic commutation, the a-c current lags the a-c voltage by greater than 180°, or

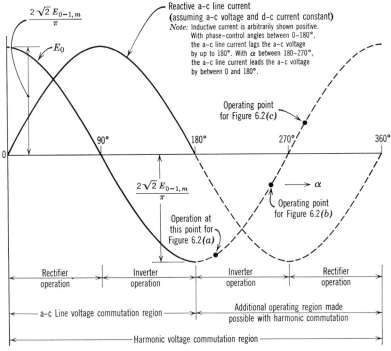

Figure 6.3 E_0 and reactive current vs. phase-control angle for the circuit in Figure 6.1.

it can then be considered leading the a-c voltage by an angle less than 180°. Therefore, the harmonic-commutated circuit permits operation in a region which places a leading load on the a-c system. For this reason, harmonic-commutated inverters have been used with zero d-c voltage and a phase-control angle slightly more than 270° to provide power-factor correction for an a-c system. The slight increase in angle beyond 270° is required to overcome the resistance and reactance regulation of the circuit. Normally, the regulation of power circuits is quite low, so that only a few degrees of phase control is required to obtain complete control of the current and the capacitive kva on the a-c system.

When a fixed source of harmonic voltage is supplied as in Figure 6.1, the harmonic-commutated inverter is capable of supplying a "dead" load. It does not require a stiff a-c bus where the a-c voltage is held relatively constant independent of the inverter operation. The dead load may have either a leading or lagging power factor. However, with such static loads, it is generally desirable to include circuits to maintain the a-c voltage waveshape reasonably constant to minimize the harmonic voltage required for proper commutation. It is also necessary to adjust the phase of the harmonic voltage when an inverter supplies a dead load. This is required to keep the harmonic voltage at the right place with respect to the commutation for the controlled-rectifying elements. If the phase position is adjusted so that commutation always occurs near the peak of the harmonic voltage wave, a minimum of externally supplied harmonic voltage is required.

6.2   FOUR-PHASE HARMONIC-COMMUTATED INVERTER WITH INTERNALLY PRODUCED HARMONIC VOLTAGE

In Figure 6.4, the harmonic voltage required to commutate the circuit is obtained by replacing the harmonic generator with a capacitive impedance. A harmonic voltage of small amplitude will commutate the circuit when operated near the phase position for a-c line commutation. When operated appreciably beyond the 0–180° interval of phase-control angles, considerable harmonic voltage is required. The increased harmonic voltage can be obtained either by using a higher capacitive impedance or greater load current. The current in Figure 6.4 operates very much like the series capacitor-commutated inverters discussed in Chapter 5, except the capacitor operates at a harmonic frequency rather than at the fundamental inverter operating frequency.

Figure 6.5 shows the important waveforms for the circuit in Figure 6.4. The assumptions upon which these waveforms are based are similar to those used for the discussions of Figure 6.1. However, the d-c current is assumed to have negligible ripple in this case to produce the triangular harmonic voltage wave forms. It is assumed that either the load current is increased or the capacitive impedance is increased to produce the greater amplitude harmonic voltage for Figures 6.5(b) and 6.5(c) as compared to Figure 6.5(a). The waveforms in Figure 6.5 are very similar to those in Figure 6.2, except the harmonic voltage has the triangular waveshape rather than a sinusoidal shape. This produces differences in the details of Figures 6.5 and 6.2, but the basic voltage waveshapes in these two figures are very nearly the same.

The phase position of the harmonic voltage is established by the transfer of current from one conducting valve to the next, which causes a reversal in the capacitor current. The capacitive impedance must be sufficient for a given load current to provide the proper amplitude harmonic voltage. The principal disadvantage of the circuit in Figure 6.4 is that the harmonic voltage amplitude is directly proportional to the load current, assuming fixed circuit constants. Thus, for very light loads, the circuit may not commutate since the harmonic voltage will approach zero. This is the

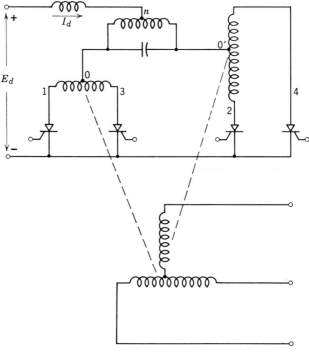

Figure 6.4 Four-phase harmonic-commutated inverter with internally produced harmonic voltage.

same disadvantage that exists for the series capacitor-commutated inverters discussed in Chapter 5.

The magnitude of the harmonic voltage required in the idealized circuit of Figure 6.4 may be determined as follows. At the commutating point

$$e_{2-n} > e_{1-n} \tag{6.1}$$

and since

$$e_{2-n} = e_{2-0'} - e_{0-n} \tag{6.2}$$

$$e_{1-n} = e_{1-0} + e_{0-n} \tag{6.3}$$

$$[e_{2-0'} - e_{0-n}] > [e_{1-0} + e_{0-n}] \tag{6.4}$$

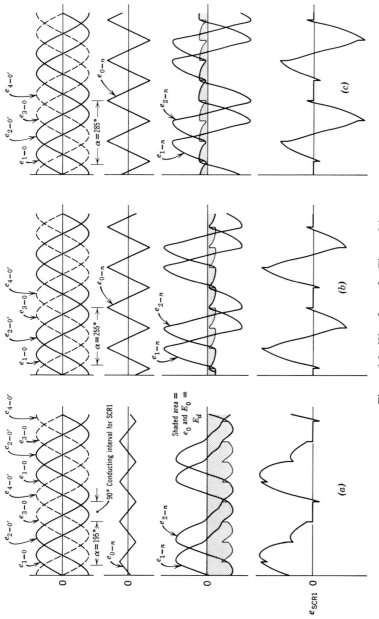

Figure 6.5   Waveforms for Figure 6.4.

but e_{0-n} is negative at the commutating point, to reduce e_{1-0} and increase $e_{2-0'}$ so that

$$[e_{2-0'} + |e_{0-n}|] > [e_{1-0} - |e_{0-n}|] \qquad (6.5)$$

or

$$2\,|e_{0-n}| > [e_{1-0} - e_{2-0'}] \qquad (6.6)$$

The amplitude of e_{0-n} for a given load current and capacitance value may be obtained as follows.

$$e_{0-n} = \frac{1}{C} \int i\, dt \qquad (6.7)$$

$$2E_{0-n,m} = \frac{1}{C} \int_0^{T/8} \frac{I_d}{2}\, dt \qquad (6.8)$$

$$2E_{0-n,m} = \frac{I_d T}{16C} = \frac{I_d}{16fC} \qquad (6.9)$$

In a practical case, the harmonic voltage may be considerably less than indicated by equation (6.9), principally because of the effects of circuit inductance. For example, with transformer leakage reactance, there is overlap of valve-conducting periods during the current transfer from one controlled rectifier to the next. During the interval when two valves are conducting, the amount of charging current for the capacitor is reduced, and this reduces the harmonic voltage amplitude. This is particularly undesirable, since more commutating voltage is required when trans-former leakage reactance is present. Thus, in general, for a practical case, the capacitance value must be less than indicated by equation (6.9) to provide for current transfer with circuit reactance and to provide negative voltage for the recovery time of the controlled-rectifying elements.

6.3   SIX-PHASE HARMONIC-COMMUTATED INVERTER

The circuit of Figure 6.6 operates very much like the six-phase star-controlled rectifier of Figure 3.10 (Chapter 3). For an idealized circuit, each valve conducts for 60°, and only one valve is conducting at any instant of time. The important waveforms for the circuit are shown in Figure 6.7, again using the same basic assumptions as were used to explain Figure 6.4. For Figure 6.6, a third harmonic of the a-c system voltage is used, and somewhat less voltage amplitude is required than for the four-phase scheme. The maximum instantaneous voltage difference between consecutive phases is less as the number of phases are increased so that less harmonic voltage is required to force commutation. Also, the harmonic frequency is greater with a greater number of phases. The fact that the harmonic voltage amplitude is somewhat less also means that the SCR voltage rating is less for a given d-c output voltage. This is indicated by comparing the e_{SCR1} waveforms in Figures 6.5 and 6.7. It should be noted that again it is assumed that the capacitive impedance is increased or the

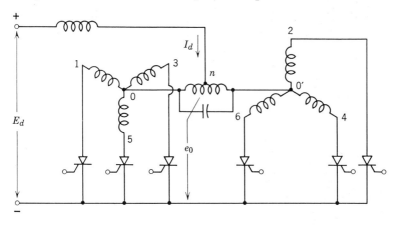

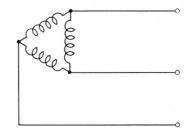

Figure 6.6   Six-phase single-way harmonic commutated inverter.

load current is increased to provide the greater harmonic voltage in Figures 6.7(b) and 6.7(c) as compared to Figure 6.7(a).

The amplitude of the harmonic voltage for the idealized circuit in Figure 6.6 is given as

$$e_{0-n} = \frac{1}{C} \int i \, dt$$

$$2E_{0-n,m} = \frac{1}{C} \int_0^{T/12} \frac{I_d}{2} \, dt$$

$$2E_{0-n,m} = \frac{I_d T}{24C} = \frac{I_d}{24fC} \tag{6.10}$$

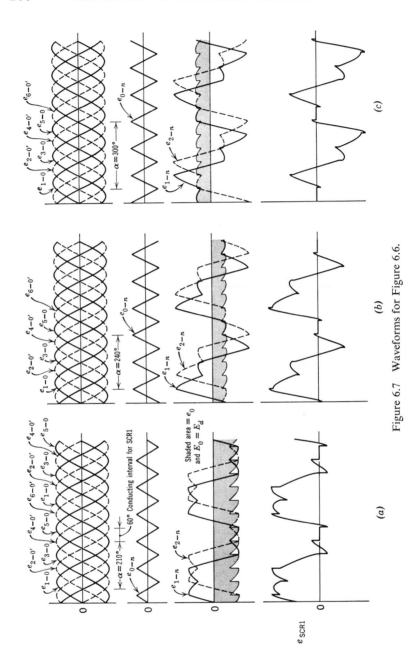

Figure 6.7   Waveforms for Figure 6.6.

6.4 OTHER HARMONIC-COMMUTATED INVERTER CIRCUITS

Figure 6.8 is a double-way version of the six-phase harmonic-commutated inverter in Figure 6.6. In general, this scheme would be more desirable than Figure 6.6 for applications involving high d-c voltage. There is still 60° conduction in each valve, as contrasted with the 120° conduction in the conventional six-phase double-way or three-phase bridge rectifier. The capacitors and mid-tapped transformer in Figure 6.8

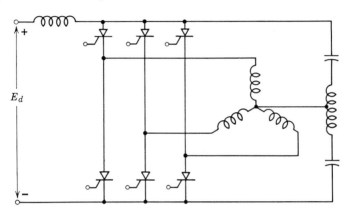

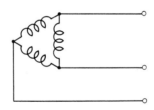

Figure 6.8   Six-phase double-way harmonic commutated inverter.

produce operation very different from that obtained in the simple phase-controlled rectifier.

Figure 6.9 is a full-wave scheme treated by Willis, which has better utility of the output transformer and improved waveshape over the schemes in Figure 6.6 or 6.8.[1,2] A circuit similar to Figure 6.9 has been used with zero d-c voltage, operating very much like a synchronous condenser supplying capacitive reactive kva for an a-c system.

Figure 6.10 is a twelve-phase arrangement containing two three-phase double-way circuits with 30° phase displacement. This arrangement has good transformer utility, 120° conduction in the valves, and the harmonic-commutating components help to cause more equal current division in the controlled rectifiers. The a-c currents to the two three-phase double-way

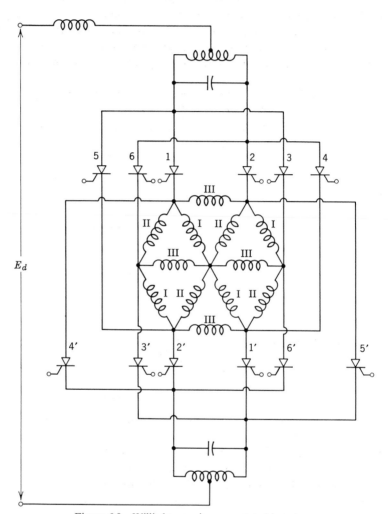

Figure 6.9   Willis harmonic-commutated inverter.

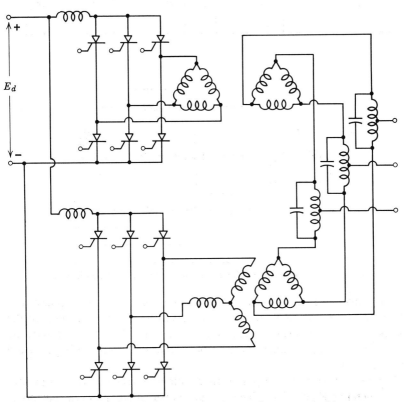

Figure 6.10   Twelve-phase double-way harmonic-commutated inverter.

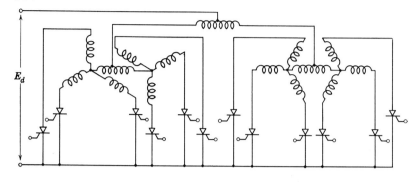

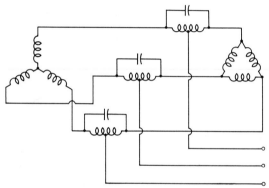

Figure 6.11   Twelve-phase single-way harmonic-commutated inverter.

circuits have fifth and seventh harmonics which are out of phase and combine to produce negligible fifth and seventh harmonic currents in the a-c system. These harmonic currents are forced to flow in the capacitors in Figure 6.10 to produce harmonics of the a-c system voltage to aid commutation.

Figure 6.11 is a twelve-phase quadruple-wye scheme which operates very much like Figure 6.10. The a-c current waveshapes and the harmonic voltage waveshapes are identical for Figures 6.10 and 6.11.

REFERENCES

1. C. H. Willis, "Harmonic Commutation for Thyratron Inverters and Rectifiers," *General Electric Review*, December 1932, p. 632.
2. C. H. Willis, "Applications of Harmonic Commutation for Thyratron Rectifiers and Inverters," *AIEE Transactions*, June 1933, pp. 701–708.
3. B. D. Bedford *et al.*, "Rectifier Characteristics with Interphase Commutation," *General Electric Review*, November 1935, pp. 499–504.

Chapter Seven

Impulse-Commutated Inverters

by W. McMurray

The term impulse commutation is applied to the use of an impulse to briefly reverse the voltage on a conducting controlled rectifier, allowing it to turn off. The magnitude of the pulse must be sufficient to extinguish the current in the controlled rectifier, and it must be long enough to provide the necessary turn-off time. Generally, the pulse is formed by means of an oscillatory inductance-capacitance (L-C) network in which the natural period is directly related to the turn-off time of the controlled rectifiers and the characteristic impedance of the network is related to the d-c supply voltage and the maximum value of load current which must be commutated. If the period of the a-c output of the inverter is long compared with the turn-off time of the controlled rectifiers, the size of the commutating components is relatively small. As the frequency of operation increases, the impulse occupies a greater portion of each half-cycle until, eventually, the commutating transient can hardly be regarded as a pulse.

The principle of impulse commutation is most clearly illustrated in inverters where the impulse is generated by auxiliary means, separate from the main power circuit. This method is termed auxiliary-impulse commutation, and an example is described in the next section. In such arrangements, loss of gating signals does not result in a commutation failure, provided that the auxiliary commutating circuit is able to turn off the last controlled rectifier that conducts.

Alternatively, the impulse may be initiated by turning on the controlled rectifier that is complementary to the controlled rectifier being turned off. That is, a pair of controlled rectifiers are gated on alternately to produce positive and negative half-cycles of the a-c output; the turning of one on starts an impulse to turn the other off. Inverters of this type, which are discussed here, are termed complementary impulse-commutated

165

inverters. The parallel inverter discussed in Chapter 4 may also be described as complementary-commutated. Removal of gating signals from this type of inverter results in a failure to commutate. The d-c power source must be disconnected to shut down the circuit.

In a third class of circuits, the impulse to commutate a particular controlled rectifier is generated as a result of turning on the controlled rectifier. That is, the circuit automatically turns itself off after a certain time and does not depend upon further action by the control circuit. Such a "self-impulse commutated" circuit is described in Chapter 10. The series inverter may also be regarded as self-commutated.

The concept of impulse commutation may be regarded as an extension of harmonic commutation, described in Chapter 6, where a harmonic voltage is introduced to provide the necessary inverse voltage to turn off the controlled rectifiers. With impulse commutation, the frequency of the harmonic is increased to the limit set by the turn-off time of the controlled rectifiers, and the harmonic oscillations between successive commutations are eliminated. However, there is a considerable difference in the operating characteristics of harmonic- and impulse-commutated inverters, as discussed on pp. 185–188.

The feedback rectifiers that are generally used with impulse-commutated inverters contribute greatly to the characteristics of the circuits. Such characteristics include good inherent voltage regulation and ability to handle wide variations in load magnitude, power factor, and frequency. They are well adapted to utilize voltage control techniques, such as discussed in Chapter 8. Thus, they have the flexibility needed in many applications. The ability of these circuits to reverse the power flow, as discussed on pp. 186–188, enhances their use as standby power supplies. Also, impulse commutated inverters have a high efficiency and relatively small size.

There are a number of different ways of producing and controlling the commutating impulse. The examples presented in this chapter by no means exhaust the possibilities, but they have been selected to convey the important principles of impulse-commutated circuits.

7.1 AUXILIARY IMPULSE-COMMUTATED INVERTER

In the circuit of Figure 7.1, which uses a center-tapped d-c supply, the main controlled rectifiers SCR1 and SCR2 are gated to conduct current to the load during alternate half-cycles of the a-c output. When the load is reactive, the feedback rectifiers D1 and D2 conduct during part of each half-cycle to return power from the load to the d-c supply. Commutation of the main controlled rectifiers is accomplished by means of the auxiliary controlled rectifiers SCR1(A) and SCR2(A) in conjunction with the capacitor C and the inductance L, which generate the impulse. To commutate main controlled rectifier SCR1, the auxiliary controlled rectifier SCR1(A) is fired, while to commutate SCR2, SCR2(A) is fired. This circuit can be used over a very wide range of frequency, up to about 5 kilocycles, or a cycle period approximately 10 times longer than the turn-off time of typical SCR's.

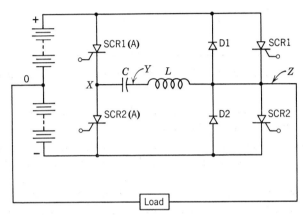

Figure 7.1   Auxiliary impulse-commutated inverter (McMurray inverter).

Theory of Operation

Suppose, initially, that main controlled rectifier SCR1 is conducting current to the load from the upper half of the d-c supply. Also, assume that the capacitor C is charged with terminal Y positive with respect to terminal X, as shown in Figure 7.2, the charge having been acquired during previous operation.

Then, to commutate the main controlled rectifier SCR1, the auxiliary controlled rectifier SCR1(A) is fired. The equivalent circuit of the inverter at this time is shown in Figure 7.2. The discharge current pulse through SCR1(A), C, and L builds up to exceed the load current I_L (assumed to flow from Z to 0 through the load at this time). In doing so, the current through SCR1 is reduced to zero. The excess of the commutating impulse current i_C over the load current I_L then flows through the feedback rectifier D1. After reaching a peak, the commutating current i_C starts to decay, and a charge of reversed polarity builds up on the capacitor C. During the time that D1 is carrying current, the forward drop of D1 appears as inverse voltage across SCR1, and turns it off.

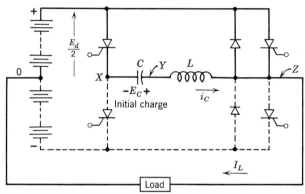

Figure 7.2   Equivalent circuit of Figure 7.1 during the first part of the commutating interval.

The second main controlled rectifier SCR2 is fired at or about the time when the capacitor current i_C returns to zero, under no load conditions, which occurs about $\pi\sqrt{LC}$ seconds, or one-half cycle of oscillation between L and C, after firing SCR1A. The equivalent circuit of the inverter now switches to the configuration of Figure 7.3. Forward voltage is reapplied to SCR1. A second and much smaller pulse of current i_C will flow from the d-c supply through SCR1(A), C, L, and SCR2 to make up the losses incurred during the first pulse and complete the charge of the capacitor to the initial magnitude, but with the opposite polarity (X positive with respect to Y). After the second pulse, SCR1(A) has reverse voltage and ceases to conduct. The capacitor is now ready to commutate SCR2 at the end of its conducting half-cycle, upon the subsequent firing of SCR2(A).

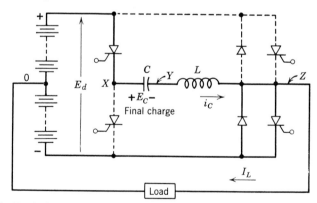

Figure 7.3   Equivalent circuit of Figure 7.1 during the second part of the commutating interval.

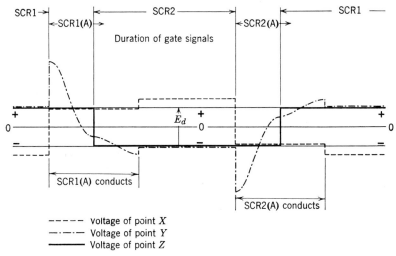

Figure 7.4 Operation of the inverter circuit (Figure 7.1) with no load (one cycle). (Drawn for a relatively high output frequency.)

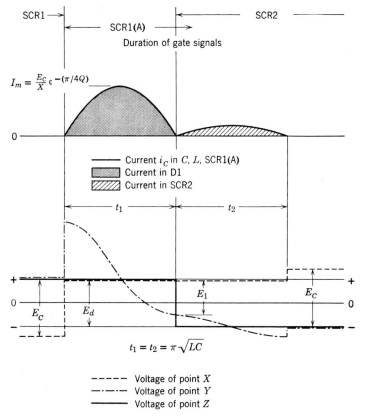

Figure 7.5 Operation of the inverter circuit (Figure 7.1) with no load; detail of the commutating interval.

169

With inductive load, D2 will conduct and switch the circuit to the configuration of Figure 7.3 before SCR2 is fired, when the commutating current i_C falls below the load current I_L. The energy stored in the commutating inductance L at this time will cause the capacitor C to charge to a higher voltage. This increase in capacitor voltage produces a larger commutating current pulse, enabling heavier load currents to be commutated. Thus, the circuit has the advantage that the commutating pulse varies with the load automatically, a desirable feature which is

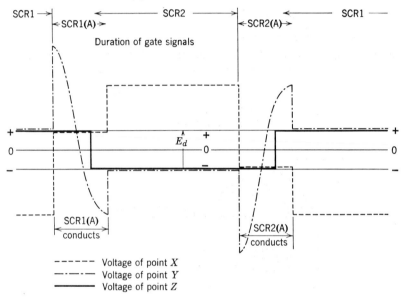

Figure 7.6   Operation of the inverter circuit (Figure 7.1) with a lagging load (one cycle).

difficult to obtain with previous inverter circuits. If the L-C commutating circuit has a high Q factor, the commutating losses are small and the inverter efficiency is high. Also, since the commutating pulse is a minimum at no load, the no-load losses are slight. The load voltage is a square wave under all conditions of loading.

The performance of the commutating circuit is analyzed in the next subsection. Voltage waveforms, depicting one cycle of steady-state operation at a relatively high frequency with no load, are sketched in Figure 7.4. The necessary durations of the gate signals applied to the controlled rectifiers are also indicated at the top of Figure 7.4. Detail of a single commutating interval for the no-load condition, including the commutating current pulses, is shown in Figure 7.5. The magnitude E_C

of the capacitor voltage between commutations is given by equation (7.28) in the next subsection, while the value E_1 at the time of switching is given by equation (7.26). It is assumed that main controlled rectifier SCR2 is gated exactly $\pi\sqrt{LC}$ seconds after auxiliary controlled rectifier SCR1(A) is gated, and SCR1 is gated $\pi\sqrt{LC}$ seconds after SCR2(A) is gated.

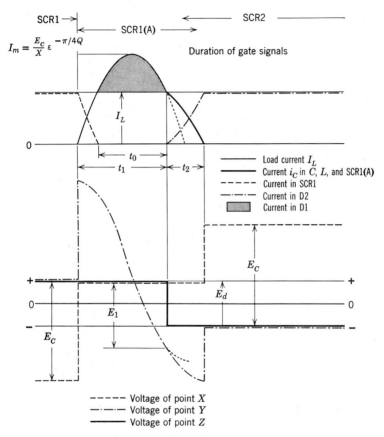

Figure 7.7 Operation of the inverter circuit (Figure 7.1) with a lagging load; detail of the commutating interval.

For the case of a lagging power factor load, Figure 7.6 illustrates the voltage waveforms for one cycle of steady-state operation. Detail of a single commutation interval is shown in Figure 7.7. It is assumed that the load has sufficient series inductance to prevent appreciable change in the magnitude of the load current during the commutating interval. With a heavy load, the magnitude of the capacitor voltage and the peak

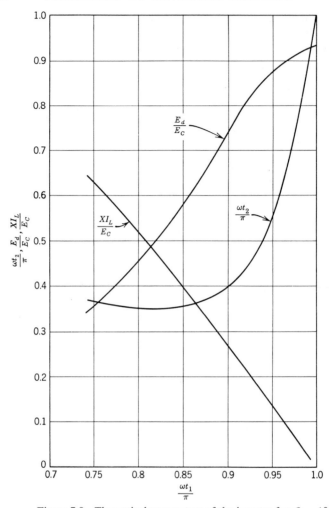

Figure 7.8   Theoretical parameters of the inverter for $Q = 10$.

commutating current I_m are considerably higher than under the no-load conditions depicted in Figure 7.5. The manner in which the values of t_1, t_2, and E_C change as I_L is increased can be predicted with the aid of a set of parameter curves such as drawn in Figure 7.8. These curves are calculated for the case $Q = 10$ by the method derived in the next subsection. However, the calculations are tedious and must be repeated for different values of Q. Also, they are subject to errors, due to the approximations involved, the nonideal nature of actual components, and the addition of cushioning to limit the rate at which forward voltage is reapplied to the controlled rectifiers. Hence, such calculations should be used only as a guide, and the actual performance determined by experiment.

The time t_0 available for recovery of the main controlled rectifier is the time during which the feedback rectifier carries the excess commutating current pulse, the cross-hatched area in Figure 7.7. The optimum commutating pulse is that which achieves the required turn-off time with the least amount of energy $\frac{1}{2}CE_C^2$. From the analysis on pp. 179–182, the optimum occurs when $I_m = 1.5I_L$.

It has been noted that the commutating pulse increases as the magnitude of the load current at the time of commutation increases. However, to take advantage of this fact, the rate of application of load must be limited.

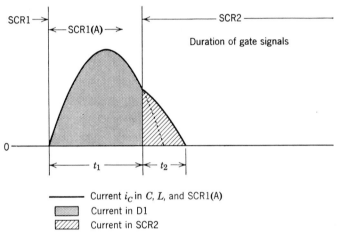

Figure 7.9   Currents during the commutation of the inverter (Figure 7.1) with no load, but with the gating of the main SCR's advanced. The voltage waveforms are the same as in Figure 7.7.

This is because the capacitor voltage E_C is boosted by an increment in load current during the second part of the commutation interval, after the pulse which turns off the main controlled rectifier. Several commutations will be required to achieve steady-state conditions after the increment in load current. Thus, an inverter designed to utilize the variable commutating pulse will be unable to commutate a large step increase in load but has the advantage of minimum commutation loss at light load.

The capacitor voltage E_C and the commutating current pulse can be built up to their maximum values even under no-load conditions, simply by advancing the gating of the main controlled rectifiers, as shown in Figure 7.9. The time delay t_1 between the gating of the auxiliary controlled rectifiers and the main controlled rectifiers is reduced to less than $\pi\sqrt{LC}$

seconds. To achieve the optimum pulse condition as determined on pp. 175–182,

$$t_1 = 0.767\pi\sqrt{LC}, \quad \text{or} \quad 2.41\sqrt{LC} \text{ seconds}$$

In this way, the equivalent circuit is always switched from the configuration of Figure 7.2 to Figure 7.3 when the current in the commutating inductance L is equal to the maximum load current which can be commutated. The voltage waveforms for this mode of operation are identical with those drawn in Figure 7.7. An inverter designed to operate in this mode will be able to take any step increase in load up to the maximum designed load without a commutation failure, since the circuit is always operating with the maximum commutating pulse.

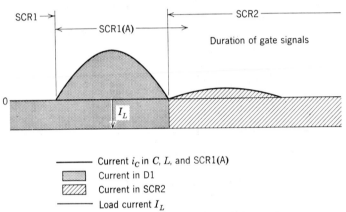

Figure 7.10   Currents during the commutation of the inverter (Figure 7.1) with a leading load. Voltage waveforms are the same as in Figure 7.5.

Operation of the inverter with purely resistive load is quite similar to the case of inductive load. The waveforms are identical during the first part of the commutating interval (time t_1), but the behavior during the second part of the commutating interval (time t_2) is slightly modified. Since the first part, t_1, includes the critical turn-off time, the design of the commutating circuit is the same.

When the load has a leading power factor, the commutation is similar to the no-load condition, as illustrated in Figure 7.10. Here, it is assumed that, while the net power factor of the load is leading, there is sufficient series inductance to prevent appreciable change in the magnitude of the load current during the commutating interval. Just before commutation, the load current is carried by a feedback rectifier, and power is being fed back into the d-c source. The first commutating pulse simply adds to the current in the feedback rectifier. The turn-off action is redundant, since the main controlled rectifier is already nonconducting. However, the pulse is necessary to reverse the charge on the capacitor. The second, or make-up commutating pulse is added to the load current which switches to the second main controlled rectifier after it is fired.

The modes of commutation that are described above as applicable to no load, lagging power-factor load, and leading power-factor load, can also occur under certain other conditions of loading. This is generally true for all impulse-commutated inverters, and examples of such conditions are discussed on pp. 198–202. Actually, the terms no load, lagging load, and leading load are here used to designate the polarity of the load current at the instant of commutation.

Analysis of Commutating Circuit

The equations for the general circuit of Figure 7.11 will first be derived. This circuit consists of a capacitor C, inductance L, and resistance R (representing losses), connected in series to a source of d-c voltage E.

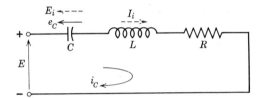

Figure 7.11   A general L-C-R circuit with a d-c source.

The intitial voltage on the capacitor is E_i, and the initial current is I_i, with polarities as indicated in Figure 7.11. The differential equation for the circuit is

$$E = E_i + \frac{1}{C} \int_0^t i_C(\tau)\, d\tau + L\frac{di_C}{dt} + Ri_C \qquad (7.1)$$

with the initial condition $i_C(+0) = I_i$. The Laplace transform of equation (7.1) is

$$\frac{E - E_i}{s} = \frac{1}{sC} i_C(s) + L[si_C(s) - I_i] + Ri_C(s) \qquad (7.2)$$

Solving equation (7.2) for $i_C(s)$, the Laplace transform of $i_C(t)$,

$$i_C(s) = \frac{\dfrac{E - E_i}{L} + sI_i}{s^2 + \dfrac{R}{L}s + \dfrac{1}{LC}} \qquad (7.3)$$

Assuming the oscillatory case, the inverse transform of equation (7.3) is

$$i_C = \frac{E - E_i}{\omega L} \epsilon^{-\alpha t} \sin \omega t - I_i \frac{\omega_0}{\omega} \epsilon^{-\alpha t} \sin (\omega t - \varphi) \qquad (7.4)$$

where

$$\left.\begin{array}{ll} \omega_0^2 = \dfrac{1}{LC}, & \alpha = \dfrac{R}{2L} \\[3mm] \omega^2 = \omega_0^2 - \alpha^2 > 0, & \varphi = \tan^{-1} \dfrac{\omega}{\alpha} \end{array}\right\} \qquad (7.5)$$

If the losses in the circuit are relatively small, then $\omega_0 \gg \alpha$ and the following approximations are valid.

$$\left.\begin{array}{l} \omega_0 \approx \omega \\[3mm] X = \sqrt{\dfrac{L}{C}} \approx \omega L \approx \dfrac{1}{\omega C} \\[3mm] \dfrac{\alpha}{\omega} = \dfrac{R}{2\omega L} \approx \dfrac{1}{2Q} \; ; \left(Q = \dfrac{X}{R}\right) \\[3mm] \varphi \approx \dfrac{\pi}{2}, \qquad \sin (\omega t - \varphi) \approx - \cos \omega t \end{array}\right\} \qquad (7.6)$$

Equation (7.4) can now be written approximately as

$$i_C \approx \left[\frac{E - E_i}{X} \sin \omega t + I_i \cos \omega t\right] \epsilon^{-(\omega t/2Q)} \qquad (7.7)$$

The expression for the voltage e_C on the capacitor C is obtained in a similar manner. The integral equation is

$$e_C = E_i + \frac{1}{C} \int_0^t i_C(\tau) \, d\tau \qquad (7.8)$$

The Laplace transform of equation (7.8) is

$$e_C(s) = \frac{E_i}{s} + \frac{1}{sC} i_C(s) \qquad (7.9)$$

$$e_C(s) = \frac{E_i}{s} + \frac{\dfrac{E - E_i}{LCs} + \dfrac{I_i}{C}}{s^2 + \dfrac{R}{L} s + \dfrac{1}{LC}} \qquad (7.10)$$

The inverse transform of (7.10) gives

$$e_C = E - (E - E_i)\frac{\omega_0}{\omega}\epsilon^{-\alpha t}\sin(\omega t + \varphi) + \frac{I_i}{\omega C}\epsilon^{-\alpha t}\sin\omega t \quad (7.11)$$

with the same approximations (7.6) as before,

$$e_C \approx E + [XI_i\sin\omega t - (E - E_i)\cos\omega t]\epsilon^{-(\omega t/2Q)} \quad (7.12)$$

The approximate general equations (7.7) and (7.12) for i_C and e_C, respectively, will now be applied to the specific conditions of the commutating circuit of the inverter of Figure 7.1, with lagging load as sketched in Figure 7.7. During the first part of the commutating interval, the equivalent circuit is as shown in Figure 7.2 with $E = 0$. If the capacitor voltage is considered positive when terminal X is positive, the initial conditions are known to be

$$E_i = -E_C, \qquad I_i = 0 \quad (7.13)$$

At the end of the first transient, after time t_1, the final conditions are therefore

$$I_L = \frac{E_C}{X}\exp\left(-\frac{\omega t_1}{2Q}\right)\sin\omega t_1 \qquad \left(\frac{\pi}{2} < \omega t_1 \le \pi\right) \quad (7.14)$$

$$E_1 = -E_C\exp\left(-\frac{\omega t_1}{2Q}\right)\cos\omega t_1 \quad (7.15)$$

Assuming particular values of Q and ωt_1, the ratios XI_L/E_C and E_1/E_C can be calculated from equations (7.14) and (7.15):

$$\frac{XI_L}{E_C} = \exp\left(-\frac{\omega t_1}{2Q}\right)\sin\omega t_1 \quad (7.16)$$

$$\frac{E_1}{E_C} = -\exp\left(-\frac{\omega t_1}{2Q}\right)\cos\omega t_1 \quad (7.17)$$

During the second part of the commutating interval, the equivalent circuit becomes as shown in Figure 7.3, with $E = E_d$. The initial conditions are

$$E_i = E_1, \qquad I_i = I_L \quad (7.18)$$

The final conditions at the end of the second transient, of duration t_2, are known to be

$$e_C = E_C, \qquad i_C = 0 \quad (7.19)$$

Therefore

$$0 = \left[\frac{E_d - E_1}{X}\sin\omega t_2 + I_L\cos\omega t_2\right]\exp\left(-\frac{\omega t_2}{2Q}\right) \quad (7.20)$$

or

$$E_1 - E_d = XI_L\cot\omega t_2 \qquad (0 < \omega t_2 \le \pi) \quad (7.21)$$

and

$$E_C = E_d + [XI_L \sin \omega t_2 - (E_d - E_1) \cos \omega t_2] \exp\left(-\frac{\omega t_2}{2Q}\right) \quad (7.22)$$

Eliminating E_d between equations (7.21) and (7.22)

$$\frac{1 - \dfrac{E_1}{E_C}}{\dfrac{XI_L}{E_C}} = \frac{\exp\left(-\dfrac{\omega t_2}{2Q}\right) - \cos \omega t_2}{\sin \omega t_2} \quad (7.23)$$

Substituting the values of XI_L/E_C and E_1/E_C obtained from equations (7.16) and (7.17) into equations (7.23), this can be solved for ωt_2. Such solutions have been obtained by graphical means for the case $Q = 10$, and values of ωt_1 between 0.75π and π. The values of E_d/E_C are then obtained from equation (7.21):

$$\frac{E_d}{E_C} = \frac{E_1}{E_C} - \frac{XI_L}{E_C} \cot \omega t_2 \quad (7.24)$$

The theoretical parameters $\omega t_2/\pi$, E_d/E_C, and XI_L/E_C of the inverter are plotted against $\omega t_1/\pi$ in Figure 7.8 for the case $Q = 10$.

For the no-load condition, a direct solution of the equations is possible, as depicted in Figure 7.5,

$$I_L = 0, \qquad \omega t_1 = \pi, \qquad \omega t_2 = \pi \quad (7.25)$$

Then, from equation (7.15),

$$E_1 = E_C \epsilon^{-(\pi/2Q)} \quad (7.26)$$

and from equation (7.22),

$$E_C = E_d + (E_d - E_1)\epsilon^{-(\pi/2Q)} \quad (7.27)$$

By eliminating E_1 between equations (7.26) and (7.27),

$$\frac{E_d}{E_C} = \frac{1 + \epsilon^{-(\pi/Q)}}{1 + \epsilon^{-(\pi/2Q)}} \approx \epsilon^{-(\pi/4Q)} \quad (7.28)$$

The ratio of the first current peak to the second at no load is

$$\frac{\dfrac{E_C}{X}\epsilon^{-(\pi/4Q)}}{\dfrac{E_C - E_d}{X}\epsilon^{+(\pi/4Q)}} = \coth\frac{\pi}{4Q} \quad (7.29)$$

$$\frac{\dfrac{E_C}{X}\epsilon^{-(\pi/4Q)}}{\dfrac{E_C - E_d}{X}\epsilon^{+(\pi/4Q)}} \approx \frac{4Q}{\pi} \qquad \text{if } Q \text{ is large} \quad (7.30)$$

The energy loss per commutation, neglecting losses during the period t_2, is

$$\tfrac{1}{2}C(E_C^2 - E_1^2) = \tfrac{1}{2}CE_C^2(1 - \epsilon^{-(\pi/Q)}) \tag{7.31}$$

Hence, the commutation loss in watts is approximately

$$fCE_C^2\left[1 - \exp\left(-\frac{\pi}{Q}\right)\right] \tag{7.32}$$

Under no-load conditions,

$$\text{commutation loss} \approx fCE_d^2\epsilon^{\pi/2Q}\left[\frac{\pi}{Q}\exp\left(-\frac{\pi}{2Q}\right)\right] = \frac{\pi}{Q}fCE_d^2 \tag{7.33}$$

Selection of Optimum Commutating Capacitance and Inductance

To achieve adequate commutation under the conditions depicted in Figure 7.7, the current i_C must exceed the load current I_L for an interval t_0, which is longer than the turn-off time of the controlled rectifiers. Three alternative pulse shapes, which satisfy this condition, are shown in Figure 7.12. The parameter χ is the ratio I_m/I_L. The optimum shape

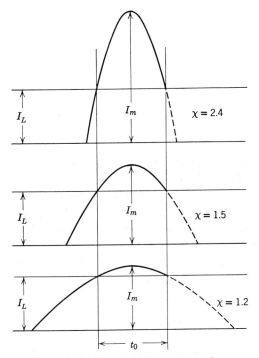

Figure 7.12 Commutating current pulses.

is judged to be the one which requires the least energy. In the analysis which follows, it is assumed that the commutating circuit has a high "Q" at its natural frequency, so that losses can be neglected. From Figure 7.12, it can be seen that

$$\cos \frac{\omega t_0}{2} = \frac{I_L}{I_m} = \frac{1}{\chi} \qquad (7.34)$$

where

$$\omega = \frac{1}{\sqrt{LC}} \qquad (7.35)$$

Therefore

$$\frac{t_0}{\sqrt{LC}} = 2 \cos^{-1} \frac{1}{\chi} = g(\chi) \qquad (7.36)$$

The energy W that the commutating circuit must provide in order to turn-off the controlled rectifiers is

$$W = \tfrac{1}{2}CE_C^2 = \tfrac{1}{2}LI_m^2 \qquad (7.37)$$

$$W = \tfrac{1}{2}\sqrt{LC}\, E_C I_m \qquad (7.38)$$

$$W = \frac{1}{2} \frac{t_0}{2 \cos^{-1} \dfrac{1}{\chi}} E_C \chi I_L \qquad (7.39)$$

Expressing equation (7.39) in normalized form as a function of the parameter χ,

$$\frac{W}{E_C I_L t_0} = \frac{\chi}{4 \cos^{-1} \dfrac{1}{\chi}} = h(\chi) \qquad (7.40)$$

The functions $g(\chi)$ and $h(\chi)$ represented by equations (7.36) and (7.40) are graphed in Figure 7.13. It is seen that the normalized commutation energy $W/E_C I_L t_0$ has a minimum value of 0.446 when $\chi = 1.5$, corresponding to the center pulse in Figure 7.12. The value of $t_0/\sqrt{LC}$ at this point is 1.68. In designing the commutating circuit, the values of E_C, I_L, t_0, and χ pertaining to maximum load and minimum supply voltage conditions must be used, which are identified by a subscript 0. The required values of C and L are then given by

$$C = \frac{\chi_0}{g(\chi_0)} \frac{I_{L0} t_{00}}{E_{C0}} = 0.893 \frac{I_{L0} t_{00}}{E_{C0}} \qquad (7.41)$$

$$L = \frac{1}{\chi_0 g(\chi_0)} \frac{E_{C0} t_{00}}{I_{L0}} = 0.397 \frac{E_{C0} t_{00}}{I_{L0}} \qquad (7.42)$$

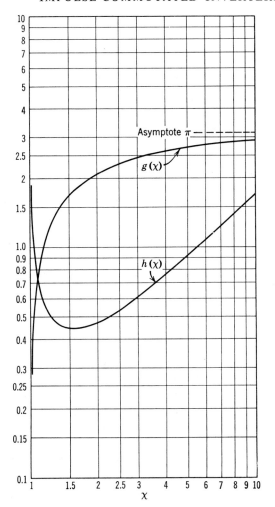

Figure 7.13 Commutation parameters $g(\chi)$, $h(\chi)$; $g(\chi) = \dfrac{t_0}{\sqrt{LC}}$ and $h(\chi) = \dfrac{W}{E_c I_L t_0}$; $\chi = I_m/I_L$.

where the numerical constants relate to the optimum condition

$$\chi_0 = 1.5, \qquad g(\chi_0) = 1.68$$

The natural frequency of the commutating circuit is

$$f' = \frac{1}{2\pi\sqrt{LC}} = \frac{g(\chi_0)}{2\pi t_{00}} = \frac{0.267}{t_{00}} \qquad (7.43)$$

The pulse width is

$$\pi\sqrt{LC} = \frac{\pi t_{00}}{g(\chi_0)} = 1.87 t_{00} \tag{7.44}$$

To allow for losses in the commutating circuit, the value of E_{C0} used in equations (7.41) and (7.42) should be multiplied by the factor $\epsilon^{-\pi/4Q}$

Alternative Circuit Configurations

Other members of the family of inverters that operate in a manner similar to that of the circuit of Figure 7.1 are shown in Figures 7.14 and

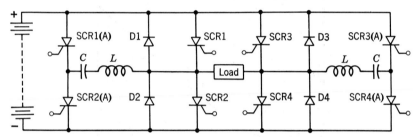

Figure 7.14   Bridge inverter, derived from Figure 7.1.

7.15. The single-phase bridge (Figure 7.14) uses the circuit of Figure 7.1 as a building block. With three such building blocks, a three-phase bridge is formed. In the center-tapped transformer arrangement of Figure 7.15, two auxiliary controlled rectifiers are required to commutate each main controlled rectifier. For example, main controlled rectifier SCR1 is turned off by firing both auxiliary controlled rectifiers SCR1(A) and SCR1(B) together. While the extra pair of auxiliary controlled

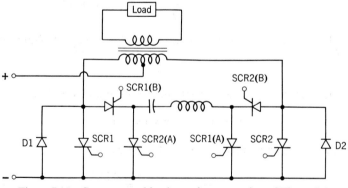

Figure 7.15   Center-tapped load transformer version of Figure 7.1.

rectifiers could be replaced by a suitable coupling transformer, the design
of such a transformer is difficult, since it would have to withstand a low-
frequency voltage and efficiently transform a heavy current pulse.

7.2   COMMUTATION WITH LIMITED INVERSE VOLTAGE

In the circuit described in Section 7.1, the voltage applied to a controlled
rectifier during its recovery interval is significantly different from the
voltage applied by all the other inverter circuits discussed thus far. With
previous circuits, an abrupt step of inverse voltage is initially applied to
the controlled rectifier that is being commutated off. The magnitude of
the step is generally of the same order as the d-c supply voltage. The
SCR voltage then rises on a ramplike curve, remaining negative for some
time longer than the necessary turn-off time. Positive, or forward voltage
is reapplied along a smooth continuation of the ramp curve, usually
having the same mathematical equation. The reverse current required to
recombine the carriers in the semiconductor controlled rectifier, which
must flow before it can support the applied inverse voltage, is limited only
by second-order effects such as lead inductance and resistance. Sometimes,
small inductances are inserted in the circuit for this purpose. Silicon-
controlled rectifiers recover their ability to block reverse current quite
abruptly, and the energy trapped in the circuit inductance can produce
a large inverse voltage spike across the controlled rectifier, in ringing with
the cell capacitance. This is generally suppressed by connecting a resistance-
capacitance filter across the controlled rectifier. The reverse recovery
current and its effects have been neglected in the circuit analyses.

In the inverter of Section 7.1, the inverse voltage applied to a controlled
rectifier is limited to the forward drop of the feedback rectifier connected
in inverse-parallel (about one volt). Also, the reverse recovery current is
limited to the excess of the commutating current pulse over the load
current. It is fortunate that the voltage across a controlled rectifier
remains slightly positive during carrier recombination, otherwise the
feedback rectifier would divert at least part of the available recovery
current. The limitation of the magnitude and slope of the recovery current
is not a drawback; in fact, it is desirable in order to avoid damaging the
cells. In practice, the charge required for carrier recombination is attained
well before the peak of the current impulse. Since the current reverses
its slope after the peak, it is necessary to keep the lead inductance of the
path through the feedback rectifier very small, so that the induced voltage
will not exceed the forward drop of the feedback rectifier and produce a
positive voltage across the controlled rectifier at this time.

While commutation with limited inverse voltage avoids the ill effects of a high peak recovery current, a new difficulty arises at the end of the commutating pulse. Forward voltage is reapplied to the controlled rectifier at a very steep rate. With the ideal components assumed in the analysis, the slope is infinite. Actually, the full d-c supply voltage is reapplied as fast as the complementary controlled rectifier can turn on. Typically, the turn-on time of a silicon-controlled rectifier varies from less than one up to several microseconds. With reactive load, a still steeper slope can arise as a result of the recovery of the feedback rectifier. The carrier recombination current required by the feedback rectifier, after it has carried the excess commutating impulse, allows the current in the commutating inductance to fall below the load current. Then, when the feedback rectifier suddenly blocks, the current, demanded by the load and commutating inductances, forces the load voltage to switch rapidly so that the complementary feedback rectifier can conduct and supply the deficiency in current. Switching can occur in a fraction of a microsecond.

Such a high rate of rise of reapplied forward voltage can cause a controlled rectifier to turn on again, resulting in a short circuit across the d-c supply. This "dv/dt" firing results from high current charging the cell capacitance. Some of the current acts upon the gate circuit, and triggers the device. The very fast switching due to inductive load can be slowed down by suitable filter circuits connected across the controlled rectifiers. Then, a special selection of controlled rectifiers, able to stand forward voltage reapplied within their own turn-on time, can be used. Alternatively, a cushioning arrangement may be added to the basic circuit. Essentially, reactance is inserted between the complementary pair of controlled rectifiers to support some of the d-c voltage just after switching, thereby limiting the rate of rise of reapplied forward voltage.

Other types of inverters can employ the limited inverse voltage mode of commutation, as well as the arrangements discussed in Section 7.1. There, no inductance is used between the d-c input and the a-c output terminals, resulting in good regulation and high efficiency. The commutating inductance carries only the commutating impulse, and can be small. However, these circuits require auxiliary controlled rectifiers and a more complex gating circuit. Arrangements in which the auxiliary controlled rectifiers are swapped for some extra reactance are described later in this chapter.

7.3   USE OF FEEDBACK RECTIFIERS

The role of the feedback rectifiers in the circuit of Section 7.1 is of such importance as to merit further discussion. When an inverter is supplying an inductive load, the load current must be provided with an alternate path after a controlled rectifier has been suddenly turned off by the commutating impulse. The conducting polarity of the complementary controlled rectifier is not appropriate, but a feedback rectifier connected (directly or effectively) in inverse parallel with such complementary controlled rectifier will provide the necessary conducting path. A similar feedback rectifier across the first controlled rectifier functions during the other half-cycle. The commutating capacitor supplies load current only during the brief commutating intervals, while the current is in the process of transferring from the outgoing controlled rectifier to the incoming feedback rectifier. During the remainder of the cycle, the commutating capacitor is inactive.

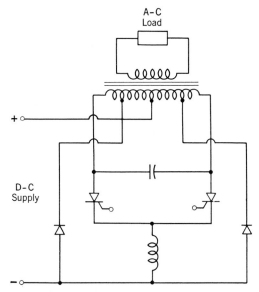

Figure 7.16   Impulse-commutated inverter derived from the parallel inverter.

Feedback rectifiers are also used in the complementary impulse-commutated inverter circuit of Figure 7.16, which is derived from the parallel inverter circuit shown in Figure 4.2 (Chapter 4). In fact, the feedback rectifiers are the only difference evident in the circuit diagram. However, in the parallel inverter, the commutating capacitor is the only alternate path for inductive load current, and must therefore be large to prevent excessive voltage buildup after commutation.

Feedback rectifiers are so termed because they return or "feed back" to the d-c supply the reactive power associated with inductive or capacitive loads. Cyclic exchange of energy between the d-c supply and the a-c load, or reactive power flow, takes place as follows. With a capacitive load, some of the energy drawn from the d-c supply through a controlled rectifier during the first part of a half-cycle of load voltage is temporarily stored in the load capacitance, and then fed back to the d-c supply through a feedback rectifier later in the same half-cycle. In the case of inductive loads, energy stored in the load inductance during the latter part of a half-cycle is returned to the d-c supply during the first part of the next half-cycle.

By handling reactive loads in this manner, the need for a large commutating capacitor and a large d-c choke is eliminated, since these elements no longer have to store large amounts of reactive energy for exchange with the load. Instead, a much smaller commutating capacitor can be used in conjunction with a small inductance, which is termed the "commutating inductance," to produce a commutating impulse. While impulse commutation and feedback rectifiers are conceptually separate, in practice they are closely linked. All of the inverters discussed in this chapter use feedback rectifiers.

The technique of using feedback rectifiers and impulse commutation to improve the performance and size of inverters is a relatively new development. The circuits described in this chapter, unlike those of previous chapters, were developed using silicon-controlled rectifiers and ordinary silicon feedback rectifiers. Hence, these circuits postdate the invention of the SCR. When only gas-tube power rectifiers were available, the size of the tubes and their accessories made the use of feedback rectifiers less attractive. Also, the longer turn-off time made impulse commutation less advantageous. Thus, there was less incentive to improve inverter circuits along the lines discussed in this chapter.

Reverse Power Flow

Another significant property imparted to inverters by using feedback rectifiers is the ability to feed back continuous power from the a-c side to the d-c side. That is, when operated in parallel with an a-c system through a suitable filter and control means, such an inverter can also act as a rectifier. This feature is useful where the inverter is employed as a standby power supply. Normally, the equipment will charge a d-c battery from the a-c power system. If the a-c source fails, the inverter can immediately reverse its direction of power transfer, and convert power from the battery into the a-c loads to maintain continuity of service. Reversal of power flow is also desirable where the a-c load includes a motor which can "overhaul." Note particularly that reversal of power flow occurs by reversal of the direct current polarity, and not by reversal of the direct voltage polarity, as in the case of a phase-controlled rectifier which is phased back beyond 90° to become an a-c line commutated inverter (see Figure 3.4, Chapter 3).

To operate in parallel with an a-c generator or as a rectifier load on an a-c generator, the inverter must be provided with several auxiliary features. An a-c filter is necessary to render its waveshape compatible with the generator sine wave. For stable operation, the filter must include some net effective series inductance. The fundamental component of the raw inverted d-c voltage corresponds to the open circuit voltage of a synchronous machine, while the inductive series reactance of the filter is analogous to the synchronous reactance of a machine. Unless both the d-c source voltage and the a-c system voltage are well regulated, the inverter must have some means of voltage control, such as discussed in chapter 8, besides the usual matching transformer. This is equivalent to field control of a synchronous machine.

Furthermore, the inverter must have some means of maintaining synchronism with the generator. The oscillator that drives the inverter can be rigidly locked to the frequency of the a-c generator, with provision for a small amount of phase variation for controlling power flow. Alternatively, the oscillator frequency can be made dependent upon the load power of the inverter, drooping in the same manner as the speed of the prime mover that drives the generator. When operating in parallel with the generator, the inverter frequency control becomes a phase or power flow control, while the voltage adjustment now controls reactive power. Other methods of synchronizing are also possible.

Operating in this manner, the inverter is a static analogue of a rotating synchronous converter or of a motor-generator set. It can convert power from dc to ac or from ac to dc, as desired, and at any power factor on the a-c side. By manipulating the controls to maintain constant a-c current, but at a varying phase angle ϕ with respect to the voltage, the diagram shown in Figure 7.17 can be obtained. This shows how the continuous power (watts) and reactive power (vars) vary with the phase angle ϕ when the volt-amperes are kept constant. Note that the d-c current is proportional to the watts. Thus, impulse-commutated inverters with feedback rectifiers can convert power over the full 360° range of a-c phase variation. No change in the commutation technique or circuit components is necessary.

The power-phase diagram of Figure 7.17 may be compared with that of Figure 6.3 (Chapter 6) for harmonic-commutated inverters. In the latter, the phase angle ϕ of the a-c current is the same as the firing angle α of the controlled rectifiers. With the circuits of this chapter, the firing angle of the controlled rectifiers relative to the a-c system voltage is very small, and corresponds to the load angle or torque angle of a synchronous

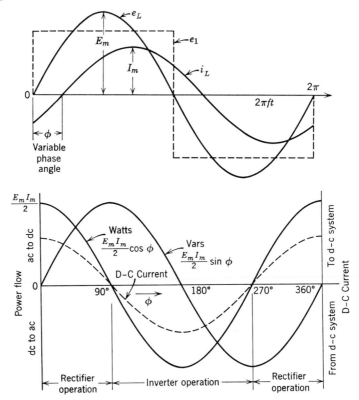

Figure 7.17   Power-phase diagram for an impulse-commutated inverter connected to an a-c system.

e_1 = square-wave output voltage of the inverter before filter.
e_L = a-c system voltage = inverter voltage after filter (the phase shift caused by the filter is neglected).
i_L = fundamental current *from* the a-c system *to* the inverter.

machine. Also, in Figure 6.3 the d-c current as well as the a-c current is kept constant, while the d-c voltage is proportional to the watts, and reverses with the direction of power transfer.

Derivation of Impulse-Commutated Inverters from Separate Inverting and Rectifying Equipments

Other types of inverters incorporating their own means of commutation, as opposed to a-c line voltage commutation, may also be operated in parallel with an a-c system. Both the parallel and series inverters can be so applied, but are able only to invert power into the a-c system. If reversal of power flow is required, a separate rectifier equipment must be added to the installation.   By certain interconnections and combinations of the inverter and rectifier components, the over-all size of the equipment can be reduced, and its performance improved.   Then, if such an equipment is operated independently of the a-c system, its performance as a pure inverter is also improved, particularly its ability to handle wide variations in load and power factor.   An example of this type of arrangement is shown in Figure 7.18.   The load seen by the parallel inverter is maintained

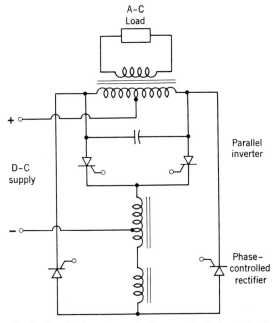

Figure 7.18   Parallel inverter with phase-controlled rectifier feedback.

approximately constant, as is necessary for its optimum performance, by suitable regulation of the phase-controlled rectifier.   The portion of the inverter output power not required by the a-c load is fed back into the d-c supply by the rectifier equipment.   However, when the true a-c load is light, a considerable amount of power circulates between the inverter and rectifier portions of the circuit.

The intimate interconnection of the inverter controlled rectifiers and the feedback rectifiers, which impulse commutation requires, may be regarded as an extreme case of inverter-rectifier combination.   It avoids circulating excessive power at light loads.   To put it another way, the close relationship of controlled rectifiers and feedback rectifiers enables efficient commutation by means of an impulse.

Simplification of Analysis

The feedback rectifiers not only are essential to the performance, but also simplify the analysis of impulse-commutated inverters. They hold the a-c output voltage to an approximate square wave proportional to the d-c input voltage, independent of the magnitude or power factor of the load. The well-known harmonic voltage components of the square wave can be applied to any given linear load impedance to obtain the harmonic components of the resultant load current. Usually, this harmonic analysis method is employed to design a filter which will reduce the harmonics to a desired limit, as discussed in Chapter 9. The commutating capacitance and inductance of the inverter circuit do not have any great effect upon the output voltage, as they do in the case of the parallel and series inverters.

Conversely, the exact nature of the load impedance need not be specified in order to perform an adequate analysis of the commutating circuit. Rather, it is necessary to know only the value of the load current during the brief interval in which commutation occurs. For design purposes, the maximum value of load current that must be commutated is the critical factor in the analysis.

Thus, the analysis of the commutating circuit can be divorced from the analysis of the load circuit, and the inverter circuits discussed in this chapter are treated in this manner. The approximations involved in this approach are practically justified when the period of the a-c output is long compared with the turn-off time of the controlled rectifiers, but introduce errors as the operating frequency increases. As seen in Chapters 4 and 5, performance calculations for the parallel and series inverters require a knowledge of the load impedance. If the load is changed, the calculations must be repeated. Without the aid of a computer, study of a wide range of complex loads is impractical, and selection of the optimum sized commutating components is difficult. For impulse-commutated inverters, performance calculations of a more general nature can be made. In particular, simple explicit equations for the optimum values of commutating capacitance and commutating inductance can be derived.

7.4   COMPLEMENTARY IMPULSE-COMMUTATED INVERTER, DERIVED FROM THE PARALLEL INVERTER

For the purposes of explanation and analysis, the circuit arrangement of Figure 7.19 is selected. This is another version of the circuit of Figure 7.16, having the same mode of operation. Without the feedback rectifiers, it reduces to a corresponding version of the parallel inverter. The d-c power supply requires a center-tap, or a neutral point established by a pair of large capacitors, which can be of the electrolytic type. By itself, this arrangement is useful where a small amount of a-c power is required from a relatively high voltage d-c source. However, it has greater importance as the building block for bridge-type circuits.

In Figures 7.16 and 7.19, the feedback rectifiers are connected to taps on the primary winding of the output transformer, rather than to the ends

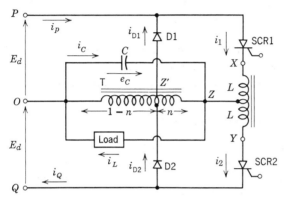

Figure 7.19   Half-bridge configuration of a complementary impulse-commutated inverter (McMurray-Bedford inverter).

of the winding, in order to enable the feedback of energy stored and otherwise "trapped" in the inductance L just after commutation, as will be described later. This results in some variation of the output voltage with the power factor of the load, but far less than is the çase with the parallel inverter.

To simplify the analysis, the following idealizing assumptions will be made.

(1) The period of the a-c output of the inverter is very much greater than the turn-off time of the silicon-controlled rectifiers.

(2) Series inductance included as part of the total load will maintain the load current substantially constant at a value I_L throughout the commutating interval. This assumption is made irrespective of whether the net power factor of the load is lagging, unity or leading. The series inductance can include the load itself, part of a filter, and the leakage inductance of the output transformer. The total effective series inductance needs to be large compared with the commutating inductance, L, but since the latter is quite small, the requirement is often met in practice.

(3) An ideal autotransformer, T, is assumed in order to establish a tap

point, Z', for connection of the feedback rectifiers. The section $Z'Z$ of the transformer winding is a small fraction "n" of the total winding OZ and is tightly coupled to it.

(4) The two halves of the commutating inductance are tightly coupled.

(5) The turn-on time of the controlled rectifiers is negligible, and their reverse current during turn-off is negligible.

(6) Losses in the inverter elements are neglected.

(7) Zero source impedance is assumed, and this is approached in practice by use of a large input capacitor of the electrolytic type. This capacitor is a necessity when the source is a rectifier equipment.

It should be noted that the practical fulfillment of assumptions 2 and 3 is made possible by assumption 1. The theory of operation developed below shows close correlation with the performance of an actual inverter.[1]

Operation with Lagging Power-Factor Load

One half-cycle of operation is illustrated in the waveforms sketched in Figure 7.20. The voltages of points P, Q, X, Y, Z, and Z', as indicated in Figure 7.19, are shown with respect to the d-c neutral point 0. Across any circuit element, the voltage is the difference between the appropriate curves. For example, the voltage across controlled rectifier SCR1 is the difference in voltage level between P and X. The locations of the currents sketched in Figure 7.20 are also specified in Figure 7.19, where the arrows indicate the direction in which the current is considered positive. The heads of the voltage arrows in Figure 7.19 indicate the polarity considered positive.

Each half-cycle is divided into five intervals, designated A, B, C, D, and E, which are bounded by changes in the state of a controlled rectifier or a feedback rectifier from a nonconducting to a conducting condition or vice versa. Such a change effectively alters the equivalent circuit of the inverter. The operation can be completely analyzed by writing the equations for each equivalent circuit and matching the boundary conditions. However, this approach would be unduly complex and would depend upon the load impedance, so that general conclusions would be difficult to extract.

From the standpoint of the designer, the most important part of the cycle is the commutating impulse, interval B in Figure 7.20. Since it is so short, it is drawn with an expanded time scale, about 10 times the scale for the other intervals. An arbitrary waveform for the load current i_L is sketched, and the approximation that i_L remains constant at a value I_L during the commutating interval B appears exaggerated, due to the

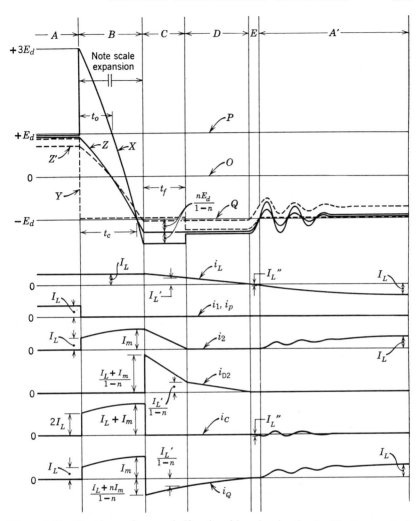

Figure 7.20 Waveforms for one half-cycle with a lagging load. The two vertical parallel lines under interval *B* indicate the duration of interval *B* to the same scale as the other intervals.

expanded time scale. With reference to Figures 7.19 and 7.20, a half-cycle of operation will be described in sequence.

INTERVAL *A*. Controlled rectifier SCR1 is conducting current to the load, drawing power from the positive line *P* of the d-c supply. If the rate of change of load current is moderate, the voltage across inductance

L is a small fraction of the d-c supply voltage E_d so that terminal Z of the load is close to the potential of the positive line P and the capacitor C is charged to the voltage E_d. At the end of interval A, the load current has attained a value I_L.

INTERVAL B.   To commutate, controlled rectifier SCR2 is gated "on," dropping the potential of point Y to the negative line Q of the d-c supply. Since the voltage on capacitor C cannot change instantaneously, a voltage of $2E_d$ must appear across winding ZY of inductance L. The same voltage is induced in winding XZ, placing an inverse voltage of $2E_d$ on controlled rectifier SCR1 and turning it off. The current I_L, originally flowing in SCR1 and winding XZ, must be transferred to winding ZY and SCR2 in order to preserve the energy stored in the inductance. Thus, a current of $2I_L$ must be drawn from capacitor C to supply the load and provide the current in SCR2.

With these initial conditions, the transient form of the impulse is calculated on pp. 198–200. The waveform is part of a sinusoid with period $2\pi\sqrt{LC}$, but the duration t_c of the commutating interval is less than one-quarter cycle, and the time t_0 available for SCR1 to recover its forward blocking ability is still shorter. This is the time for which point X remains above the potential of line P, placing inverse voltage on SCR1.

At the time when load terminal Z reaches the potential of the negative d-c line Q, the voltage across the commutating inductance is zero and the current i_2 has built up to a maximum value I_m. If the feedback rectifiers were connected to terminal Z instead of the tap Z', D2 would now conduct, and the current I_m would circulate through winding ZY, SCR2, and D2, thereby "trapping" the energy $\frac{1}{2}LI_m^2$. Feedback rectifier D2 would also carry the load current, feeding energy from the load back to the negative side of the d-c supply, until such time as the load current reversed. However, the trapped energy would eventually be dissipated as losses in winding ZY, SCR2, and D2 so that the size of these components would be greatly increased, and the circuit efficiency would be low.

INTERVAL C.   By connecting the feedback rectifiers to the tap Z', the trapped energy can also be fed back to the d-c supply. The onset of interval C is delayed until point Z' reaches the potential of the negative d-c line Q, bringing rectifier D2 into conduction. The current i_2 will have dropped from its maximum value I_m, but if the fraction n is small (0.1 to 0.2), the drop in i_2 will be slight. Rectifier D2 clamps the magnitude of the voltage impressed on winding OZ' to the d-c supply voltage E_d. The voltage $nE_d/(1 - n)$, induced in the winding section $Z'Z$, is applied to section YZ of the commutating inductance and reduces the trapped circulating current linearly from the value I_m to zero in time t_f

$$t_f = \frac{LI_m(1 - n)}{E_d n} \tag{7.45}$$

This current, circulating through winding $Z'Z$, is transformed into winding $Z'O$ and fed back into the d-c source. During this interval, the magnitude of the load voltage is $E_d/(1 - n)$, and the voltage across the commutating inductance raises the reapplied forward voltage on controlled rectifier SCR1 to the value $2E_d/(1 - n)$.

INTERVAL D. This interval occurs after the trapped circulating current has been reduced to zero, but the continuing feedback of energy from the inductive load keeps rectifier D2 in conduction. The load voltage remains at the value $E_d/(1 - n)$. The voltage $nE_d/(1 - n)$ induced in winding section $Z'Z$ now appears as inverse voltage across SCR2.

INTERVAL E. When the load current reverses its polarity, rectifier D2 blocks, and the increasing load current in the new direction is supplied by discharge of the capacitor C until the potential of terminal Z reaches the level of the d-c line Q.

INTERVAL A'. Controlled rectifier SCR2 now commences to conduct again. In order to do so, after being reverse biased during intervals D and E, a positive gating signal must be reapplied at this time, or else the gating voltage applied at the start of interval B must have been maintained. It is found more convenient to use square-wave gating signals.

Since the load current built up during interval E cannot immediately pass through the inductance L, the capacitor C initially continues to discharge, starting a period of "ringing" between L and C. The peak ringing voltage is less than $nE_d/(1 - n)$. Eventually, the ringing is damped out by the load and by losses. At the end of interval A', the load current has reached a magnitude I_L, as in the corresponding interval A when SCR1 was conducting.

A similar sequence of events follows in the half-cycle following commutation from SCR2 to SCR1 again. With high power-factor loads, intervals D and E are absent. The load current reverses during interval C, and rectifier D2 blocks when the increasing load current in the new direction meets the decreasing trapped circulating current. Thus, interval C merges directly into interval A', and SCR2 maintains continuous conduction. The no-load condition can be considered as a special case in which i_L, including I_L, is set equal to zero.

Operation With Leading Power Factor Load

Waveforms, illustrating the operation of the inverter circuit of Figure 7.19 with leading power factor load, are drawn in Figure 7.21. One-half cycle is shown, which is divided into six intervals, a, b, c, d, e, and f. The commutating intervals b and c have an expanded time scale, about 10

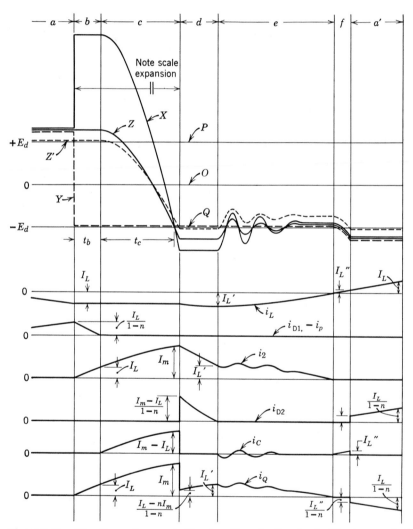

Figure 7.21. Waveforms for one half-cycle with a leading load. The two vertical parallel lines under interval C indicate the duration of intervals b and c to the same scale as the other intervals.

times the scale for the other intervals. An arbitrary load current waveform is sketched and, during intervals b and c, an approximately constant magnitude I_L is assumed. The sequence of operation in a half-cycle will now be analyzed, with reference to Figures 7.19 and 7.21.

INTERVAL a.   The load current is negative and rectifier D1 is conducting, feeding energy from the load back into the positive side of the d-c supply. The voltage across the load and capacitor C is raised to the value $E_d/(1 - n)$ by the action of the autotransformer T. Controlled rectifier SCR1 is inversely biased by the voltage $nE_d/(1 - n)$ induced in winding $Z'Z$. At the end of interval a, the load current is $-I_L$, and the current in D1 is $I_L/(1 - n)$.

INTERVAL b.   Commutation is initiated by gating "on" controlled rectifier SCR2, effectively connecting point Y to the negative line Q of the d-c supply. The capacitor C maintains point Z at its former potential, so that a voltage of $E_d(2 - n)/(1 - n)$ is impressed across winding ZY of the commutating inductance. The same voltage induced in winding XZ increases the inverse voltage on SCR1 to $2E_d/(1 - n)$. However, in the case of leading loads, this turn-off action is redundant.

The current i_2 increases linearly from zero to the value I_L in a time t_b given by

$$t_b = \frac{LI_L(1 - n)}{E_d(2 - n)} \tag{7.46}$$

During the same period, the current in D1 is reduced to zero, after which D1 blocks. Essentially, the load current has been transferred from the path through D1 to the path through SCR2.

INTERVAL c.   The capacitor C now enters into the commutating transient, producing an impulse which increases the current i_2 from I_L to a peak value I_m as C reverses its charge. Interval c is analyzed on pp. 201–202 and lasts for a little over one-quarter cycle of the natural frequency of L and C.

INTERVAL d.   This interval is quite similar to interval C with lagging load. Rectifier D2 conducts, and the tapped transformer permits recovery of the excess energy trapped in the inductance L after commutation. The current i_2 falls linearly from the peak value I_m at the rate of $nE_d/(1 - n)L$ amperes per second until i_2 is equal to the load current, assumed to have reached a value I_L' by the end of interval d.

INTERVAL e.   Rectifier D2 blocks, and the magnitude of the load voltage falls from $E_d/(1 - n)$ to approximately E_d, after a period of "ringing" between capacitor C and inductance L. This is similar to interval A' with lagging load.

INTERVAL f.   When the load current reverses, controlled rectifier SCR2 blocks, and the increasing load current in the new direction charges capacitor C back to the level $E_d/(1 - n)$ volts.

INTERVAL a'.   Rectifier D2 starts conducting again, and feeds energy from the load back into the negative side of the d-c supply. The conditions are now reversed from the original state in interval a.

Analysis of Commutating Impulse, Interval B, with Lagging Power-Factor Load

Referring to Figures 7.19 and 7.20, time is measured from the instant when controlled rectifier SCR2 is turned "on." The current i_2 flowing in controlled rectifier SCR2 is chosen as the independent variable. The initial conditions at time $t = +0$ are

$$\text{Current in inductance } L: \ i_2(+0) = I_L$$

$$\text{Voltage on capacitor } C: \ e_c(+0) = E_d$$

The differential equation describing the commutating transient may be written in terms of the voltages around the loop $OZYQ$, Figure 7.19:

$$E_d = L\frac{di_2}{dt} - e_C(+0) + \int_0^t \frac{i_2 + I_L}{C}\, d\tau \tag{7.47}$$

The Laplace transform of equation (7.47) is, with the initial conditions,

$$\frac{2E_d}{s} = Lsi_2(s) - LI_L + \frac{i_2(s)}{sC} + \frac{I_L}{s^2 C} \tag{7.48}$$

Solving equation (7.48) for $i_2(s)$, the transform of $i_2(t)$,

$$i_2(s) = \frac{\dfrac{2E_d}{L} + sI_L - \dfrac{I_L}{sLC}}{s^2 + \dfrac{1}{LC}} \tag{7.49}$$

The inverse transform of equation (7.49) gives

$$i_2 = \frac{2E_d}{\omega L}\sin \omega t + I_L(2\cos \omega t - 1) \tag{7.50}$$

where

$$\omega = \frac{1}{\sqrt{LC}} \tag{7.51}$$

If the fraction n is small, the commutating interval B ends very soon after the voltage on the inductance L passes through zero, that is, when

$$e_{YZ} = L\frac{di_2}{dt} = 0 \tag{7.52}$$

This occurs after a time t_c, where

$$\tan \omega t_c = \frac{E_d}{\omega L I_L} \tag{7.53}$$

or

$$t_c = \sqrt{LC} \tan^{-1} x \tag{7.54}$$

where the parameter x is defined as

$$x = \frac{E_d}{I_L}\sqrt{\frac{C}{L}} \tag{7.55}$$

This may be expressed as the ratio of the "instantaneous" load impedance at commutation, E_d/I_L, to the characteristic impedance of the commutating circuit, $\sqrt{L/C}$.

The maximum value I_m attained by the current i_2 is obtained by putting $t = t_c$ in equation (7.50)

$$I_m = 2\sqrt{\frac{C}{L}E_d{}^2 + I_L{}^2} - I_d \tag{7.56}$$

The ratio I_m/I_L may be stated as a function $f(x)$ of the parameter x:

$$\frac{I_m}{I_L} = f(x) = 2\sqrt{x^2 + 1} - 1 \tag{7.57}$$

The time t_0 during which controlled rectifier SCR1 has reverse voltage, as a fraction of the natural period of the commutating circuit ($\sqrt{LC}$ seconds per radian), may also be derived in terms of x.

$$\frac{t_0}{\sqrt{LC}} = g(x) = \sin^{-1}\frac{x}{\sqrt{x^2 + 1}} - \sin^{-1}\frac{x}{2\sqrt{x^2 + 1}} \tag{7.58}$$

The energy W, trapped in the inductance L after commutation, is

$$W = \tfrac{1}{2}LI_m{}^2 = \tfrac{1}{2}LI_L{}^2[f(x)]^2 \tag{7.59}$$

$$\frac{W}{E_d I_L t_0} = h(x) = \frac{[f(x)]^2}{2xg(x)} = \frac{(2\sqrt{x^2 + 1} - 1)^2}{2x\left(\sin^{-1}\dfrac{x}{\sqrt{x^2 + 1}} - \sin^{-1}\dfrac{x}{2\sqrt{x^2 + 1}}\right)} \tag{7.60}$$

In equation (7.60), the quantity $E_d I_L t_0$ is the energy which is diverted from flowing through SCR1 during the period when it is reverse biased, while W is the total energy developed by the commutating impulse in order to accomplish this diversion. The commutation parameters defined by equations (7.57), (7.58), and (7.60) are graphed in Figure 7.22.

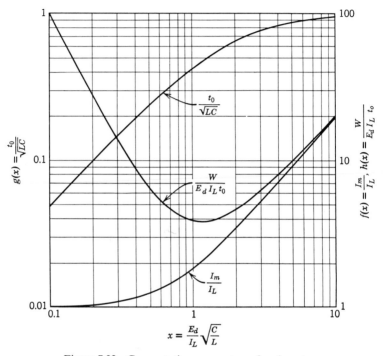

Figure 7.22   Commutation parameters of an inverter.

The analysis in this subsection applies in general to any case where a controlled rectifier is carrying current just before commutation, as well as to actual lagging loads. For example, the case of pure resistive loading is similar.

Commutating Impulse With No Load

The equations describing the commutating interval when the inverter is unloaded may be derived from the equations for the case of lagging power-factor load setting I_L equal to zero, or x infinite. Equation (7.50) becomes

$$i_2 = \frac{2E_d}{\omega L} \sin \omega t \tag{7.61}$$

and

$$t_c = \frac{\pi}{2}\sqrt{LC} \tag{7.62}$$

$$t_0 = \frac{2}{3}t_c = \frac{\pi}{3}\sqrt{LC} \tag{7.63}$$

$$I_m = 2E_d\sqrt{\frac{C}{L}} \tag{7.64}$$

$$W = 2CE_d^2 \tag{7.65}$$

The above equations apply not only to the case of no load, but to any condition where the load current is substantially zero during the commutating interval. For example, a resistance load with a series L-C filter tuned to the fundamental a-c component draws a current in phase with the fundamental a-c voltage. Thus, the current is zero when the a-c voltage is zero, that is, during commutation.

Analysis of Commutating Impulse, Interval c, with Leading Power-Factor Load

With reference to Figures 7.19 and 7.21, time will be measured from the instant when rectifier D1 blocks.

The initial conditions at time $t = +0$ are

Current in inductance L: $i_2(+0) = I_L$
Voltage on capacitor C: $e_c(+0) = E_d/(1 - n)$

The differential equation describing the commutating transient in terms of i_2 as the independent variable may be written in terms of the voltages around the loop $OZYQ$, Figure 7.19:

$$E_d = L\frac{di_2}{dt} - \frac{E_d}{1 - n} + \int_0^t \frac{i_2 - I_L}{C}\, d\tau \tag{7.66}$$

The Laplace transform of equation (7.66) is, with the initial conditions,

$$\frac{E_d}{s}\left(\frac{2 - n}{1 - n}\right) = Lsi_2(s) - LI_L \frac{i_2(s)}{sC} - \frac{I_L}{s^2 C} \tag{7.67}$$

Solving equation (7.67) for $i_2(s)$, the transform of $i_2(t)$,

$$i_2(s) = \frac{\dfrac{E_d}{L}\left(\dfrac{2 - n}{1 - n}\right) + sI_L + \dfrac{I_L}{sLC}}{s^2 + \dfrac{1}{LC}} \tag{7.68}$$

Applying the inverse transformation to equation (7.68),

$$i_2 = \frac{E_d}{\omega L}\left(\frac{2-n}{1-n}\right)\sin \omega t + I_L \tag{7.69}$$

It follows that

$$t_c = \frac{\pi}{2\omega} = \frac{\pi}{2}\sqrt{LC} \tag{7.70}$$

and

$$I_m = I_L + E_d\left(\frac{2-n}{1-n}\right)\sqrt{\frac{C}{L}} \tag{7.71}$$

These results apply generally to any case where a feedback rectifier is carrying current just before commutation, as well as to true leading power-factor loads. For example, pronounced harmonic currents of suitable phase can produce this condition, even though the fundamental component of the load current has unity or lagging power factor.

Selection of Commutating Capacitance and Inductance

The most severe commutation duty occurs under the conditions analyzed on pp. 198–200, when the current I_L through the controlled rectifiers just before commutation has a maximum specified value I_{L0}. If the d-c supply voltage $2E_d$ varies, its minimum value is critical. Here, the reverse voltage time t_0 must not be less than the maximum turn-off time t_{00} specified for the controlled rectifiers. The parameter x, equation (7.55) will have a minimum value x_0 which the designer may select. Then, substituting these particular values in equations (7.55) and (7.58) and manipulating, normalized equations for the capacitance C and inductance L are obtained:

$$C\frac{2E_d}{I_{L0}t_{00}} = \frac{2x_0}{g(x_0)} \tag{7.72}$$

$$L\frac{I_{L0}}{2E_d t_{00}} = \frac{1}{2x_0 g(x_0)} \tag{7.73}$$

The functions represented by equations (7.72) and (7.73) are plotted in Figure 7.23. Any pair of values of C and L obtained from Figure 7.23 will provide a sufficient commutating impulse.

The minimum value of capacitance is readily estimated by noting that it must supply a current at least equal to twice the load current throughout the turn-off time, during which the capacitor is discharged from a voltage magnitude E_d to zero. Then, equating CE_d to $2I_{L0}t_{00}$ and manipulating into the form of equation (7.72), the minimum value of four seen in Figure 7.23 is verified. This must be increased to allow for the additional current built up in the inductance during commutation. The smaller the inductance is, the larger is the current increment, and the larger must be the capacitance. The parameter x_0 is a measure of the relative size of C and L. It is desired to determine an optimum choice of x_0 which results in most efficient operation.

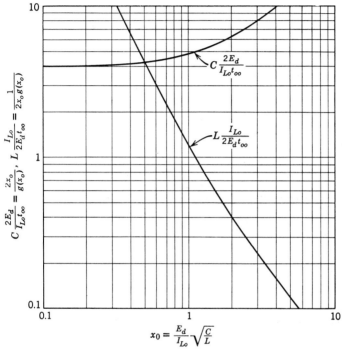

Figure 7.23   Variation of C and L with parameter x_0.

The major criterion for selecting x_0 is taken to be the minimization of the energy W trapped in the inductance L after commutation. This energy not only determines the rating of the inductance, but also increases the circuit losses because of imperfect efficiency in feeding the energy back to the d-c supply. The circulating power involved is $2Wf$ watts. It is seen from Figure 7.22 that the selection $x_0 = 1.15$ will minimize the trapped energy W_0 under the most severe commutation condition, specified above,

$$\frac{W_0}{E_d I_{Lo} t_{00}} = h(x_0) = 3.87 \text{ minimum} \tag{7.74}$$

However, this choice of x_0 does not minimize the trapped energy W when the load current at commutation I_L is less than the maximum value I_{L0}. With no load, for instance, we obtain from equations (7.65) and (7.72):

$$\frac{W}{E_d I_{L0} t_{00}} = \frac{2x_0}{g(x_0)} \tag{7.75}$$

This is the same function of x_0 as given by equation (7.72) and plotted in Figure 7.23, which is a minimum when x_0 is small. For intermediate

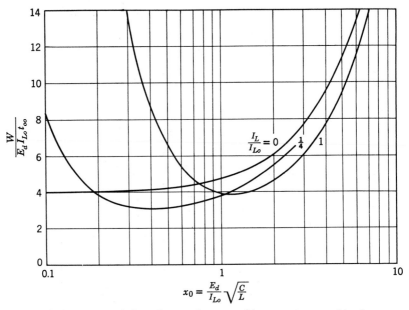

Figure 7.24   Variation of trapped energy with parameter x_0 and load.

levels of loading, the energy W can be related to the fixed quantity $E_d I_{L0} t_{00}$ with the aid of equations (7.59) and (7.74):

$$\frac{W}{E_d I_{L0} t_{00}} = h(x_0) \frac{W}{W_0} = h(x_0) \left[\frac{I_L f(x)}{I_{L0} f(x_0)} \right]^2 = h(x_0) \left[\frac{x_0 f(x)}{x f(x_0)} \right]^2 \tag{7.76}$$

In Figure 7.24, the variation of $W/E_d I_{L0} t_{00}$ with x_0 is shown for each of the following load conditions.

(1) $I_L = I_{L0}$, $x = x_0$, maximum load, equation (7.74).
(2) $I_L = I_{L0}/4$, $x = 4x_0$, calculated from equation (7.76).
(3) $I_L = 0$, $x \to \infty$, no load, equation (7.75).

It is seen that if x_0 is within the range 0.75 to 1.15, the value of W changes little with load, and is close to a minimum. For most inverter designs, the selection of $x_0 = 1$ is judged to be the optimum, for which $g(1) = 0.425$ (Figure 7.22). Then, from equations (7.72) and (7.73), the optimum values of C and L are given by

$$C = \frac{t_{00}I_{L0}}{0.425E_d} \tag{7.77}$$

$$L = \frac{t_{00}E_d}{0.425I_{L0}} \tag{7.78}$$

The highest peak current in a controlled rectifier is the greatest value of I_m, which occurs with a leading power-factor load, when the d-c supply voltage is a maximum, according to equation (7.71). These conditions also give the maximum value of W.

Another parameter of interest is the rate of rise of reapplied forward voltage on a controlled rectifier after it has been turned off. The steepest slope occurs when $I_L = I_{L0}$, and is nearly uniform. A forward voltage of $2E_d$ is reapplied in the time

$$t_{c0} - t_{00} = \left[\frac{\tan^{-1} x_0}{g(x_0)} - 1\right]t_{00} \tag{7.79}$$

$$= 0.85t_{00} \quad \text{when} \quad x_0 = 1$$

The average rate of voltage rise is then $2E_d/0.85t_{00}$, which must not exceed the maximum allowable rate of rise specified for the controlled rectifiers.

Selection of the Tap Fraction n

The average current in the controlled rectifier during interval C is $I_m/2$. If a constant value E_a is assumed for the forward drop of the controlled rectifier, the energy dissipated in the SCR during interval C is

$$\tfrac{1}{2}I_mE_at_f = \frac{LI_m^2E_a(1-n)}{2E_dn} = W\frac{E_a(1-n)}{E_dn} \tag{7.80}$$

where the value of t_f has been obtained from equation (7.45). Thus, a certain fraction of the trapped energy W is lost in the controlled rectifier, and more is lost in the feedback rectifier and the windings of the inductance and transformer.

For efficient recovery of the trapped energy, a large value of n is desirable, especially when the d-c supply voltage is low. However, a small value of n is desired to minimize the peak voltage $2E_d/(1-n)$ applied to the controlled rectifiers, and to reduce the variation of load voltage with power factor. The best compromise appears to be the selection of a value of n in the range 0.1 to 0.2, tending toward the upper limit when the d-c supply voltage is low and the inverter frequency is high.

The relatively large magnitude of the trapped circulating current and energy is one disadvantage of this circuit arrangement, compared with other impulse commutated inverters. However, at a frequency of 60 cps where the cycle period is about 800 times longer than the turn-off time of typical silicon controlled rectifiers, the performance is generally satisfactory. In small size inverters, for example 1 kva or less, it may be expedient to eliminate the taps on the transformer and dissipate the trapped energy in resistors connected in series with the feedback rectifiers. At a frequency of 400 cps, where the cycle period is typically 120 times longer than the turn-off time, the handling of the trapped energy may become a problem in high performance inverters. Above 1000 cps, it is generally better to select an alternative circuit arrangement.

Bridge-Circuit Configurations

A single-phase bridge version of the circuit discussed above is shown in Figure 7.25. This is essentially two circuits of the type shown in Figure

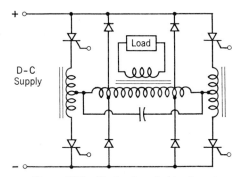

Figure 7.25   Single-phase bridge inverter.

7.19 which are operated 180° out of phase, with a common transformer and commutating capacitor, making a d-c neutral point unnecessary.

The three-phase bridge inverter in Figure 7.26 is also derived from the circuit of Figure 7.19. In each phase, the commutating capacitance is split into two halves, connected from the a-c line to each side of the d-c supply. Alternatively, one capacitor per phase could be connected to a d-c neutral point. In either case, the mode of operation is the same. Each of the a-c line voltages (points A, B, and C) exhibit an approximate square wave with respect to the midpotential of the d-c supply. However, the three-phase connection eliminates third harmonics from the a-c load voltages, both line-to-line and line to a-c neutral. Hence, there is a third harmonic voltage between the neutral points of the a-c and d-c systems. These waveshapes will be discussed further in Chapters 8 and 9.

Unless the primary windings of the output transformers are wye-connected to the d-c neutral point, the potential of a line in which a

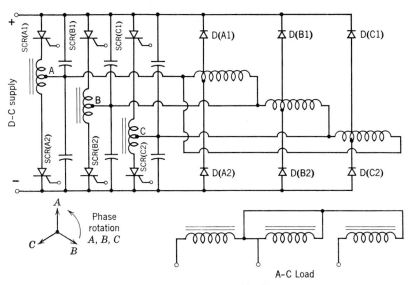

Figure 7.26 Three-phase bridge inverter.

feedback rectifier is conducting depends upon the state of the other two lines as well as the transformer tap fraction n. With delta-connected primary windings, as shown in Figure 7.26, the location of the taps must be appropriate to the phase rotation in order to force feedback of the energy trapped in the commutating inductances. For example, just after controlled rectifier SCR(A1) is gated on, current will circulate through SCR(A1) and feedback rectifier D(A1). To force decay of this current, induced voltage must raise the potential of point A above the positive d-c line. Hence, line B must be negative and either SCR(B2) or D(B2) must be conducting.

As in the case of the circuits in Figures 7.16 and 7.19, the bridge circuits of Figures 7.25 and 7.26 also revert to corresponding versions of the parallel inverter if the feedback rectifiers are omitted, and the size of the commutating capacitance and inductance is increased.

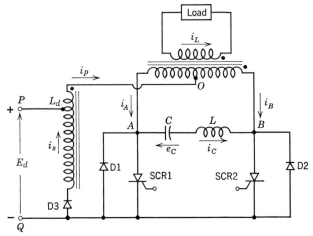

Figure 7.27 Inverter with center-tapped ac (limited inverse voltage).

7.5 COMPLEMENTARY IMPULSE-COMMUTATED INVERTER WITH LIMITED INVERSE VOLTAGE

Like other families of inverters, this type of circuit can exist in two basic configurations. One is the center-tapped a-c arrangement of Figure 7.27, using a single-way transformer. This is the prototype of polyphase single-way, or star circuits. The other basic configuration is the center-tapped d-c circuit of Figure 7.28, using a double-way transformer, which is the building block for bridge circuits.

Depending upon the particular type of inverter, it is generally easier to describe and understand the mode of operation with reference to one,

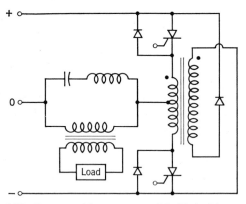

Figure 7.28 Inverter with center-tapped dc (limited inverse voltage).

rather than the other, of these basic configurations. In the present case, it is believed that the single-way version of Figure 7.27 is easier to understand. Therefore, the theory of operation will be developed for this arrangement.

In principle, each controlled rectifier performs a dual role, combining the two functions which were assigned to separate main and auxiliary controlled rectifiers in the inverter of Section 7.1. When first gated on, a controlled rectifier serves as an "auxiliary" unit to impress a turn-off commutating pulse upon its complementary controlled rectifier. At the end of the commutating interval, it assumes the role of a "main" controlled rectifier and carries load current when so required. During commutation, the transformer primary windings in Figure 7.27 are short-circuited, thereby collapsing the load voltage to zero, while the d-c supply voltage is held off by a linear reactor L_d provided for this purpose. To limit the overshoot of load voltage which occurs after commutation, while the magnetic flux in reactor L_d is resetting, a secondary winding on L_d is connected across the d-c supply through an auxiliary feedback rectifier D3.

Theory of Operation

To simplify the analysis, all the pertinent assumptions listed in Section 7.4 will be made here, too. In addition, it will be assumed that all the windings on the transformer in Figure 7.27 are tightly coupled, and that the secondary winding has the same number of turns as either half of the primary winding. Also, the secondary winding of the choke L_d has N times as many turns as the primary winding and is closely coupled to it.

It should be borne in mind that many of the results presented are approximate solutions of approximate equations. A rough approximation may suffice for qualitative understanding of the circuit, but better accuracy is desirable for design work. In the following analysis, the implications of some of the approximations, and the difference between certain degrees of accuracy, will be studied with reference to parameter ratios typical of practical design. More particularly, the conditions $L_d = 6L$, $N = 5$, and $Q = 10$ will be taken as an example. The waveform illustrations are drawn for these values.

Typical waveforms of the load voltage developed by this circuit are shown in Figure 7.29, for both no load and inductive load conditions. In each half-cycle, four distinct intervals bounded by some discontinuity are present, as indicated in Figure 7.29. The region during and immediately following commutation (intervals *1*, *2*, and the beginning of *3*) will be studied more closely. In a given half-cycle, the difference in magnitude

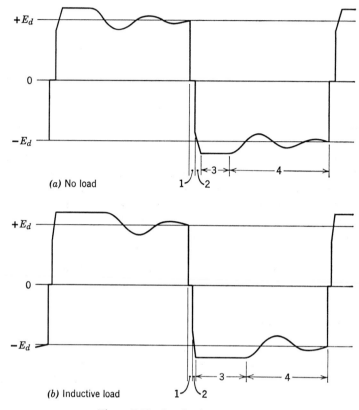

Figure 7.29   Load voltage waveforms.

between the load voltage and the d-c supply voltage E_d appears across the choke L_d. Under steady-state conditions, this choke voltage must integrate to zero average value, so that the average magnitude of the load voltage is E_d.

Analysis of No Load Condition

In Figure 7.27, the current arrows indicate the direction in which current flow will be considered positive. The heads of the voltage arrows indicate the polarity considered positive in the equations.

At the end of the half-cycle in which controlled rectifier SCR1 conducts, it is assumed that all previous oscillations have died out. Therefore, the transformer center tap O (Figure 7.27) is at the potential E_d of the positive d-c line P and, since end A is connected to the negative line Q through

SCR1, the induced potential at end B is $2E_d$. The commutating capacitor C is charged to the voltage $2E_d$, with terminal A negative.

INTERVAL 1. To start interval 1 of the next half-cycle, controlled rectifier SCR2 is gated on, connecting point B to the negative line Q. Capacitor C discharges through inductance L, SCR2, and feedback rectifier D1. The forward voltage drop across D1 appears as inverse voltage across SCR1, turning it off. The equivalent discharge circuit is drawn in Figure 7.30(a), on which the initial conditions are indicated. The arrows in Figures 7.30(a) to 7.30(d) indicate the same conventionally positive polarities as shown in Figure 7.27. Substituting these initial conditions into the general equations (7.7) and (7.12), derived in the

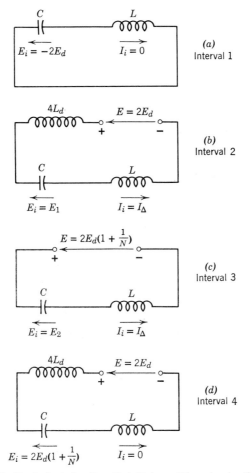

Figure 7.30   Equivalent circuits with initial conditions (no load).

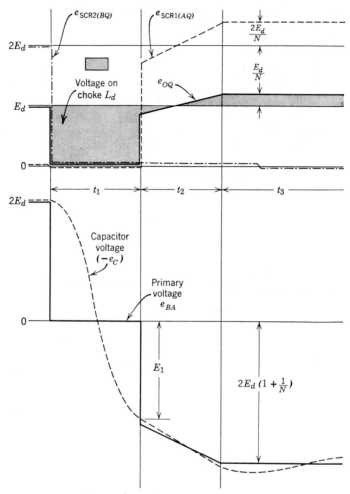

Figure 7.31   Voltage waveforms (no load).

analysis on pp. 175–179, the approximate equations for the damped sinusoidal commutating transient are

$$i_C = \frac{2E_d}{X} \epsilon^{-(\omega t/2Q)} \sin \omega t \qquad (7.81)$$

$$e_C = -2E_d \epsilon^{-(\omega t/2Q)} \cos \omega t \qquad (7.82)$$

where

$$\omega = \frac{1}{\sqrt{LC}} \quad \text{and} \quad X = \sqrt{\frac{L}{C}}$$

The waveforms are sketched in Figure 7.31 and 7.32, interval *1*.

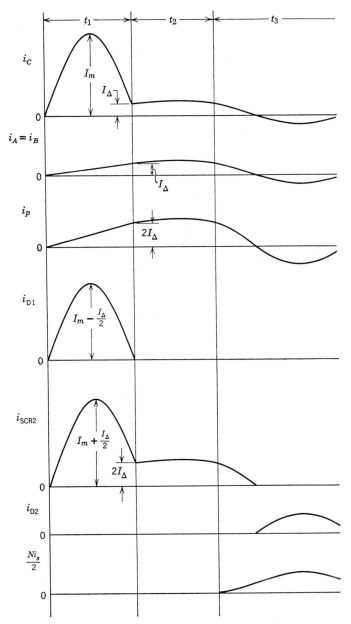

Figure 7.32   Current waveforms (no load).

During interval *1*, when both SCR2 and D1 are conducting, points *A*, *B* and therefore *O* are all at the negative potential of *Q*, short-circuiting the transformer primary windings so that the output voltage is zero. The d-c supply voltage E_d is supported by the inductance L_d, which draws a linearly rising current from the source:

$$i_p = \frac{E_d t}{L_d} \tag{7.83}$$

This current splits equally between the two halves of the transformer windings:

$$i_A = i_B = \tfrac{1}{2} i_p = \frac{E_d t}{2L_d} \tag{7.84}$$

The current in controlled rectifier SCR2 is $(i_C + i_B)$, while the current in feedback rectifier D1 is $(i_C - i_A)$. Interval *1* ends when the rising current i_A meets the commutating current i_C as it falls toward the end of its first half-cycle. This causes rectifier D1 to block, and the equivalent circuit is changed. If I_Δ is the value of currents i_C, i_A at the end of interval *1* of duration t_1, then by equations (7.81) and (7.84)

$$I_\Delta = \frac{2E_d}{X} \exp\left(-\frac{\omega t_1}{2Q}\right) \sin \omega t_1 \tag{7.85}$$

$$I_\Delta = \frac{E_d t_1}{2L_d} \tag{7.86}$$

The simultaneous solution of (7.85) and (7.86) gives quite accurate values of I_Δ and t_1, but cannot be obtained algebraically. As a first approximation, we may assume that

$$t_1 \approx \pi\sqrt{LC} \tag{7.87}$$

which, put into equation (7.86), gives

$$I_\Delta \approx \frac{E_d \pi \sqrt{LC}}{2L_d} \tag{7.88}$$

A more accurate solution may be obtained by noting that

$$\sin \omega t_1 = \sin(\pi - \omega t_1) \approx \pi - \omega t_1 \tag{7.89}$$

Substituting approximations (7.87) and (7.89) in equation (7.85),

$$I_\Delta \approx \frac{2E_d}{X} \epsilon^{-(\pi/2Q)}(\pi - \omega t_1)$$

$$= \frac{2E_d}{L} \epsilon^{-(\pi/2Q)}(\pi\sqrt{LC} - t_1) \tag{7.90}$$

Solving equations (7.86) and (7.90) for t_1 and I_Δ, we obtain

$$t_1 = \gamma \pi \sqrt{LC} \tag{7.91}$$

$$I_\Delta = \gamma \frac{E_d \pi \sqrt{LC}}{2L_d} \tag{7.92}$$

where

$$\gamma = \frac{1}{1 + \dfrac{L}{4L_d} \epsilon^{\pi/2Q}} \tag{7.93}$$

For the typical values $L_d = 6L$, $Q = 10$,

$$\gamma = \frac{1}{1 + \dfrac{1.17}{24}} = 0.955 \tag{7.94}$$

Thus, the first approximations (7.87) and (7.88) are in error by about 5 per cent. The value of I_Δ, relative to the peak value I_m of the commutating current i_C, is

$$\frac{I_\Delta}{I_m} = \frac{\gamma \dfrac{E_d \pi \sqrt{LC}}{2L_d}}{2E_d \sqrt{C/L} \exp\left(-\dfrac{\pi}{4Q}\right)} = \gamma \pi \frac{L}{4L_d} \epsilon^{\pi/4Q} = 0.135 \tag{7.95}$$

To obtain the value E_1 of the capacitor voltage e_C at the end of interval *1*, we may note that

$$\cos \omega t_1 = -\cos(\pi - \omega t_1) \approx -1 \tag{7.96}$$

and, hence, from equation (7.82)

$$E_1 = -2E_d \exp\left(-\frac{\omega t_1}{2Q}\right) \cos \omega t_1$$

$$\approx 2E_d \exp\left(-\frac{\pi}{2Q}\right)$$

$$= 2E_d \times 0.855 \quad \text{for} \quad Q = 10 \tag{7.97}$$

It should be noted that the inverse voltage applied across the auxiliary feedback rectifier D3 is $(N + 1)E_d$ during interval *1*. With $N = 5$ typically, the PIV rating of D3 must exceed 6 times the d-c supply voltage.

INTERVAL 2. When the feedback rectifier D1 blocks, the equivalent circuit switches to the configuration of Figure 7.30(*b*), where the d-c supply voltage and the inductance L_d, both referred to the total primary winding of the transformer, are in series with the commutating L-C circuit. The final conditions of interval 1 [equations (7.92) and (7.97)] become the initial conditions of interval 2, and are indicated on Figure 7.30(b). Applying equations (7.7) and (7.12) to the new equivalent circuit,

$$i_C = \left[\frac{2E_d - E_1}{X'} \sin \omega' t + I_\Delta \cos \omega' t \right] \epsilon^{-(\omega' t/2Q)} \tag{7.98}$$

$$e_C = 2E_d + [X' I_\Delta \sin \omega' t - (2E_d - E_1) \cos \omega' t] \epsilon^{-(\omega' t/2Q)} \tag{7.99}$$

where

$$X' = \sqrt{\frac{L + 4L_d}{C}}, \qquad \omega' = \frac{1}{\sqrt{(L + 4L_d)C}} \tag{7.100}$$

and time t is now measured from the start of interval 2. Thus, the response is again a damped sinusoid, but of a lower frequency than in interval 1. With the typical condition $L_d = 6L$, we obtain $\omega' = \omega/5$ and $X' = 5X$.

The voltage reapplied to controlled rectifier SCR1 and diode D1 is the voltage from point A to B [see Figures 7.27 and 7.30(b)] which equals $2E_d$ less the voltage across $4L_d$. Since the two series inductances share the difference between the applied voltage $2E_d$ and the capacitor voltage e_C in proportion to their inductance values,

$$e_{\text{SCR1}} = 2E_d - \frac{4L_d}{4L_d + L} (2E_d - e_C) \tag{7.101}$$

$$= \frac{e_C + 2E_d \dfrac{L}{4L_d}}{1 + \dfrac{L}{4L_d}} \tag{7.102}$$

The initial reapplied forward voltage step is then

$$e_{\text{SCR1,initial}} = \frac{E_1 + 2E_d \dfrac{L}{4L_d}}{1 + \dfrac{L}{4L_d}}$$

$$= 2E_d \frac{\epsilon^{-(\pi/2Q)} + \dfrac{L}{4L_d}}{1 + \dfrac{L}{4L_d}}$$

$$= 2E_d \times 0.860 \quad \text{for} \quad Q = 10, L_d = 6L \tag{7.103}$$

which is practically the same as E_1, the initial capacitor voltage. In general, the rate of rise of voltage must be limited by accessory means to avoid failure of the controlled rectifiers. This modifies the transition between intervals *1* and *2*, but such will not be considered here.

Interval *2* ends when the voltage induced in the secondary winding of choke L_d becomes equal to the d-c supply voltage, causing auxiliary feedback rectifier D3 to conduct and limit any further voltage rise. The voltage at point O (and the load voltage) is clamped at the level $E_d(1 + 1/N)$, and the voltage on SCR1 is twice as much. The voltage E_2 on the capacitor C at the end of interval *2* can be found by using equation (7.102):

$$2E_d\left(1 + \frac{1}{N}\right) = \frac{E_2 + 2E_d\dfrac{L}{4L_d}}{1 + \dfrac{L}{4L_d}} \tag{7.104}$$

$$E_2 = 2E_d\left[1 + \frac{1}{N}\left(1 + \frac{L}{4L_d}\right)\right] \tag{7.105}$$

Now, equating the right-hand sides of equations (7.99) and (7.105), a transcendental equation for t_2, the duration of interval 2, is obtained. An approximate solution can be reached by assuming that $\omega' t_2$ is a small angle, so that we may put $\sin \omega' t_2 \approx \omega' t_2$ and $\epsilon^{-(\omega' t_2/2Q)} \approx 1$ in equation (7.99). Thus,

or
$$E_2 \approx 2E_d + X'I_\Lambda \omega' t_2 - (2E_d - E_1) \tag{7.106}$$

$$t_2 \approx \frac{C(E_2 - E_1)}{I_\Lambda} \tag{7.107}$$

From equation (7.107), it is seen that we have, in effect, assumed that the current in the circuit remains constant at I_Λ throughout interval 2, so that the capacitor is charged linearly from voltage E_1 to voltage E_2. The closeness of the approximation can be checked by estimating the value of $\omega' t_2$ with the aid of equations (7.107), (7.100), (7.105), (7.97), and (7.92):

$$\omega' t_2 = \frac{2E_d C\left[1 + \dfrac{1}{N}\left(1 + \dfrac{L}{4L_d}\right) - \epsilon^{-(\pi/2Q)}\right]}{\sqrt{(L + 4L_d)C}\,\dfrac{E_d \pi \sqrt{LC}}{2L_d}}$$

$$= \frac{1 + \dfrac{1}{N}\left(1 + \dfrac{L}{4L_d}\right) - \epsilon^{-(\pi/2Q)}}{\gamma\pi\sqrt{\left(1 + \dfrac{L}{4L_d}\right)\dfrac{L}{4L_d}}} \tag{7.108}$$

$= 0.566$ radians or $32.5°$ with $N = 5$, $L_d = 6L$, $Q = 10$.

Thus, the accuracy of the assumption that $\omega' t_2$ be small is fair.

INTERVAL 3. With the diode D3 conducting to limit the transformer primary voltage to $2E_d(1 + 1/N)$, the equivalent circuit for the commutating components is as shown in Figure 7.30(c). The initial conditions for interval 3 are the final conditions for interval 2: the capacitor voltage is E_2, given by equation (7.105), and the current is still approximately I_Δ [equation (7.92)]. These initial conditions may be substituted into equations (7.7) and (7.12) to obtain equations for the commutating circuit transient. The capacitor voltage oscillates about the level

$$2E_d(1 + 1/N)$$

at a frequency $\omega = 1/\sqrt{LC}$. Generally, the duration of interval 3 is long enough for the oscillations to die out. It should be noted that the oscillating current is carried by SCR2 and D2 alternately.

The energy dissipated by the oscillations depends on the difference between the initial and final conditions of the components:

$$W_3 = \tfrac{1}{2}C\left[E_2 - 2E_d\left(1 + \frac{1}{N}\right)\right]^2 + \tfrac{1}{2}LI_\Delta{}^2 \tag{7.109}$$

$$\approx \tfrac{1}{2}C(2E_d)^2\left(\frac{L}{4L_dN}\right)^2 + \tfrac{1}{2}C(2E_d)^2\left(\gamma\pi\frac{L}{4L_d}\right)^2 \tag{7.110}$$

$$= \tfrac{1}{2}C(2E_d)^2 \times 0.00007 + \tfrac{1}{2}C(2E_d)^2 \times 0.0156 \tag{7.111}$$

where the values are calculated for $N = 5$, $L_d = 6L$. This is a small loss, even compared with the loss involved in producing the commutating pulse during interval 1, which is (with $Q = 10$ as an example):

$$W_1 = \tfrac{1}{2}C(2E_d)^2 - \tfrac{1}{2}CE_1{}^2 - \tfrac{1}{2}LI_\Delta{}^2 \tag{7.112}$$

$$\approx \tfrac{1}{2}C(2E_d)^2(1 - \epsilon^{-(\pi/Q)}) - \tfrac{1}{2}LI_\Delta{}^2 \tag{7.113}$$

$$= \tfrac{1}{2}C(2E_d)^2 \times 0.27 - \tfrac{1}{2}LI_\Delta{}^2 \tag{7.114}$$

The voltage across the primary winding of the choke L_d during interval 3 is E_d/N, and the equation for the currents referred to the primary side is

$$Ni_s + i_p = i_{p,\,\text{initial}} - \frac{E_dt}{NL_d} \tag{7.115}$$

where $i_p = 2i_C$, i_p initial $\approx 2I_\Delta$ and time is measured from the start of interval 3. Thus, the secondary current is given by

$$\frac{N}{2}i_s \approx I_\Delta - \frac{E_dt}{2NL_d} - i_C \tag{7.116}$$

Interval *3* ends when i_s becomes zero and rectifier D3 blocks. If it is assumed that the commutating circuit current i_C has decayed to zero by this time, the duration t_3 of interval *3* can be obtained from equation (7.116):

$$t_3 \approx \frac{2I_\Delta L_d N}{E_d} \qquad (7.117)$$

Comparing equations (7.117) and (7.86), it is seen that

$$t_3 = Nt_1 \qquad (7.118)$$

The physical interpretation of these results is that the flux established in the choke L_d during interval *1* is reset to zero during interval *3*. Also, the energy drawn from the d-c supply and stored in L_d during interval *1* is fed back into the d-c supply during interval *3* (less some losses, of course).

INTERVAL *4*. When rectifier D3 blocks, the equivalent circuit reverts to the same form as during interval *2*, but with initial conditions as shown in Figure 7.30(*d*). The capacitor voltage oscillates about the level $2E_d$, at the lower frequency $\omega' = 1/\sqrt{(L + 4L_d)C}$. As in interval *3*, the oscillating current is carried alternately by D2 and SCR2. If the output frequency is low enough, the oscillations decay before the next commutation. The energy dissipated by the oscillations is then

$$W_4 = \tfrac{1}{2}C(2E_d)^2 \left[\left(1 + \frac{1}{N} \right) - 1 \right] \qquad (7.119)$$

$$= \tfrac{1}{2}C(2E_d)^2 \frac{1}{N^2} \qquad (7.120)$$

$$= \tfrac{1}{2}C(2E_d)^2 \times 0.04 \quad \text{for} \quad N = 5 \qquad (7.121)$$

Comparing equations (7.111), (7.114), and (7.121), it is seen that most of the no-load losses occur during intervals *1* and *4*, the division being dependent upon the values of Q and N. Neglecting the relatively small losses in intervals *2* and *3*, the total no-load power loss at low frequency f is

$$p_{\text{loss}} \approx 2f(W_1 + W_4) \qquad (7.122)$$

$$= fC(2E_d)^2 \left(1 - \epsilon^{-(\pi/Q)} + \frac{1}{N^2} \right) \qquad (7.123)$$

$$= fC(2E_d)^2 \times 0.31 \text{ for our sample conditions}$$

As the output frequency increases, the next commutation occurs before all the energy has been dissipated in interval *4*. The initial conditions for interval *1* must then include a different voltage on capacitor C or current in L and L_d. Analysis becomes complicated and dependent upon frequency. The magnitude of the commutating pulse generated during interval *1* can be higher or lower than under low-frequency conditions. This makes the circuit unsuitable for variable high-frequency operation. However, for a fixed frequency, the circuit can be designed to give a maximum commutating pulse. The theoretical upper frequency limit f_m is reached when interval *4* has shrunk to nothing. Any further increase in frequency will encroach upon the resetting time t_3 of reactor L_d.

$$f_m \leq \frac{1}{2\pi\sqrt{LC}(N+1)} \qquad (7.124)$$

At the upper limit frequency, the initial capacitor voltage is $2E_d(1 + 1/N)$, and the no-load loss is approximately

$$p_{\text{loss},\,m} \approx f_m C (2E_d)^2 \left(1 + \frac{1}{N}\right)^2 [1 - \epsilon^{-(\pi/Q)}] \qquad (7.125)$$

$$= f_m C (2E_d)^2 \times 0.39 \text{ for our sample conditions}$$

Thus, the high frequency loss per cycle is close to that at low frequency. The magnitude of the commutating pulse is increased by the factor $(N+1)/N$. This circuit is therefore useful up to fairly high frequencies, about 3 kilocycles.

Analysis with Lagging Load

Detail of the first part of a half-cycle of operation with lagging load is shown in the voltage waveforms of Figure 7.33 and the current waveforms of Figure 7.34. These may be compared with the no-load waveform sketches in Figures 7.31 and 7.32, respectively. With a low operating frequency, the assumption that the load current remains constant at a value I_L during the period shown is reasonable, when the series inductance of the load is large compared with the inductance L_d.

INTERVAL *1*.   It is assumed that the initial charge on the capacitor C is still $2E_d$, so that the equivalent commutating circuit and initial conditions remain as shown in Figure 7.30(*a*). Also, equations (7.81) and (7.82) are still valid. However, the initial d-c current i_p, flowing through the choke L_d, transformer winding OA, and controlled rectifier SCR1, is equal to the load current I_L. During interval *1*, additional currents as given by

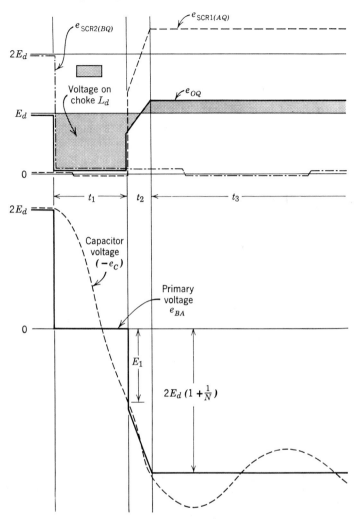

Figure 7.33   Voltage waveforms (lagging load).

equations (7.83) and (7.84) begin to flow, so that the total currents are

$$i_p = I_L + \frac{E_d t}{L_d} \tag{7.126}$$

$$i_A = I_L + \frac{E_d t}{2L_d} \tag{7.127}$$

$$i_B = \frac{E_d t}{2L_d} \tag{7.128}$$

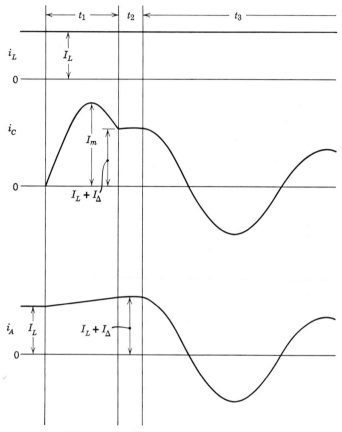

Figure 7.34   Current waveforms (lagging load).

As the commutating current i_C rises above the current i_A, conduction in SCR1 is extinguished, and the excess of i_C over i_A flows through feedback rectifier D1. Interval *1* ends when i_C falls back to equal i_A again, causing D1 to block. The time t_1 may be found by equating (7.127) to (7.81) and solving the transcendental equation

$$I_L + \frac{E_d t_1}{2L_d} = \frac{2E_d}{X} \exp\left(-\frac{\omega t_1}{2Q}\right) \sin \omega t_1 \qquad (\pi/2 < \omega t_1 < \pi) \quad (7.129)$$

Having determined t_1, the current $I_L + I_\Delta$ and the capacitor voltage E_1 at the end of interval *1* may be obtained from equations (7.81) and (7.82). However, it is not important to know these particular values. The inverse bias time t_0 available for turn-off of SCR1 is of prime interest. A fairly accurate estimate of t_0 may be obtained by replacing the damped sinusoidal

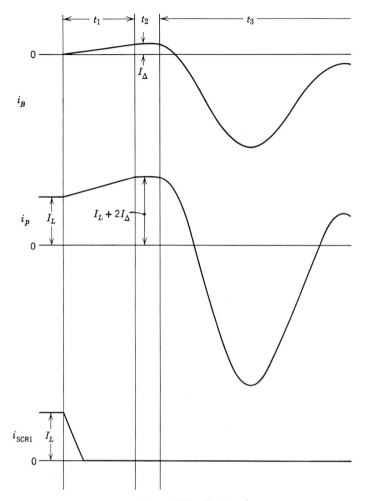

Figure 7.34   *Continued.*

current i_C [equation (7.81)] with an undamped sinusoid having the same peak value, and replacing the linearly rising current i_A [equation (7.127)] with a constant current equal to the true value at the time $(\pi/2)\sqrt{LC}$ (see Figure 7.35):

$$i_C = \frac{2E_d}{X} \epsilon^{-(\pi/4Q)} \sin \omega t \qquad (7.130)$$

$$i_A = I_L + \frac{E_d \pi \sqrt{LC}}{4L_d} \qquad (7.131)$$

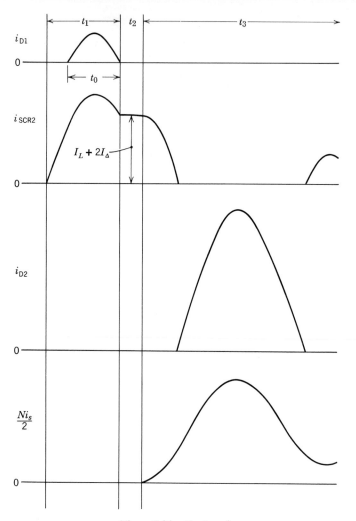

Figure 7.34   *Continued.*

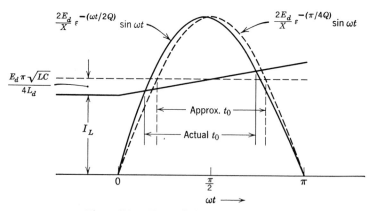

Figure 7.35   Turn-off time approximation.

224

Then, from Figure 7.35, it is seen that

$$t_0 \approx 2\sqrt{LC} \cos^{-1} \frac{I_L + \dfrac{E_d \pi \sqrt{LC}}{4L_d}}{\dfrac{2E_d}{X} \exp\left(-\dfrac{\pi}{4Q}\right)} \tag{7.132}$$

The optimum design of the commutating circuit is similar to that for the inverter circuit discussed on pp. 179–182 except that allowance must be made for an increment in current above the load current, as given by equation (7.131).

INTERVAL 2. The equivalent circuit is the same as for no-load conditions [Figure 7.30(b)], but the initial conditions are different. However, the final capacitor voltage E_2 is the same as before [equation (7.105)]. The capacitor C is charged from a voltage E_1, the end condition of interval 1, to the voltage E_2 by an approximately constant current $I_L + I_\Delta$.

INTERVAL 3. Here again, the equivalent circuit of Figure 7.30(c) applies, with the same initial capacitor voltage but with an initial current of approximately $I_L + I_\Delta$. As at no load, the capacitor voltage oscillates about the level $2E_d(1 + 1/N)$ at a frequency

$$\omega = \frac{1}{\sqrt{LC}}$$

but the amplitude is greater because more energy is trapped in the inductance L. In both of the basic dissipation equations (7.109) and (7.112), the last term becomes $\frac{1}{2}L(I_L + I_\Delta)^2$. However, with or without load, the total dissipation in intervals 1 and 3 is given by the expression

$$W_1 + W_3 \approx \tfrac{1}{2}C(2E_d)^2\left(1 + \frac{L}{4L_dN}\right)^2 - \tfrac{1}{2}CE_1^2 \tag{7.133}$$

$$\approx \tfrac{1}{2}C[(2E_d)^2 - E_1^2] \tag{7.134}$$

Since E_1 becomes smaller as the load increases, the losses are greater. The current in the secondary winding of the choke L_d is given by

$$\frac{Ni_s}{2} = i_L + I_\Delta - \frac{E_d t}{2NL_d} - i_C \tag{7.135}$$

where the load current i_L decreases from its initial value I_L. Interval 3 ends when i_s becomes zero and D3 blocks. Comparing equations (7.116) and (7.135), it is seen that the duration of interval 3 depends upon the load current, and is longer than with no load.

INTERVAL *4*. The equivalent circuit is similar to Figure 7.30(*d*), with modification due to loading. The capacitor voltage oscillates about the level $2E_d$, at the frequency ω', and dissipates about the same energy as at no load [equation (7.120)]. However, with inductive load, the oscillations persist longer than at no load. The load is a high impedance at the frequency ω', and the load current maintains either SCR2 or D2 in conduction. Hence, the oscillations see only the low incremental resistance of a diode, rather than the full forward drop. With resistive loading, the oscillations are more rapidly damped.

Comparison with Previous Circuits

Operation of the circuit has been analyzed in detail for the conditions of no load and lagging power-factor load. As in the case of the other impulse-commutated inverters, the circuit can also operate with purely resistive loads and with leading power-factor loads.

This inverter was conceived as a compromise between the two circuits previously discussed in Sections 7.1 and 7.4. It attempts to incorporate the better features and avoid some of the disadvantages of each of these previous circuits. In particular, the *L-C* circuit that generates the commutating pulse is allowed almost one half-cycle of natural oscillation through a local low-impedance path, so that most of the energy stored in the commutating capacitor before commutation is returned to the capacitor after commutation, and only the losses have to be made up from the d-c supply, as in the inverter of Section 7.1. With the circuit of Section 7.4, substantial energy is drawn from the d-c supply during commutation. This energy becomes "trapped" after commutation and results in a heavy circulating current through the controlled rectifiers and feedback rectifiers while it is being fed back into the d-c supply. The present circuit involves less trapped energy and employs a better method of feeding it back. In general, to circumvent the inefficient interchange of energy between the commutating components and the d-c supply, the commutating pulse must not pass through the d-c supply.

7.6 HIGH-FREQUENCY INVERTER

As noted in the opening paragraph on p. 165 the commutating impulse occupies a greater portion of each half-cycle as the operating frequency increases. Eventually, a frequency is reached where the next commutating impulse starts immediately upon completion of the recovery from the last commutation. At this frequency, those parts of certain inverter circuits that are used to block or limit oscillation of the commutating circuit between successive commutations can be removed. Although the imagination must be stretched to recognize such reduced circuits as impulse-commutated, they exhibit similar output characteristics.

An example of this type of circuit that provides a simple, but efficient, high-frequency inverter is shown in the half-bridge configuration of Figure 7.36. The commutating means is simply a series L-C circuit connected in parallel with the load. Its mode of operation is illustrated by the waveforms in Figure 7.37. For simplicity, the load current is

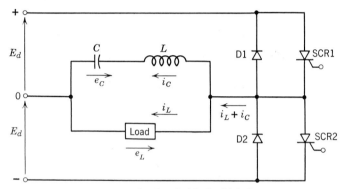

Figure 7.36   Inverter circuit suitable for high frequency.

assumed to be a lagging square wave, such as obtained with a phase-controlled rectifier or a series saturable reactor. In another form of the circuit, a pair of back-to-back auxiliary-controlled rectifiers are connected in series with the L-C circuit, and are gated in the same manner as the auxiliary-controlled rectifiers in the circuit of Section 7.1. Essentially, oscillation of the commutating circuit is interrupted by the auxiliary-controlled rectifiers at the times A and B, Figure 7.37, until the desired half-cycle duration has elapsed. Thus, it is another type of auxiliary impulse-commutated inverter with limited inverse voltage, but is less efficient than the arrangement of Section 7.1, since the commutating pulse passes through the d-c supply.

The reduced commutating circuit in Figure 7.36 can also be regarded as a filter, tuned to have a net leading power factor at the operating frequency of the inverter, and designed to over compensate the lagging power factor of the heaviest load. In this way, the inverter always sees a leading power factor load, that is, current switches from a controlled rectifier to a feedback rectifier before the end of each half-cycle (see Figure 7.37). Hence, the controlled rectifiers are turned off before the next half-cycle begins with turn-on of the complementary controlled

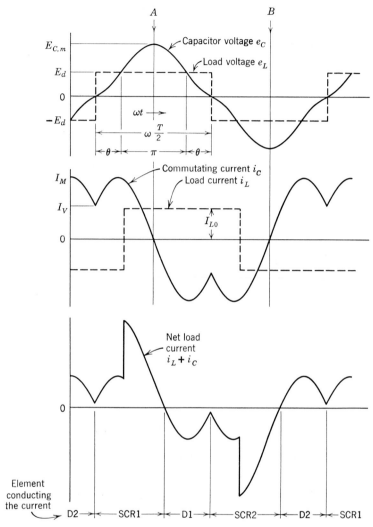

Figure 7.37   Waveforms in high frequency inverter (drawn for $\theta = 60°$).

rectifiers. It has previously been noted in Sections 7.1 and 7.4 that the commutating impulse is redundant when the load has a leading power factor.

In general, inverters that supply only leading power-factor loads do not require special commutating means. Where filters are used to make the load have a leading power factor, such filters can be regarded as the commutating circuit. More complex filters can provide harmonic attenuation in addition to commutation.[2]   However, power-factor correcting filters are quite large unless the operating frequency is high. Approaching impulse commutation from this point of view, the commutating circuit is a special kind of filter designed to reduce the size by producing a leading current with an extremely distorted waveform that "corrects" the power factor only during the critical commutating intervals.

Analysis of Commutating Circuit

With the square-wave load voltage generated by the circuit of Figure 7.36, the L-C circuit draws a double-peaked current having a cusped valley at the time of commutation, as seen in Figure 7.37. To preserve the assumed square-wave voltage, the valley current I_V must equal or exceed the maximum load current I_{L0} in magnitude. Denoting the natural angular frequency and characteristic impedance of the L-C circuit by $\omega = 1/\sqrt{LC}$ and $X = \sqrt{L/C}$ respectively, equations (7.7) and (7.12) derived by the general analysis on pp. 175–179 may be applied. At the beginning of a positive half-cycle, the conditions are

$$E = E_d, \qquad E_i = 0, \qquad I_i = I_V \qquad (7.136)$$

Then, the commutating circuit equations during a positive half-cycle are (neglecting the exponential damping factor):

$$i_C = \frac{E_d}{X} \sin \omega t + I_V \cos \omega t \qquad (7.137)$$

$$e_C = E_d + X I_V \sin \omega t - E_d \cos \omega t \qquad (7.138)$$

The period T of the square-wave output may be related to the value of ω through the angular parameter θ, as defined in Figure 7.37.

$$\omega T = 2\pi + 4\theta \qquad (7.139)$$

Equations (7.137), (7.138), and (7.139) can now be used to derive expressions for the peak capacitor voltage $E_{C,m}$, peak current I_M, and valley current I_V in terms of E_d, X, and the parameter θ. Substituting the conditions at the end of the positive half-cycle, $\omega t = \omega T/2 = \pi + 2\theta$, and $i_C = -I_V$, into equation (7.137)

$$-I_V = -\frac{E_d}{X} \sin 2\theta - I_V \cos 2\theta \qquad (7.140)$$

from which

$$\frac{X I_V}{E_d} = \cot \theta \qquad (7.141)$$

The peak current I_M occurs when $\omega t = \theta$ in equation (7.137)

$$I_M = \frac{E_d}{X} \sin \theta + I_V \cos \theta \qquad (7.142)$$

Equation (7.142) with (7.141) gives

$$\frac{I_M}{I_V} = \sec \theta \qquad (7.143)$$

The peak voltage $E_{C,m}$ occurs when $\omega t = (\pi/2) + \theta$ in equation (7.138).

$$E_{C,m} = E_d + XI_V \cos \theta + E_d \sin \theta \qquad (7.144)$$

From equations (7.144) and (7.141)

$$\frac{E_{C,m}}{E_d} = 1 + \operatorname{cosec} \theta \qquad (7.145)$$

By squaring equation (7.145) and dividing by the product of (7.139) and (7.141), we obtain

$$\frac{CE_{C,m}^2}{TI_V E_d} = \frac{(1 + \operatorname{cosec} \theta)^2}{(2\pi + 4\theta) \cot \theta} \qquad (7.146)$$

Since the quantity $TI_V E_d$ is a constant determined by the load specifications, equation (7.146) is proportional to the energy storage rating of the capacitor. Similarly, squaring equation (7.143), multiplying by (7.141), and dividing by (7.139), we obtain

$$\frac{LI_M^2}{TI_V E_d} = \frac{\sec^2 \theta \cot \theta}{2\pi + 4\theta} \qquad (7.147)$$

which is proportional to the size of the inductance. The total size of the commutating components is proportional to the sum of equations (7.146) and (7.147), which is a minimum when $\theta \approx 45°$. Hence, this is the optimum design choice of the parameter θ. With the load current waveshape shown in Figure 7.37, the time t_0 available for turning off the controlled rectifiers is

$$t_0 = \frac{2\theta}{\omega} = \frac{\theta T}{\pi + 2\theta} \qquad (7.148)$$

When $\theta = 45° = \pi/4$ radians, $t_0 = T/6$. This determines the upper frequency limit at which the circuit can be used.

REFERENCES

1. W. McMurray and D. P. Shattuck, "A Silicon-Controlled Rectifier Inverter with Improved Commutation," *AIEE Transactions*, Volume 80, Part I, 1961, pp. 531–42.
2. R. R. Ott, "A Filter for Silicon-Controlled Rectifier Commutation and Harmonic Attenuation in High Power Inverters," AIEE Conference Paper, CP 62–222, New York, January 28–February 2, 1962.

Chapter Eight

Inverter Voltage Control

Most inverter applications require a means of voltage control. This control may be required because of variations in the inverter source voltage, regulation within the inverter, or because it is desired to provide stepless adjustment of the inverter output voltage. The methods of control can be grouped into three broad categories.

(1) Control of voltage supplied to the inverter.
(2) Control of voltage within the inverter.
(3) Control of voltage delivered by the inverter.

There are a number of well-known methods of controlling the d-c voltage supplied to an inverter or the a-c voltage delivered by an inverter. These include the use of saturable reactors, magnetic amplifiers, induction regulators, phase-controlled rectifiers, and transistor-series or shunt regulators. With the introduction of high-speed, efficient, and extremely reliable solid-state switching devices, including transistors and silicon-controlled rectifiers, considerable effort has been expended to develop new methods of voltage control. In general, these improved controls involve switching techniques where voltage control is achieved by some form of switching time-ratio control. Chapter 10 includes a discussion of several important time-ratio controls to provide d-c output voltage control from a relatively fixed or independently variable source of d-c voltage. These approaches may be used to provide efficient control of the d-c voltage supplied to an inverter.

One of the most advantageous means of controlling inverter output voltage is to incorporate switching time-ratio controls within the inverter circuit. This basic form of inverter voltage control is the principal emphasis of this chapter. With this technique, it is often possible to

231

include inverter output voltage control without significantly adding to the total number of circuit components or individual component ratings. A single-phase pulse-width control technique is discussed to illustrate the important principles of this means of control. By properly gating the inverter-controlled rectifying devices, it is possible to vary the fundamental component of the inverter output voltage. Polyphase versions of this same technique are also presented. In addition, pulse-width control techniques are discussed where the inverter valves are switched "on" and "off" a number of times during each half-cycle of the inverter operating frequency. With this method of control, it is possible to substantially reduce or eliminate lower-frequency harmonics. Therefore, with a minimum of filtering, a good output voltage waveform is produced over a wide inverter voltage control range.

Another very efficient method of control involves using multiple inverters, and summing their output voltages. The resulting total voltage is controlled by varying the phase angle between the individual inverters. In certain cases, this method of control is really another form of pulse-width modulation, and the waveforms are the same as with other pulse-width modulation circuits. Circuits using multiple inverters with controlled phase displacement are particularly advantageous when sufficient power rating is involved to require multiple valves.

8.1   CONTROL OF VOLTAGE SUPPLIED TO INVERTER

There are a number of possible techniques for controlling the d-c voltage supply to an inverter and thereby controlling the inverter a-c output voltage. The principal d-c voltage control techniques used for this purpose are as follows.

(1) Induction regulators.
(2) Saturable reactors.
(3) Magnetic amplifiers.
(4) Phase-controlled rectifiers.
(5) Transistor-series or shunt regulators.
(6) Semiconductor switching-type d-c voltage controls.

Items 1 to 4 may be used only when the inverter d-c voltage is obtained from an a-c source of power, while items 5 and 6 may be used when either rectified ac or dc is available. The techniques of items 1 to 5 are quite well known and have been extensively covered in previous literature. The semiconductor switching types of voltage control are the most recent of those mentioned above. These generally involve some form of pulse-width modulation or other forms of switching time-ratio control. These switching techniques can generally provide more efficient and faster response d-c voltage controls than the other techniques listed above. A discussion of several switching-type d-c voltage controls is included in Chapter 10.

The principal advantages of voltage-control schemes, where the d-c inverter supply voltage is controlled, are stated below.

(1) The inverter output voltage waveshape and it harmonic content are not significantly changed as the voltage is controlled.

(2) In certain applications, the principal reason for voltage control is to compensate for source-voltage fluctuations. The inverter can be designed for a very limited voltage range if a control is provided to hold the inverter d-c supply voltage relatively constant in spite of wide variations in the source voltage. Inverters designed for small input voltage variations are more efficient, both in terms of power loss and component utilization.

(3) A number of well-known and reliable techniques, as mentioned earlier in this section, are available for controlling the d-c voltage delivered to an inverter.

The principal disadvantages of controlling the d-c voltage to an inverter are the following ones.

(1) The commutating voltage in many inverters is proportional to the d-c input voltage (see Chapter 7). This makes the inverter current capability decrease as the d-c voltage is reduced. Therefore, the controlling of the d-c voltage to such inverters is not desirable when a large variation in the output voltage is required and when high load current is necessary at the reduced load voltages. If these inverters are designed to reliably commutate the highest currents at a reduced d-c voltage then, at the high d-c input voltages, there is excessive commutating voltage which, generally, produces increased circulating currents resulting in higher circuit losses.

(2) The power delivered by the inverter is handled twice, once by the d-c voltage control, and once by the inverter. This generally involves more equipment than is required if the voltage control function can be within the inverter itself. This may not be the case where the supply voltage varies over a wide range while relatively constant inverter output voltage and commutation capability are required.

(3) Most efficient d-c voltage control schemes require filtering in the d-c circuit. This may cause slower response time in a complete closed-loop voltage-regulated inverter power supply.

8.2   CONTROL OF VOLTAGE WITHIN INVERTER

Parallel Inverter-Commutation Angle Control

As described in Chapter 4, the commutating angle of a parallel inverter is approximately equal to the power-factor angle of the total a-c circuit; that is, the power factor of the a-c impedance, including the commutating capacitance, output transformer, and load. With a fixed d-c voltage, the inverter output voltage varies with the commutating angle or a-c circuit power factor. Thus, it is possible to control the output voltage of a parallel inverter by varying the load power factor. This may be done in one of several different ways. These include the control of saturable reactors in shunt with the load, phase-controlling electric valves to vary the current through inductors in parallel with the load, and using feedback-rectifier circuits. These feedback circuits contain rectifiers to feed back current to the d-c source whenever the load voltage rises above a certain value, thereby tending to clamp the inverter load voltage to the specified value. Usually, inductance is connected in series with the feedback-rectifier circuits to limit the circulating currents. The inductive load produced by these circulating currents is most effective in limiting the inverter output voltage.

In general, commutation angle control is suitable for applications requiring only a small amount of voltage control. When a large control range is required, the kva rating of the components used to provide the necessary control range may become excessive.

Parallel or Series Inverter Frequency Control

The inverter operating frequency may be varied to control the output voltage of either the parallel inverter or the series inverter. In the parallel inverter this is essentially just another method of varying the power factor angle of the total a-c circuit. In the series inverter, as discussed in Chapter 5, the output voltage is changed whenever the inverter operating frequency is varied with respect to the resonant frequency of the inverter circuit. Possibly the most practical way of changing this relationship is to vary the inverter operating frequency, although it is also possible to use saturable reactors to change the resonant frequency of the circuit. Of course, inverter frequency control is applicable only when the output frequency need not be constant, and it is the most suitable for applications which require only a small amount of voltage control. For situations requiring a large amount of voltage control under heavy load, the kva rating of the components to provide the necessary control will generally become excessive.

Pulse Width Voltage Control

SINGLE PHASE. An excellent method of controlling the voltage within an inverter involves the use of pulse modulation techniques. In essence, with this technique the inverter output voltage involves a pulse width modulated wave, and the voltage is controlled by varying the duration of the output voltage pulses.[1] Figure 8.1 shows the elementary form of two inverter circuits which can be used to provide pulse width voltage control. The SCR commutating circuit components are not shown.

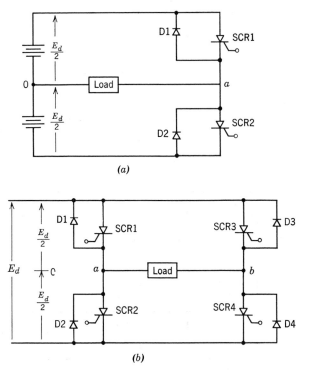

(a)

(b)

Figure 8.1   (a) Single-phase inverter with center-tapped d-c source. (b) Single-phase bridge inverter. (*Note.* The SCR commutating circuits are not shown.)

In Figure 8.1(a), for the full-voltage condition, SCR1 and SCR2 are controlled to connect point a alternately to the positive line for a complete half-cycle and to the negative d-c line for a complete half-cycle. The output voltage e_{a-0} is a square wave with amplitude $E_d/2$. Reduced output voltage may be obtained when SCR1 and SCR2 are turned on for shorter time intervals so that both valves are open for an appreciable time. For resistive loads the circuit operates in the ideal manner producing a width modulated output voltage wave with the output voltage equal to zero for the time intervals when both SCR1 and SCR2 are off. With inductive loads, the operation of the circuit of Figure 8.1(a) in this mode is more

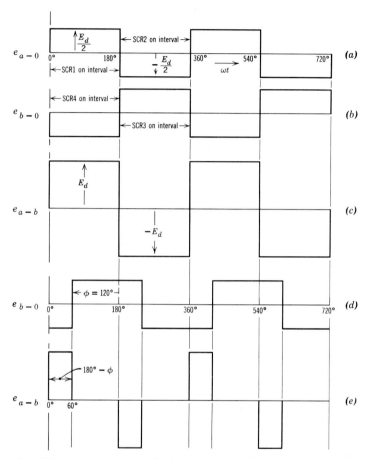

Figure 8.2   Waveforms for the circuit of Figure 8.1(b) with no phase displacement and with $\phi = 120°$.

complicated since, when both valves are off, the load current must continue to flow at least for some interval of time. In this case, the output voltage is no longer the simple width-modulated pulsating wave but, rather, there is a time during each half-cycle when the inductive load feeds back current to the d-c source and connects point a to the d-c source through one of the feedback diodes D1 or D2.

Because of the different modes of operation of Figure 8.1(a) with or without reactive load, Figure 8.1(b) will be used as the vehicle to present the concepts of pulse-width voltage control. For the present discussion, the circuit of Figure 8.1(b) is assumed to operate in the following manner. Each SCR is gated on for a time interval equal to a half-cycle of the output voltage wave. In addition, it is assumed that SCR1 and SCR2 are never on simultaneously, which means that they may be operated from a single square-wave gating supply. The same restrictions are assumed for SCR3 and SCR4. As indicated previously, the commutating circuits are not shown in Figure 8.1(b), and it is assumed that one of the commutating schemes discussed in Chapter 7 is used to turn off the SCR's after a 180°-conduction interval.

The waveforms for the full output voltage condition are thus, as shown in Figures 8.2(a), 8.2(b), and 8.2(c). Figures 8.2(a) and 8.2(b) show the voltage points a and b, respectively, to the theoretical neutral of the d-c source. The term "theoretical neutral" is used, since an actual center tap is not required on the d-c supply. However, it is convenient to discuss the voltages with respect to a center tap on the voltage source. The voltages e_{a-0} and e_{b-0} are simply square waves, since points a and b are alternately connected to one end of the d-c source and then to the other.

The load voltage is

$$e_{a-b} = e_{a-o} - e_{b-o} \qquad (8.1)$$

Figure 8.2(c) shows the load voltage for this maximum voltage condition. The load voltage will deviate from a square wave only by a small amount caused by the method of commutation actually used. The resulting a-c voltage is practically a square wave for many inverter applications, as with present SCR devices the commutation interval is relatively short for inverter operating frequencies up to 1000 cps. (see Chapter 7).

Voltage control is achieved by varying the phase of the conduction intervals of SCR1 and SCR2 with respect to SCR3 and SCR4. Figures 8.2(d) and 8.2(e) show the conditions when the conduction intervals of SCR3 and SCR4 are advanced by an angle $\phi = 120°$. The load voltage is alternating pulses of controlled width, as shown in Figure 8.2(e). This method of pulse-width control is obtained by adding two square-wave voltages, which are shifted in phase with respect to each other. As shown in Figure 8.2, the pulse width is $180° - \phi$, where ϕ is the phase displacement angle between the two square-wave voltages e_{a-o} and e_{b-o}. Thus, in this circuit, pulse-width control is achieved by the vector addition of a-c voltages e_{a-o} and e_{b-o}. The inverter output voltage can be smoothly adjusted from a maximum to zero by either phase-advancing or retarding the control signals for one pair of SCR's with respect to the other. Where

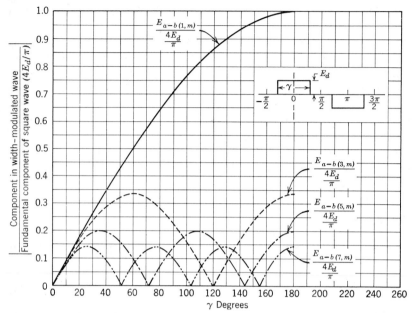

Figure 8.3   Pulse width-modulated wave harmonic content, expressed as a fraction of the fundamental component in a square wave.

voltage control is not required, the phase displacement can be fixed to produce a pulse length that will minimize certain harmonics in the output waveform.

The harmonic content of the inverter voltage, e_{a-b} in Figure 8.2, can be expressed as a function of the phase displacement angle as

$$e_{a-b} = \sum_n A_n \cos n\omega t \qquad (8.2)$$

$n = 1, 3, 5, \ldots$, (considering symmetrical pulses about 0 and π, as shown in Figure 8.3)

where

$$A_n = \frac{2}{\pi} \int_0^\pi e_{a-b} \cos n\omega t \, d(\omega t) \qquad (8.3)$$

$$e_{a-b} = E_d, \qquad \text{for} \quad 0 < \omega t < \frac{\gamma}{2} \qquad (8.4)$$

$$e_{a-b} = 0, \qquad \text{for} \quad \frac{\gamma}{2} < \omega t < \frac{\pi}{2} \qquad (8.5)$$

where

$$\gamma = 180 - \phi$$

Thus,

$$A_n = \frac{4}{\pi} \int_0^{\gamma/2} E_d \cos n\omega t \; d(\omega t) \tag{8.6}$$

$$= \frac{4}{\pi} E_d \left[\frac{\sin n\omega t}{n} \right]_0^{\gamma/2}$$

$$= \frac{4E_d}{n\pi} \sin n\frac{\gamma}{2}$$

And,

$$e_{a-b} = \sum_n \frac{4E_d}{n\pi} \left(\sin n\frac{\gamma}{2} \right) (\cos n\omega t) \tag{8.7}$$

$$n = 1, 3, 5, \ldots\ldots$$

or

$$E_{a-b(n,m)} = \frac{4E_d}{n\pi} \sin n\frac{\gamma}{2} \tag{8.8}$$

The maximum value of the fundamental component occurs when

$$\frac{\gamma}{2} = \frac{\pi}{2}, \quad \text{or} \quad e_{a-b} \text{ is a square wave, and is}$$

$$E_{a-b(1,m)} \bigg|_{\gamma/2=\pi/2} = \frac{4E_d}{\pi} \tag{8.9}$$

Figure 8.3 is a plot of the fundamental and first three harmonic components of the pulse-width modulated wave, expressed as a fraction of the maximum fundamental voltage, $4E_d/\pi$.

As shown in Figure 8.3, the harmonic voltages vary cyclically in amplitude, as the fundamental voltage is reduced by narrowing the pulse width. With very narrow pulse widths, the amplitudes of the harmonic voltages can approach the magnitude of the fundamental voltage. The harmonic voltages, expressed as a per cent of the pulse-width controlled fundamental voltage, are shown in Figure 8.4.

This is a very good method of voltage control for a reasonable output voltage range. However, when the output voltage is reduced to a low value, the harmonics become large as compared to the fundamental component.

In a practical case, in addition to the harmonic content in the output voltage, it is important to know the conduction intervals for each SCR and diode as a function of load power factor and pulse length. This information is required to select devices of the proper rating, to determine the maximum current to be commutated and to calculate the ripple current

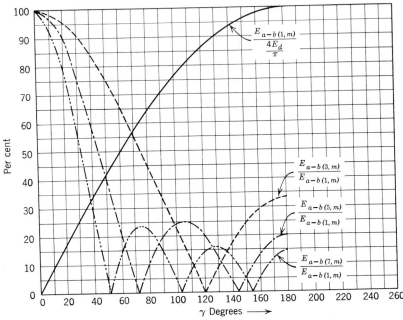

Figure 8.4   Pulse width-modulated wave harmonic content, expressed as a percentage of the fundamental component.

drawn from the d-c supply. The conduction angles for the semiconductor devices in the circuit of Figure 8.1(b) are shown in Figures 8.5 to 8.9, for several different load power factors and phase-displacement angles. The following assumptions are made for Figures 8.5 to 8.9.

(1) The SCR's and rectifiers are ideal elements; that is, they have negligible forward drop when conducting, infinite reverse resistance, and the SCR's have infinite forward resistance when off.

(2) The inverter load current is sinusoidal. In addition, it is assumed to be directly proportional to the fundamental component of the inverter output voltage, as would be the case with a fixed load impedance.

(3) The commutating currents are neglected, and the effect of commutation on the voltage waveforms is considered negligible.

With unity power-factor load and maximum inverter voltage, the fundamental components of load voltage and current are identical and in phase. The fundamental component of the square wave of voltage is a sinusoidal wave with amplitude $4E_d/\pi$, as shown in Figure 8.5. Furthermore, it is apparent that for this phase-displacement angle, $\phi = 0°$, each SCR conducts for 180°. For a reduced output voltage with 50 per cent

$4E_d/\pi$ fundamental component, a phase-displacement angle $\phi = 120°$ is required, as indicated in Figure 8.6. The fundamental component of load voltage, or current, is shifted in phase with respect to both the fundamental components of e_{a-o} and e_{b-o} by an angle equal to one half the phase

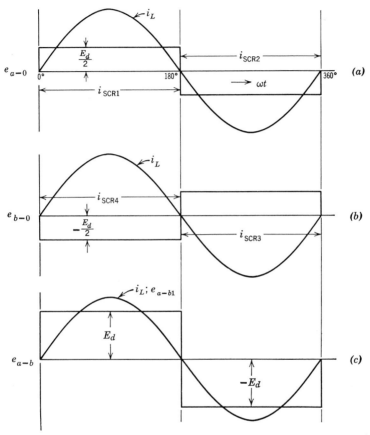

Figure 8.5   Waveforms for the circuit in Figure 8.1(b) with load power factor $= 1.0$ and $\phi = 0°$.

displacement, or $\phi/2$. The load current is leading the fundamental component of voltage e_{a-o} by $\phi/2$, and is lagging the fundamental of e_{b-o} by $\phi/2$ in Figure 8.6.

The following definitions will be used throughout these discussions.

(a) A "power-current" flow is when current is flowing through two SCR's from the d-c source to the load.

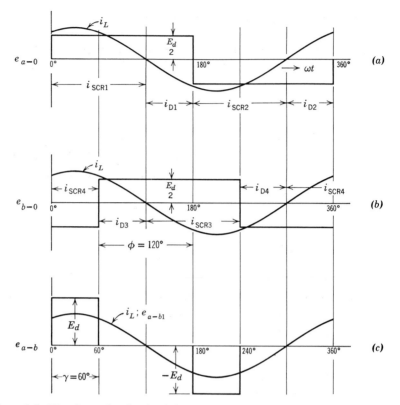

Figure 8.6   Waveforms for the circuit in Figure 8.1(*b*) with load power factor = 1.0 and $\phi = 120°$.

(*b*) A "circulating-current" flow is when the load current is flowing through one SCR and a rectifier. This circulating current does not flow through the d-c power source.

(*c*) A "feedback-current" flow is when current is flowing through two rectifiers: the load, and the d-c source. This operating mode returns power to the d-c source.

Hence, Figure 8.6 indicates a power-current flow for the time intervals from zero to 60° and 180 to 240° during each cycle, and a circulating current for all other intervals of time. No feedback current flows with a unity power-factor load.

For a lagging power-factor load, the SCR and rectifier currents are a function of the power-factor angle θ, in addition to the phase-displacement angle ϕ. Figures 8.7 and 8.8 show the conditions for $\phi = 0°$ and $\phi = 120°$ for a 0.5 lagging power factor load. In Figure 8.7, the current is no longer in phase with the fundamental component of e_{a-0}. The point in the cycle where load current starts to flow through an SCR, after the SCR is switched on, is delayed by the load power-factor angle θ. For example, the load current starts to flow through SCR1, in Figure 8.7, 60° after e_{a-o} becomes positive, and this same sort of thing occurs for each SCR. When the phase-displacement angle is 120°, as in Figure 8.8, the load current is shifted in phase with respect to the fundamental of e_{b-o} by an amount equal to $\phi/2$ plus the load power-factor angle. The load current is in phase with the fundamental component of e_{a-o}, since the phase lag produced by the lagging power factor is cancelled by the phase lead produced by the phase displacement.

For the conditions in Figure 8.5, the diodes do not conduct, so that only a "power current" flows. In Figures 8.6 and 8.8, during a portion of the

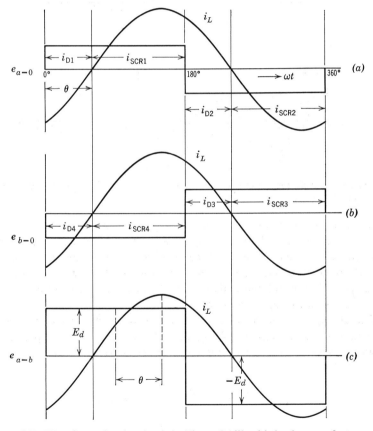

Figure 8.7   Waveforms for the circuit in Figure 8.1(b) with load power factor = 0.5 lagging and $\phi = 0°$.

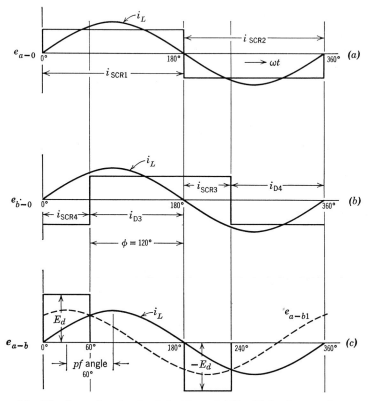

Figure 8.8   Waveforms for the circuit in Figure 8.1(b) with load power factor $= 0.5$ lagging and $\phi = 120°$.

cycle, a power current flows and during another part of the cycle a "circulating current" flows. In Figure 8.7 a power current flows during part of the cycle and a "feedback current" flows during the time intervals when two diodes conduct simultaneously. Figure 8.9 shows a condition of load power factor and phase displacement which results in all three modes of current flow during each cycle.

The average current flowing in the SCR's and rectifiers can be readily determined for the circuit of Figure 8.1(b) for all phase displacement angles and power factors by referring to the waveforms in Figures 8.5 through 8.9.

$$I_{\text{SCR1}} = I_{\text{SCR2}} = \frac{I_m}{2\pi} \int_0^{\pi - |\theta + \phi/2|} \sin \omega t \, d(\omega t) \tag{8.10}$$

$$I_{\text{SCR3}} = I_{\text{SCR4}} = \frac{I_m}{2\pi} \int_0^{\pi - |\theta - \phi/2|} \sin \omega t \, d(\omega t) \tag{8.11}$$

Here I_m is the maximum value of the sinusoidal load current and

$$\left| \theta + \frac{\phi}{2} \right| \quad \text{or} \quad \left| \theta - \frac{\phi}{2} \right|$$

is the magnitude of the sum of or difference between the power-factor angle and one half the phase-displacement angle. The sign convention is

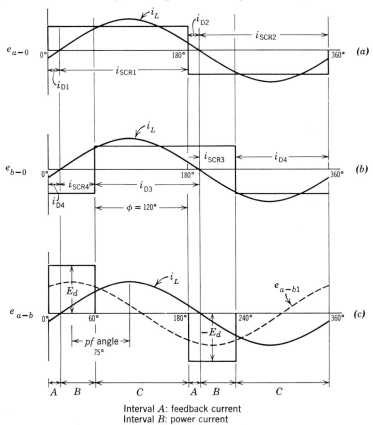

Interval A: feedback current
Interval B: power current
Interval C: circulating current

Figure 8.9 Waveforms for the circuit in Figure 8.1(b) with load power factor = 0.259 lagging and $\phi = 120°$.

that θ is positive when the power factor is leading, and ϕ is positive when e_{b-o} is advanced with respect to e_{a-o}.

The average rectifier currents are

$$I_{D1} = I_{D2} = \frac{I_m}{2\pi} \int_0^{|\theta + \phi/2|} \sin \omega t \; d(\omega t) \tag{8.12}$$

$$I_{D3} = I_{D4} = \frac{I_m}{2\pi} \int_0^{|\theta - \phi/2|} \sin \omega t \; d(\omega t) \tag{8.13}$$

The current drawn from the d-c source then becomes the difference between the SCR currents and the feedback rectifier currents.

$$I_d = \frac{I_m}{2\pi}\left[\int_0^{\pi-|\theta+\phi/2|} \sin\omega t \, d(\omega t) + \int_0^{\pi-|\theta-\phi/2|} \sin\omega t \, d(\omega t) \right.$$
$$\left. - \int_0^{\theta+\phi/2} \sin\omega t \, d(\omega t) - \int_0^{\theta-\phi/2} \sin\omega t \, d(\omega t) \right] \quad (8.14)$$

$$I_{d,e} = \sqrt{\frac{I_m^2}{2\pi}\left[\int_0^{-|\theta+\phi/2|} \sin^2\omega t \, d(\omega t) + \int_0^{-|\theta-\phi/2|} \sin^2\omega t \, d(\omega t) \right.}$$
$$\left. - \int_0^{|\theta+\phi/2|} \sin^2\omega t \, d(\omega t) - \int_0^{|\theta-\phi/2|} \sin^2\omega t \, d(\omega t) \right] \quad (8.15)$$

The same procedure can be followed to determine the waveform and conduction angles for leading power-factor loads.

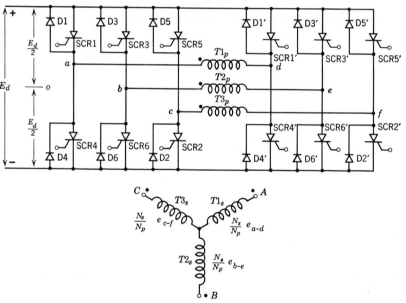

Figure 8.10   Twelve-phase double-way inverter.

In summary, for Figure 8.1(b), pulse-width voltage control is achieved by varying the phase position of the gating signals to one side of the inverter with respect to the other. The inverter output voltage e_{a-b} is independent of the load power factor. This is an excellent means of voltage control for a reasonable control range, since no additional power-circuit components are required to achieve voltage control. Where a wide voltage-control range is required, the percentage harmonics become large when the voltage is reduced to a low value. In addition, the utilization of the circuit components is reduced at low inverter output voltages.

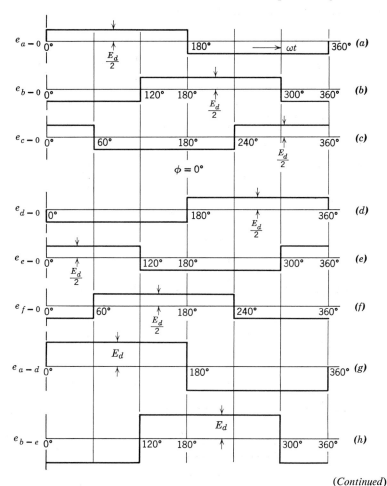

(*Continued*)

Figure 8.11 Waveforms for the circuit in Figure 8.10.

POLYPHASE. This same basic method of voltage control can be extended to polyphase systems simply by combining several of the bridge arrangements of the type shown in Figure 8.1(*b*). The basic power circuit of an inverter with a three-phase output is shown in Figure 8.10. This inverter is made up of three bridge circuits of the type shown in Figure 8.1(*b*). In this configuration, each single-phase bridge supplies one line-to-neutral voltage of the wye-connected output. The inverter voltage is again controlled by phase displacement of one half of each bridge with respect to the other half. The phase-shifted portion of each bridge is shifted the same amount, for a given output voltage, so as to maintain the three-phase relationship for the complete circuit. The three-phase arrangement

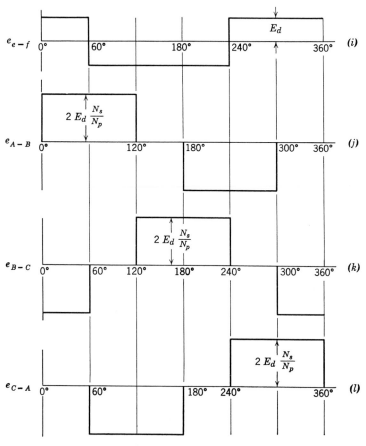

Figure 8.11   *Continued.*

eliminates the third harmonic in the line-to-line output voltage waveform for any pulse length of voltage applied to the transformer primaries. (The wye-connected secondary is used, as a delta connection would result in circulation of undesired third-harmonic currents).

When the remaining harmonics—fifth, seventh, eleventh, thirteenth, and so on—are added vectorially for a given phase displacement while retaining the phase relations to produce three phase output, the ratio of each harmonic to the fundamental remains constant. Hence, the magnitudes of the harmonics can again be obtained from Figure 8.3, except the lowest frequency component is the fifth harmonic. Figure 8.11 shows the

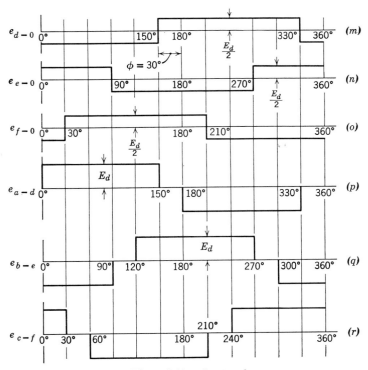

Figure 8.11   *Continued.*

important voltage waveforms for this three-phase circuit. These waveforms are indicated for several different phase-displacement angles ϕ. The line to line voltage e_{A-B} is simply determined as follows.

$$e_{a-d} = e_{a-0} - e_{d-0} \qquad (8.16)$$

$$e_{b-e} = e_{b-0} - e_{e-0} \qquad (8.17)$$

$$e_{A-B} = (e_{a-d} - e_{b-e})\frac{N_S}{N_P} \qquad (8.18)$$

More valves can be used with conduction intervals starting at different phase positions to further reduce harmonics in pulse-length voltage controls. In higher power inverters, this is a very practical technique, since multiple valves are required to deliver the rated power output. Figure 8.12 shows how six single-phase bridge inverters of the same basic type as in Figure 8.1(*b*) can be interconnected to produce a three-phase output voltage containing no fifth or seventh harmonics. Since the three-phase arrangement eliminates third harmonics and multiples thereof, the line-to-line output voltages from the circuit in Figure 8.12 contain no harmonics

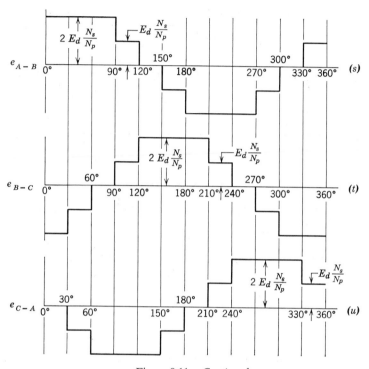

Figure 8.11   *Continued.*

below the eleventh for any pulse length of voltage applied to the transformer primaries.

The *I* and *I'* inverters in Figure 8.12 have single secondary windings on their single-phase output transformers. This portion of the circuit is operated in the same manner as in Figure 8.10. However, the secondary windings of the *I* and *I'* transformers are not connected directly to the output neutral. The *II* and *II'* inverters provide the center portion of the wye-connected output, as shown in the vector diagram of Figure 8.12. The valves in the *II* and *II'* inverters are operated 30° out of phase from the corresponding *I* and *I'* valves. Each output transformer for the *II* and

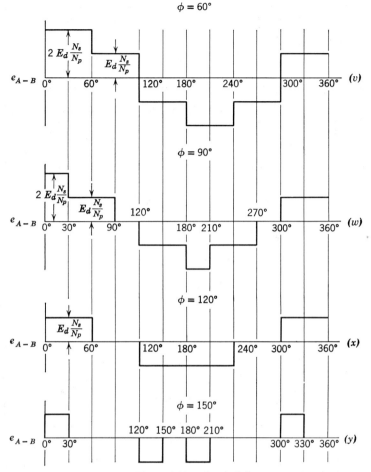

Figure 8.11   *Concluded.*

II' inverters has two equal-turn secondary windings. The turns ratio from the primary winding to one secondary winding is $1/\sqrt{3}$ that for the I and I' transformers. The line-to-line output voltages for the circuit in Figure 8.12 are then

$$e_{A-B} = e_{a-d} - 2e_{h-k} + e_{g-j} + e_{i-l} - e_{b-e} \qquad (8.19)$$

$$e_{B-C} = e_{b-e} - 2e_{i-l} + e_{h-k} + e_{g-j} - e_{c-f} \qquad (8.20)$$

$$e_{C-A} = e_{c-f} - 2e_{g-j} + e_{i-l} + e_{h-k} - e_{a-d} \qquad (8.21)$$

where the relationships between the voltage magnitudes are as shown in Figure 8.12.

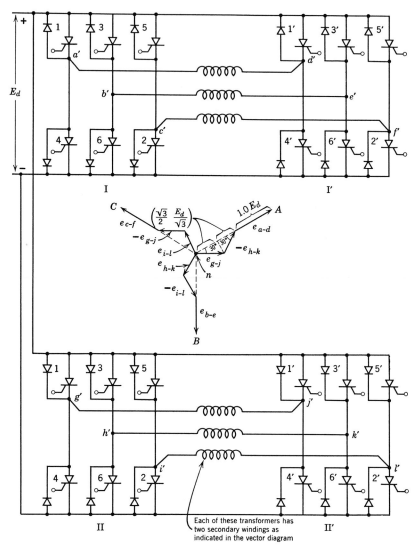

Figure 8.12   Twenty-four phase, double-way inverter. Assuming unity-turns ratio on *I-I′* transformers, the secondary voltages, at full output, are

$$|e_{a-d}| = |e_{b-e}| = |e_{c-f}| = 1.0E_d$$

and secondary voltages from *II-II′* are

$$|e_{g-j}| = |e_{h-k}| = |e_{i-l}| = \frac{E_d}{\sqrt{3}}$$

The vector relationships for the fifth and seventh harmonics will show that these harmonics do not appear in the line-to-neutral and, therefore, the line-to-line output voltages for the circuit in Figure 8.12. This may be shown also in the following manner.   Consider only the fundamental, fifth, and seventh harmonic components in the voltages that combine to produce the line-to-neutral voltage e_{A-n}.

$$e_{a-d} = E_1 \cos \omega t + E_5 \cos 5\omega t + E_7 \cos 7\omega t \qquad (8.22)$$

$$e_{h-k} = \frac{E_1}{\sqrt{3}} \cos(\omega t - 150) + \frac{E_5}{\sqrt{3}} \cos(5\omega t - 30)$$

$$+ \frac{E_7}{\sqrt{3}} \cos(7\omega t - 330) \quad (8.23)$$

$$e_{g-j} = \frac{E_1}{\sqrt{3}} \cos(\omega t - 30) + \frac{E_5}{\sqrt{3}} \cos(5\omega t - 150)$$

$$+ \frac{E_7}{\sqrt{3}} \cos(7\omega t - 210) \quad (8.24)$$

Now, the combining of these to determine e_{A-n} gives

$$e_{A-n} = e_{a-d} - e_{h-k} + e_{g-j} \qquad (8.25)$$

$$e_{A-n}\Big|_{\substack{\text{fifth}\\ \text{harmonic}}} = E_5 \cos 5\omega t - \frac{E_5}{\sqrt{3}} \cos(5\omega t - 30)$$

$$+ \frac{E_5}{\sqrt{3}} \cos(5\omega t - 150) \quad (8.26)$$

and, using the relation $\cos(x + y) = \cos x \cos y - \sin x \sin y$,

$$e_{A-n}\Big|_{\substack{\text{fifth}\\ \text{harmonic}}} = E_5 \cos 5\omega t - \frac{E_5}{\sqrt{3}}\left(\frac{\sqrt{3}}{2} \cos 5\omega t + \frac{\sin 5\omega t}{2}\right)$$

$$+ \frac{E_5}{\sqrt{3}}\left(-\frac{\sqrt{3}}{2} \cos 5\omega t + \frac{\sin 5\omega t}{2}\right)$$

$$= E_5 \cos 5\omega t - \frac{E_5}{2} \cos 5\omega t$$

$$- \frac{E_5}{2\sqrt{3}} \sin 5\omega t - \frac{E_5}{2} \cos 5\omega t$$

$$+ \frac{E_5}{2\sqrt{3}} \sin 5\omega t = 0 \qquad (8.27)$$

$$e_{A-n}\bigg|_{\substack{\text{seventh}\\ \text{harmonic}}} = E_7 \cos 7\omega t - \frac{E_7}{\sqrt{3}} \cos (7\omega t - 330)$$

$$+ \frac{E_7}{\sqrt{3}} \cos (7\omega t - 210)$$

$$= E_7 \cos 7\omega t - \frac{E_7}{\sqrt{3}}\left(\frac{\sqrt{3}}{2} \cos 7\omega t - \frac{\sin 7\omega t}{2}\right)$$

$$+ \frac{E_7}{\sqrt{3}}\left(-\frac{\sqrt{3}}{2} \cos 7\omega t - \frac{\sin 7\omega t}{2}\right)$$

$$= E_7 \cos 7\omega t - \frac{E_7}{2} \cos 7\omega t$$

$$+ \frac{E_7}{2} \frac{\sin 7\omega t}{\sqrt{3}} - \frac{E_7}{2} \cos 7\omega t$$

$$- \frac{E_7}{2\sqrt{3}} \sin 7\omega t = 0 \qquad (8.28)$$

A harmonic analysis of the output voltage from the circuit of Figure 8.12 yields the same results as previously obtained by equation (8.8), except that now the third, fifth, and seventh harmonics, and multiples thereof, do not exist. The lowest harmonics present in the output voltage are the eleventh and thirteenth. Hence, for applications with sufficiently large power requirements to justify the use of 24 valves and the additional control circuit complexity, the circuit of Figure 8.12 provides rather good output voltage waveshape over a relatively wide voltage control range.

Multiple Pulse-Width Voltage Control—Four Added Commutations per Half Cycle[2]

As previously discussed, the circuits in Figure 8.1 may be used with pulse-width voltage control over a limited range. At the maximum output voltage condition, a square-wave voltage is delivered to the load. The square wave contains $33\frac{1}{3}$ per cent third harmonics, 20 per cent fifth harmonics, and $14\frac{2}{7}$ per cent seventh harmonics, for example. When the fundamental component of output voltage is reduced, by pulse-width control, the percentage of harmonics is increased. Various polyphase arrangements may be used to eliminate certain harmonics as discussed in the previous section.

Particularly, in applications where a minimum number of valves are required to handle the power involved, it is desirable to use other techniques to further reduce the harmonics in the inverter output voltage over the voltage control range. One such technique is illustrated by the waveform in Figure 8.13. This inverter output voltage waveform may be obtained by operating the circuit in Figure 8.1(a) with a form of complementary impulse commutation with four extra commutations per half-cycle. The values of the angles α_1 and α_2 will be such that certain

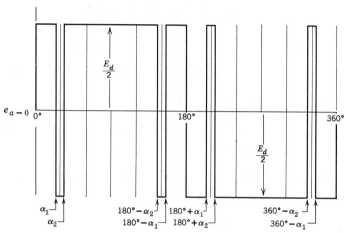

Figure 8.13   Output voltage waveform of the circuit in Figure 8.1(a) with four added commutations per half-cycle.

harmonics in the square-wave form are reduced or eliminated. The Fourier analysis for the waveform given in Figure 8.13 is

$$E_{a-o(n,m)} = \frac{4}{\pi} \int_0^{\alpha_1} \frac{E_d}{2} \sin n\omega t \, d(\omega t) \tag{8.29}$$

$$- \frac{4}{\pi} \int_{\alpha_1}^{\alpha_2} \frac{E_d}{2} \sin n\omega t \, d(\omega t)$$

$$+ \frac{4}{\pi} \int_{\alpha_2}^{90°} \frac{E_d}{2} \sin n\omega t \, d(\omega t)$$

$$E_{a-o(n,m)} = \frac{4}{\pi} \frac{E_d}{2} \left(\frac{1 - 2 \cos n\alpha_1 + 2 \cos n\alpha_2}{n} \right) \tag{8.30}$$

The third and the fifth harmonics are the largest present in the square-wave voltage, and are also the most difficult to filter because they are relatively close to the fundamental frequency. If the numerator of the expression in parenthesis in equation (8.30) could be made equal to zero for a given value of n, then the particular harmonic would be eliminated from the load voltage waveshape. For the third harmonic, the expression is

$$E_{a-o(3,m)} = \frac{4}{\pi} \frac{E_d}{2} \left(\frac{1 - 2 \cos 3\alpha_1 + 2 \cos 3\alpha_2}{3} \right) \tag{8.31}$$

For the fifth harmonic, the expression is

$$E_{a-o(5,m)} = \frac{4}{\pi} \frac{E_d}{2} \left(\frac{1 - 2 \cos 5\alpha_1 + 2 \cos 5\alpha_2}{5} \right) \tag{8.32}$$

For both the third and fifth harmonics to be equal to zero,

$$E_{a-o(3,m)} = E_{a-o(5,m)} = 0 \tag{8.33}$$

And, substituting from (8.31) and (8.32),

$$1 - 2 \cos 3\alpha_1 + 2 \cos 3\alpha_2 = 0 \tag{8.34}$$

$$1 - 2 \cos 5\alpha_1 + 2 \cos 5\alpha_2 = 0 \tag{8.35}$$

Equations (8.34) and (8.35) are subject to the following constraints.

$$0° < \alpha_2 < 90°$$

$$0° < \alpha_1 < 90°$$

$$\alpha_1 < \alpha_2$$

The approximate solutions for the values of α_1 and α_2 are

$$\alpha_1 = 23.62°$$

$$\alpha_2 = 33.30°$$

The approximate solution of the simultaneous trigonometric equations (8.34) and (8.35) for the values of α_1 and α_2 can be substituted in equation (8.30) to determine the fundamental and harmonic voltage content of the load voltage waveform. Table 8.1 gives the value of the components in the low harmonic waveform and the 180° square wave for comparison purposes.

The information in Table 8.1 indicates that the fundamental is reduced to 83.9 per cent of the fundamental frequency component in the 180° square wave. A comparison of the percentage harmonics present in both waveforms indicates that the third and fifth harmonic have been reduced to zero. The percentages of the seventh, ninth, and eleventh harmonics have been increased. However, these higher-frequency harmonics are easier to filter.

With only two valves as in the circuit in Figure 8.1(a), the angles α_1 and α_2 can be controlled to vary the output voltage. However, it is then possible only to eliminate either the third or the fifth harmonic, over the complete control range. This method of voltage control can be more effectively employed when four valves are used, as in the circuit in Figure 8.1(b). This circuit may be operated with four added commutations per half-cycle to eliminate third and fifth harmonics in three different ways. Each pair of valves can provide voltage control by varying α_1 and α_2 while

Table 8.1

Order of harmonic (n)	Value of component in square wave	Value of component in low harmonic wave relative to fundamental in square wave	Value of component in low harmonic wave as percentage of fundamental in low harmonic wave
1	1.000	0.839	100.0
3	0.333	0	0
5	0.200	0	0
7	0.143	0.248	29.6
9	0.111	0.408	48.6
11	0.091	0.306	36.4

maintaining zero third harmonic. The phase displacement between the two pairs of valves can be fixed to eliminate the fifth harmonic. The second way of operating is the opposite of this. The voltage is controlled by varying α_1 and α_2 for each pair of valves while maintaining zero fifth harmonic, and the two pairs of valves are operated with the correct phase displacement to eliminate third harmonics.

The third way of operating Figure 8.1(b), to achieve voltage control with no third or fifth harmonics, is the most logical extension of the operation previously discussed for Figure 8.1(a). In this mode of operation, each pair of valves is operated with fixed angles α_1 and α_2 to produce e_{a-o} and e_{b-o} waveforms, as in Figure 8.13, with no third and fifth harmonics. The output voltage is controlled by phase-shifting the gating of one pair of valves with respect to the other. The waveforms for various phase displacement angles considering this mode of operation are shown in Figure 8.14. The output voltage waveform is similar for any of the three ways of operating Figure 8.1(b), which have been discussed.

The general relationship to determine the harmonic content of the load voltage waveform as a function of the order of the harmonic and the

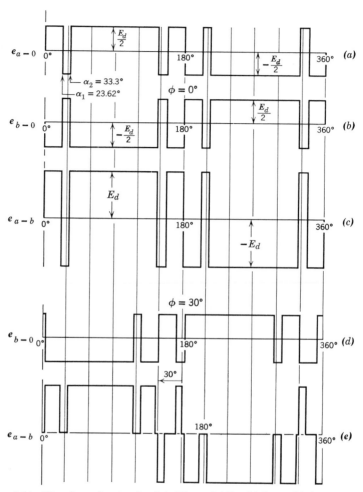

Figure 8.14   Waveforms for the circuit in Figure 8.1(b) with four added commutations per half-cycle.

fundamental frequency phase-shift angle can be determined in a similar manner to that used to obtain equation (8.8). The expression is

$$E_{a-b(n,m)} = \frac{4E_d}{\pi} \left[\frac{1 - 2 \cos n(23.62°) + 2 \cos n(33.3°)}{n} \right] \cos \frac{n\phi}{2} \quad (8.36)$$

In equation (8.36), n is the order of the harmonic 1, 3, 5, 7, 9, 11 . . . ∞, ϕ is the fundamental frequency phase-shift angle between the two sides of the inverter, and E_d is the magnitude of the d-c voltage.

The substitution of the value $n = 1$ into equation (8.36) determines the

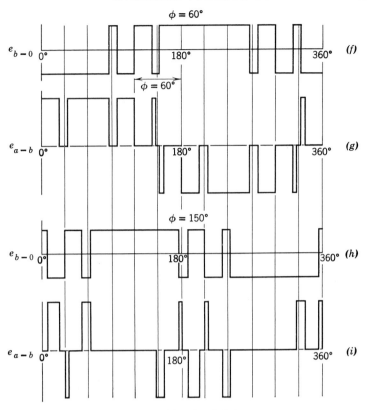

Figure 8.14 *Concluded.*

magnitude of the fundamental frequency voltage as a function of phase shift.

$$E_{a-b(1,m)} = \frac{4E_d}{\pi} [0.839]\left[\cos \frac{n\phi}{2}\right] \tag{8.37}$$

Equation (8.37) shows that the peak fundamental voltage varies as the cosine of one half of the phase shift angle as before.

The substitution of $n = 3, 5, 7, 9,$ and 11 into equation (8.36) gives the variation in harmonic voltages as a function of phase shift angle.

$$E_{a-b(3,m)} = 0 \tag{8.38}$$

$$E_{a-b(5,m)} = 0 \tag{8.39}$$

$$E_{a-b(7,m)} = \frac{4E_d}{\pi} [0.248]\left[\cos \frac{7\phi}{2}\right] \tag{8.40}$$

$$E_{a-b(9,m)} = \frac{4E_d}{\pi} [0.408]\left[\cos \frac{9\phi}{2}\right] \tag{8.41}$$

$$E_{a-b(11,m)} = \frac{4E_d}{\pi} [0.306]\left[\cos \frac{11\phi}{2}\right] \tag{8.42}$$

Table 8.2 indicates the magnitude of the harmonics determined from equations (8.37) through (8.42). The harmonic magnitudes are expressed as a fraction of $4E_d/\pi$ for seven values of phase-shift angle from 0 to 180°.

Table 8.2

ϕ	0°	30°	60°	90°	120°	150°	180°
Order of harmonic:							
1	0.839	0.811	0.727	0.593	0.423	0.217	0
3	0	0	0	0	0	0	0
5	0	0	0	0	0	0	0
7	0.248	0.064	0.215	0.176	0.124	0.240	0
9	0.408	0.288	0	0.288	0.408	0.288	0
11	0.306	0.296	0.265	0.216	0.153	0.076	0

It is frequently not necessary to eliminate the third harmonic, since it is not present in many polyphase connections. Therefore, it is preferable to locate the "notches" so as to eliminate the fifth and seventh harmonic voltages so that the lowest harmonic is the eleventh. Using a similar analysis to that used to obtain equations (8.31) and (8.32), the fifth and seventh harmonic voltages in the waveform shown in Figure 8.13 are

$$E_{a-0(5,m)} = \frac{4}{\pi}\frac{E_d}{2}\left(\frac{1 - 2\cos 5\alpha_1 + 2\cos 5\alpha_2}{5}\right) \tag{8.38}$$

$$E_{a-0(7,m)} = \frac{4}{\pi}\frac{E_d}{2}\left(\frac{1 - 2\cos 7\alpha_1 + 2\cos 7\alpha_2}{7}\right) \tag{8.39}$$

For both the fifth and seventh harmonic voltages to be equal to zero, the following two equations must be solved.

$$1 - 2\cos 5\alpha_1 + 2\cos 5\alpha_2 = 0 \tag{8.40}$$

$$1 - 2\cos 7\alpha_1 + 2\cos 7\alpha_2 = 0 \tag{8.41}$$

The approximate solution of these two simultaneous trigonometric equations is

$$\alpha_1 = 16.25° \tag{8.42}$$

$$\alpha_2 = 22.07° \tag{8.43}$$

Therefore, the harmonic content of the load voltage for any harmonic is

$$E_{a-0(n,m)} = \frac{4}{\pi}\frac{E_d}{2}\left[\frac{1 - 2\cos n(16.25°) + 2\cos n(22.07°)}{n}\right] \tag{8.44}$$

When the circuit in Figure 8.1(b) is operated with this α_1 and α_2, the similar expression to equation (8.36) for output voltage is

$$E_{a-b(n,m)} = \frac{4E_d}{\pi} \left[\frac{1 - 2 \cos n(16.25°) + 2 \cos n(22.07°)}{n} \right] \cos \frac{n\phi}{2} \quad (8.45)$$

In most polyphase circuits the third harmonics, and multiples thereof, will be eliminated by the transformer connection. The addition of the two "notches" in the voltage waveform can eliminate the fifth and seventh harmonics. Therefore, with the use of twelve valves in a three phase circuit as shown in Figure 8.10, and the proper α_1 and α_2, it is possible to obtain 100 per cent voltage control with no harmonics below the eleventh in the inverter output voltage waveform.

Multiple Pulse-Width Voltage Control—Many Commutations per Half-Cycle

The circuits discussed in the previous section may also be operated to provide multiple pulse-width voltage control where many pulses occur on each half-cycle of the fundamental frequency. By this means it is possible to further reduce the harmonics in the output voltage waveform. In fact, it is generally possible to eliminate all harmonics below the frequency of the pulses which occur in the output voltage.

There are numerous forms of this type of voltage control; that is, the pulse frequency can be synchronized with the fundamental inverter operating frequency, the pulse frequency can be independent of the inverter frequency, the pulse width of consecutive pulses during each half-cycle can be controlled in a sinusoidal or approximately sinusoidal fashion, and the like. It is interesting to note that if the pulse widths are controlled according to a trapezoidal program on each half-cycle, the inverter output voltage will very closely approximate a sine wave.

Figures 8.15 and 8.16 show two examples of the load voltage waveforms that may be obtained for this type of operation, using the circuit in Figure 8.1(b). In Figure 8.15 consecutive pulses have equal width. One method of controlling the voltage would be to simultaneously vary the width of all pulses while maintaining the pulse repetition frequency constant and "locked" to the fundamental operating frequency. In Figure 8.16 the pulse widths of consecutive pulses vary in a sinusoidal fashion. The voltage could be controlled by varying the width of all pulses while still retaining the sinusoidal relationship and a constant repetition rate for the voltage pulses.

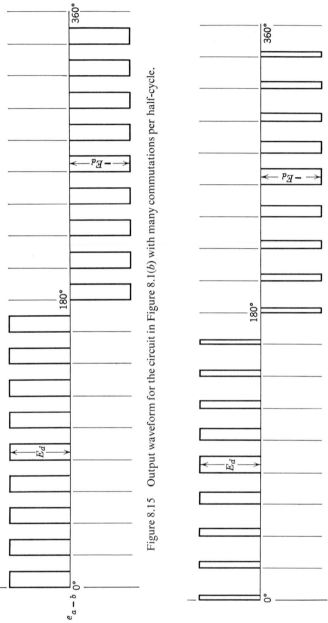

Figure 8.15   Output waveform for the circuit in Figure 8.1(b) with many commutations per half-cycle.

Figure 8.16   Output waveform of the circuit in Figure 8.1(b) with many "programmed" commutations per half-cycle.

8.3 CONTROL OF VOLTAGE DELIVERED BY INVERTER

The a-c voltage delivered by an inverter can be controlled by the use of all of the conventional a-c voltage regulation techniques. These include saturable reactors, magnetic amplifiers, induction regulators, or a-c phase-controlled rectifiers. These schemes have the disadvantage that where a wide control range is required, the kva rating of the control equipment is

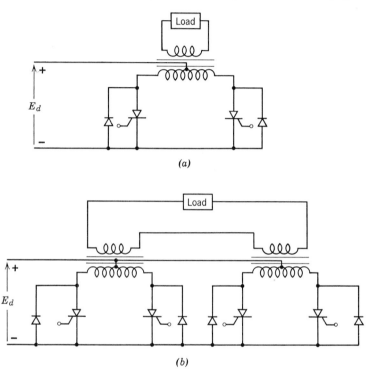

(a)

(b)

Figure 8.17 (a) Single-phase single-way inverter. (b) Double single-phase single-way inverter.

quite large. A basically different method of voltage control that is quite advantageous for many applications is the use of multiple inverters and phase-shifting the output of one inverter with respect to the other to control the voltage.[3,4] Possibly, the simplest form of this voltage control scheme is illustrated in Figure 8.17. The basic inverter circuit is shown in Figure 8.17(a). The output voltage from two of these circuits is added by connecting the transformer secondary windings in series to provide the load voltage. The load voltage is therefore the vector sum of the two inverter voltages and, thus, it can be smoothly controlled from zero to a maximum as the phase relationship between the inverters is controlled. It is important to point out that Figure 8.1(*b*) may also be considered a form of this mode of voltage control. This is particularly true when each half of the bridge is considered as one inverter; so that the phase shifting of one half of the bridge with respect to the other is similar to the process involved for Figure 8.17(*b*). In fact, in many cases, it is fairly difficult to distinguish between pulse-width control and phase-shift control of multiple inverters.

A polyphase form of this method of voltage control is discussed in the Appendix (pp. 264–278).[5]

REFERENCES

1. D. V. Jones, "Variable Pulse Width Parallel Inverters," U.S. Patent 3,075,136, January 22, 1963.
2. F. G. Turnbull, "Selected Harmonic Reduction in Static DC–AC Inverters," *IEEE Transactions*, Paper 63–1011, 1963.
3. P. D. Corey, "Methods for Optimizing the Waveform of Stepped-Wave Static Inverters," *AIEE Transactions*, CP 62–1147, Denver, June 17–22, 1962.
4. D. L. Anderson, A. E. Willis, and C. E. Winkler, "Advanced Static Inverter Utilizing Digital Techniques and Harmonic Cancellation," NASA Technical Note D-602, Washington, May 1962.
5. C. W. Flairty, "A 50 KVA Adjustable-Frequency 24-Phase Controlled Rectifier Inverter," AIEE Industrial Electronics Symposium, Boston, September 20–21, 1961.

APPENDIX.   A 50 kva ADJUSTABLE-FREQUENCY 24-PHASE CONTROLLED-RECTIFIER INVERTER*

C. W. FLAIRTY, *Manager—Inverter Circuits Engineering, Low Voltage Switchgear Department, General Electric Company, Philadelphia, Pennsylvania.*

INTRODUCTION

The present low cost of converting ac to dc in large blocks of power, as well as the availability of increasing amounts of d-c power from unconventional sources, demand that circuits be developed that are capable of inverting not only at power frequencies but at higher and lower frequencies. A majority of the industrial applications can be covered with frequencies

* A paper that was presented at the AIEE Industrial Electronics Symposium, Boston, September 20–21, 1961.

ranging from a few cycles per second to around 10 kc. High conversion efficiency is required in addition to circuits requiring a minimum kva of equipment per kilowatt of output power. Another requirement is high reliability to cut maintenance cost, to enable the use of power in unattended locations, or in areas where rotating machinery with slip rings or brushes cannot be used. The rapidly decreasing prices of silicon-controlled rectifiers, along with their increased power ratings make the solid-state inverter a strong competitor even today for rotating machinery in many areas.

One important application example for the solid-state inverter is the synchronous motor drive where precise speed regulation is required over a wide range. The static inverter offers this adjustable frequency and adjustable voltage power supply with the precision of solid-state, tuning forks or crystal-controlled oscillators. Speed regulation is then essentially independent of supply voltage or load variations, and dependent only on the low power-adjustable frequency oscillator. Since the motor requires a constant volts per cycle for best torque capacity without overheating and excessive current drain from the source, voltage control proportional to frequency control is required of its power supply. The inverter with proper connections and control circuits can supply this adjustable output voltage either independent of, or in conjunction with the change in frequency. In addition, the inverter eliminates the costly maintenance problems that accompany the a-c or d-c motors with slip rings or commutators since the SCR offers long-life potentialities just as provided by plain silicon rectifiers.

The objectives in this inverter development were to provide not only adjustable frequency and voltage control, but a sinusoidal output with low harmonic distortion. In short, this power supply was to provide, with solid state (no moving parts except for fans), the adjustable frequency and voltage power previously obtained from motor-generator sets. Specifications for the equipment were as follows.

Input   125 v dc $\pm$ 10%
Output
(1) 50 kva isolated output.
(2) 240 v ac, 3ϕ regulated and adjustable over $\pm$10% range and unregulated down to zero voltage.
(3) Include 120 v a-c output from a 2:1 step-down autotransformer.
(4) $\pm$ 1% voltage regulations from 0 to 100% load and $\pm$ 10% input voltage variation.
(5) Continuously adjustable frequency from 50 to 500 cps.
(6) $\pm$ $\frac{1}{2}$% frequency regulation.

(7) Less than 5% total harmonic content when filtered.

(8) Operable over the full range of leading and lagging power factors at reduced loads, and full kva from 0.7 lagging to unity.

The equipment was also designed for an over-all conversion efficiency of 90%, quiet operation, and portability.

INVERTER APPROACH

The basic approach to this inverter development was to utilize a multiphase configuration. Obvious advantages of a multiphase approach are as follows.

(1) Decreased harmonic content in the output voltage.

(2) Little or no paralleling of SCR's without sacrificing device utilization.

(3) Lower harmonic current requirements from the source or energy storage devices.

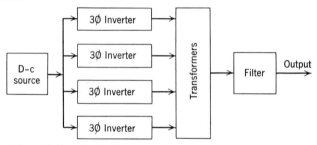

Figure 8.18   Block diagram of the 24-phase inverter configuration.

The specific version of the multiphase inverter used was a 24-phase configuration. A block diagram of this circuit configuration is shown in Figure 8.18 and consists of four identical three-phase bridge inverters, each with separate output transformers having common series-connected secondaries in the appropriate sequence. All four inverters are supplied from a common d-c bus; in this case, 125 v ± 10%. The power-switching devices used in each bridge inverter were the General Electric Company 200-v SCR's. The use of these devices enables the equipment to deliver 50-kva output without paralleling SCR's.

In this circuit approach, each of the four three-phase bridge inverters operates in a square-wave mode, such that each of the SCR'S has a conduction interval of 180° rather than the conventional 120°-conduction period for plain rectifiers in a bridge configuration. Since the four inverters are identical, the waveshapes associated with only one inverter

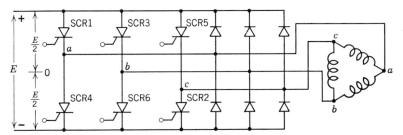

Figure 8.19 Basic three-phase bridge inverter with feedback rectifiers and transformer primary connections.

(Figure 8.19) require detailed discussion. In Figure 8.19 the SCR's are numbered in their natural firing sequence, and a d-c neutral point o is shown for discussion purposes. The three output terminal points of the inverter are labeled a, b, and c respectively, and are tied to the delta-connected primary of one output transformer.

With a three-phase gating signal available to the SCR's in the firing sequence 1 through 6, the output voltage of one inverter is readily determined as shown in Figure 8.20, and can be explained as follows. Since a 180°-conduction mode is used in the inverters, each output terminal point

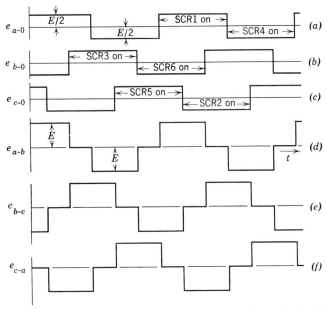

Figure 8.20 Theoretical line-to-dc neutral and line-to-line voltages for the three-phase bridge inverter shown in Figure 8.19.

in Figure 8.19 is alternately connected to the positive or negative d-c bus if we neglect rectifier drop. Therefore, a plot of the line to d-c neutral voltages e_{a-0}, e_{b-0} and e_{c-0} is useful in determining output voltages. For example, the voltage e_{a-0} [Figure 8.20(a)] is positive $E/2$ during the 180° interval when SCR1 is conducting and negative $E/2$ during the conduction interval of SCR4. The voltages e_{b-0} and e_{c-0} are identical to that of e_{a-0}, except phase displaced by 120° and 240° respectively, as shown in Figures 8.20(b) and 8.20(c).

The line-to-line voltages e_{a-b}, e_{b-0}, and e_{c-0}, and become

$$e_{a-b} = e_{a-0} - e_{b-0} \tag{8.46}$$

$$e_{b-c} = e_{b-0} - e_{c-0} \tag{8.47}$$

$$e_{c-a} = e_{c-0} - e_{a-0} \tag{8.48}$$

as shown in Figure 8.20. The line-to-line voltages are the familiar 120° waves that contain no third harmonic voltage, as expressed by the equation,

$$e_{(\theta)} = \frac{4E}{\pi}\left(\frac{\sqrt{3}}{2}\right)\left[\cos\theta - \frac{1}{5}\cos 5\theta + \frac{1}{7}\cos 7\theta - \frac{1}{11}\cos 11\theta + \cdots\right] \tag{8.48}$$

and the lowest theoretical harmonic present in the output voltage waveform of this bridge inverter is the fifth harmonic.

In larger power applications that require a lower harmonic content in the output wave, a filter to remove fifth and seventh harmonics becomes large and bulky. Therefore, it is highly desirable to improve the unfiltered waveshape by means of other inverter configurations if this improvement does not come at the expense of poorer utilization of the SCR's. In addition, the larger power applications normally require more than six SCR's, which means a paralleling of devices in the inverter if only one three-phase bridge is used. One method of obtaining an improved voltage waveform is to operate a second inverter in parallel from the same d-c source, but having its output phase displaced by 30° from that of inverter number 1. With proper transformer connections, a twelve-phase inverter output is obtained as shown in Figure 8.21. The lowest harmonics present in this output waveform are the eleventh and thirteenth as shown in equation (8.49).

$$e_{(\theta)} = \frac{4E}{\pi}\sum_{n}[K_1 + K_2\cos n2\gamma + K_3\cos n4\gamma]\frac{\sin n\theta}{n} \tag{8.49}$$

$$n = 1, 3, 5, 7, \ldots$$

where $\gamma = 15°$ and the constants are determined by the transformer turns ratio. For many applications, such as supplies for motors, this inverter

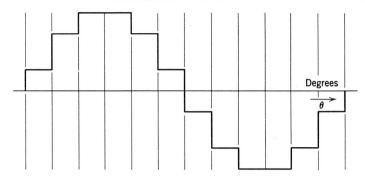

Figure 8.21   A twelve-phase inverter output waveform.

output waveshape is adequate without additional improvement. However, a further step can be taken to improve the unfiltered output waveshape and, at the same time, double the power-handling capabilities of the circuit by using a second twelve-phase inverter configuration with outputs phase displaced 15° from the output of the first twelve-phase inverter. The complete configuration then consists of four three-phase bridge inverters

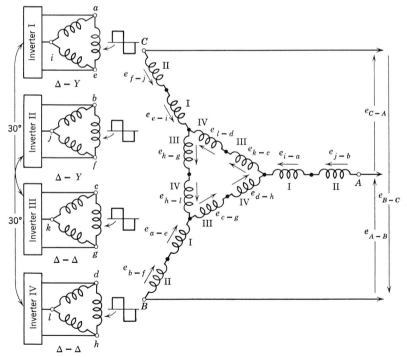

Figure 8.22   Diagram of transformer configuration for a 24-phase inverter.

identical to that shown in Figure 8.19 and is called a 24-phase inverter. Gate signals to each of the four inverters are phase displaced, such that the corresponding line-to-line voltage appearing on each of the transformer primaries is phased as shown in Figure 8.22. Hence, inverters *I* and *III*, with outputs phase displaced by 30°, make up one twelve-phase inverter while inverters *II* and *IV*, also having outputs displaced by 30°, make up the second twelve-phase inverter. If the two twelve-phase inverter outputs are so phased that *I* leads *II* by 15°, *II* leads *III* by 15°, and *III* leads *IV* by 15°, a 24-phase output voltage waveform is obtained from line to line of the secondaries, as shown in Figure 8.23(*b*). This waveform requires only a small amount of filtering to get total harmonics to 5%, since the lowest frequency significant harmonics present are the twenty-third and twenty-fifth with 4.35% and 4.0% of each respectively. The Fourier series for this output wave becomes

$$e_{(\theta)} = \frac{4E}{\pi} \sum_{n} \left[K_1 \cos \frac{n\gamma}{2} + K_2 \cos \frac{n3\gamma}{2} + K_3 \cos \frac{n5\gamma}{2} \right.$$
$$\left. + K_4 \cos \frac{n7\gamma}{2} + K_5 \cos \frac{n9\gamma}{2} \right] \frac{\sin n\theta}{n} \quad (8.50)$$

$$n = 1, 3, 5, 7, \ldots$$

Therefore, by the use of a 24-phase inverter configuration the lowest significant harmonic present in the output has been raised from the fifth in one bridge output to the twenty-third. The total number of SCR's required has increased by a factor of 4, but the total kva capability of the equipment has also increased by the same factor. Hence, the SCR's are utilized equally as well as in the three-phase bridge inverter.

VOLTAGE CONTROL

Control of the output voltage is essential in most inverter applications, either to compensate for variations of the input supply and internal regulation, or because the load demands adjustable voltage, or both. This requirement, when coupled with the additional necessity of maintaining low harmonic content in the output, provides further reason for use of a multiphase approach. Voltage control is obtained in this type of inverter by changing the phase displacement between the two twelve phase inverters (see Figure 8.22). For example, if the twelve-phase inverter made up of *I* and *III* is phase-advanced from the twelve-phase inverter *II* and *IV*, the output line-to-line voltage will be reduced. A 165° phase advance of *I* and *III* from the 24-phase position reduces the output voltage to zero. On the other hand, a 15° phase retard of *I* and *III* from the 24-phase position gives maximum output voltage and the in-phase condition with inverters *II* and *IV*. Since line-to-line voltages e_{A-B}, e_{B-C}, and e_{C-A} are determined by summing the instantaneous secondary voltages, output waveforms can readily be obtained for any phase position by equations (8.51) through (8.52).

$$e_{A-B} = (-e_{j-b} - e_{i-a} + e_{d-h} + e_{c-g} + e_{a-e} + e_{b-f}) \qquad (8.51)$$

$$e_{B-C} = (-e_{b-f} - e_{a-e} + e_{h-l} + e_{k-g} + e_{e-i} + e_{f-j}) \qquad (8.52)$$

$$e_{C-A} = (-e_{f-j} - e_{e-i} + e_{l-d} + e_{k-c} + e_{i-a} + e_{j-b}) \qquad (8.53)$$

The theoretical waveforms of the unfiltered output voltage are shown in Figure 8.23 for several positions. For comparison purposes, oscillograms of measured output voltage waveforms are shown in Figure 8.24.

Since the 24-phase inverter waveshape occurs only at one phase position [see Figure 8.23(*b*)] it is obvious that the principal harmonics of a twelve-phase inverter (the eleventh and thirteenth) are introduced back into the output voltage as the fundamental is reduced. However, the lowest theoretical harmonic present, even at low values of fundamental voltage, is the eleventh. Therefore, the inverter filter must be designed to reduce eleventh and thirteenth harmonics in addition to the higher order harmonics when phase displacement is used for for voltage control. Figure 8.25 shows a plot of percentage, fundamental, and harmonic voltages present in the output waveform as a function of phase displacement between the twelve-phase inverters. For still further improvement in the unfiltered waveshape of the lower output voltages, a 48-phase configuration could be used containing two 24-phase inverters with phase displacement between their outputs for voltage control. However, the further improvement in output waveshape before filtering will be overshadowed by the increased circuit complexity unless the total kva rating of the system is quite large.

FREQUENCY CONTROL

One of the outstanding features of the solid-state approach to adjustable frequency power supplies is the accuracy of, and the simplicity of obtaining the frequency reference, and the ability to obtain precise frequency control with an open-loop system. In this equipment, the desired output frequency is determined by the use of a simple unijunction transistor oscillator operated from a regulated d-c supply. The unijunction oscillator provides timing pulses for the gating circuitry that, in turn, supplies signals in proper sequence to all of the inverter-controlled rectifiers. A simple

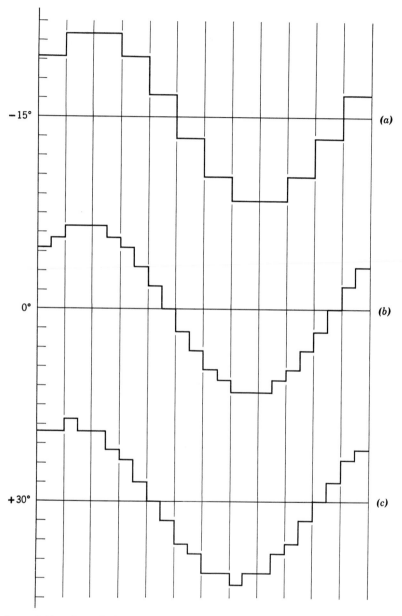

Figure 8.23   Theoretical output voltage waveforms at several phase displacement positions (0° refers to the 24ϕ output condition).

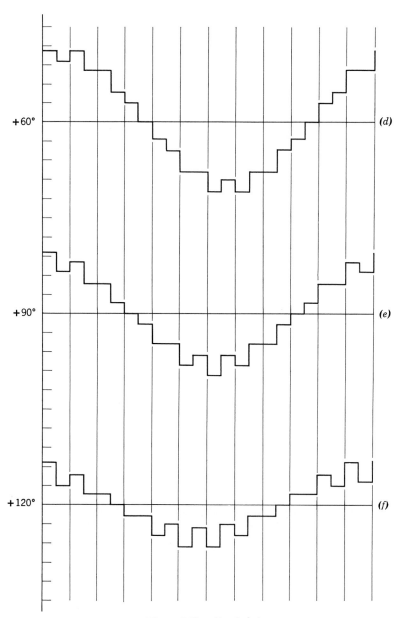

Figure 8.23   *Concluded.*

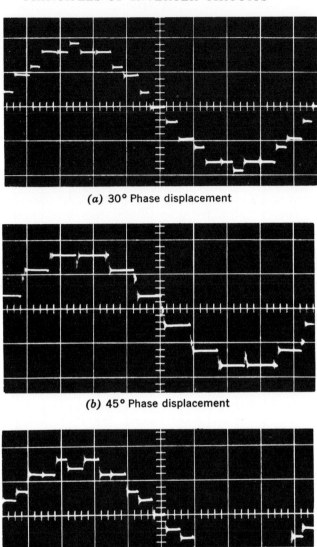

(a) 30° Phase displacement

(b) 45° Phase displacement

(c) 60° Phase displacement

Figure 8.24 Measured output voltage waveforms at various displacement positions (0° represents the 24ϕ output condition). (a) 30° Phase displacement. (b) 45° Phase displacement. (c) 60° Phase displacement. (d) 90° Phase displacement. (e) 120° Phase displacement. (f) 150° Phase displacement.

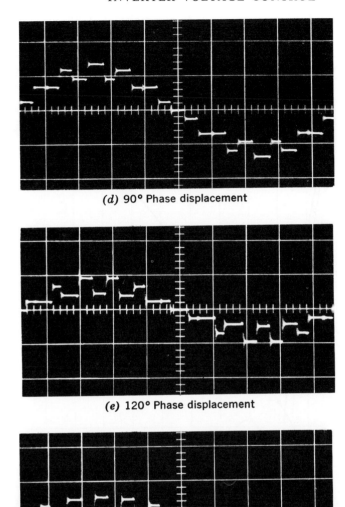

(d) 90° Phase displacement

(e) 120° Phase displacement

(f) 150° Phase displacement

Figure 8.24 *Concluded*.

oscillator of this type can provide accuracies of $\pm \frac{1}{2}\%$ over industrial ambient temperatures without the use of special temperature compensation. Under controlled temperature conditions, the unijunction transistor oscillator accuracy is greatly improved. For systems requiring

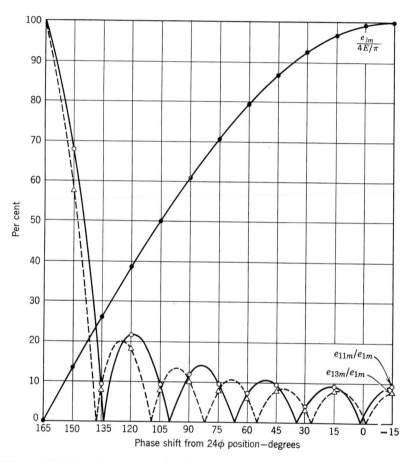

Figure 8.25 Plot of percentage fundamental and harmonic output voltages as a function of phase displacement.

extreme accuracies, tuning-fork oscillators or crystal-controlled oscillators are frequently used. However, of primary importance is the fact that this master oscillator alone determines the output frequency of the inverter. Inverter loading, even during transient conditions such as starting a motor, does not affect the output frequency.

Table 8.3.   Over-all Circuit Efficiency vs. Frequency and Percent Load

Frequency	Percentage of full load		
(cps)	(20 %)	(60 %)	(100 %)
50	90	92	88.3*
100	89	91.7	87.8*
300	83.5	89.3	86*
500	75.3	86	85.3*

* It should be pointed out that the efficiency at full load is less than at 60% loading because of excessive source regulation near full load resulting in heavier currents and, hence, more circuit losses for a given output power. In fact, all measurements listed were for input d-c voltages less than the nominal design value of 125 v. Therefore, all efficiencies listed would be higher when measured at the nominal supply of 125 v dc.

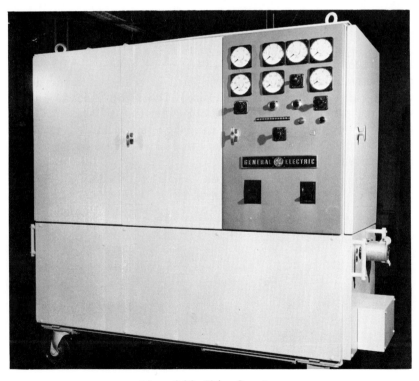

Figure 8.26   50-kva Inverter.

PERFORMANCE

Evaluation of this 24-phase inverter has been completed under all loading, frequency, and power-factor conditions of the specifications. All of these specifications were met, and some were exceeded. Efficiency of the equipment was measured over extremes of load and frequency range, and calculated on the basis of watts out and total watts input, including firing circuits and fans, in addition to the inverter losses. Typical measured efficiencies are listed in Table 8.3 for various percentage loads and different frequencies.

CONCLUSION

This 50-kva inverter equipment (Figure 8.26) announced by the General Electric Company in February 1961, along with other types and ratings under development, promises to be very competitive with conventional power supplies that provide adjustable or fixed frequency and voltage power. Rapidly decreasing prices of SCR's, and the development of larger device power ratings, along with improved inverter techniques, are all important factors towards making this type of inverter economically competitive with the rotating machine. The higher efficiency, quiet operation, precision frequency control independent of load, and static operation with potentially long life make the inverter approach attractive even, at present, in some applications that require these general supply characteristics.

Chapter Nine

Improving the Inverter Output Waveshape

by D. P. Shattuck

In most inverters, a square-wave voltage is produced at some point in the circuit because of the switching action that occurs during the inversion process. For the simple mechanical-switch or transistor-switch inverters, the d-c source voltage is switched alternately from one polarity to the other, producing a square-wave voltage output to the load (see Chapter 2). With series capacitor commutation, a square-wave voltage is impressed upon an *L-C* resonant circuit in series with the load. The *L-C* circuit provides the means for commutation and also accomplishes filtering action. Thus, the load voltage may approach a sinusoidal waveform in series inverters. In the parallel inverter, the d-c reactor and commutating capacitor also serve the dual function of providing commutation and some load voltage filtering. With a large commutating capacitor, the output voltage may approach a sinusoidal waveform in parallel capacitor-commutated inverters.

The simplest forms of the very important new family of inverters discussed in Chapter 7 produce square-wave output voltage. These impulse-commutated inverters have low regulation, and are capable of operating over wide load current and load power factor ranges. With fast solid-state switching devices, the commutating impulse can have a very short duration. Thus, the losses associated with commutation are minimized so that modern impulse-commutated inverters provide extremely efficient and highly reliable inversion systems. In a great many inverter applications, the harmonic content in the load voltage must be less than a specified amount. For a typical a-c electric power system, the magnitude of a single harmonic may be specified not to exceed 5 per cent of the fundamental component with a limit of 10 per cent total harmonic distortion. The total distortion is the square root of the sum of the squares of all of the

279

harmonics present. One of the important reasons for harmonic distortion limits is heating in a-c motor loads. The harmonics present usually result in increased motor losses without corresponding improvement in torque output. In many instrumentation circuits and components, inaccuracies may result from harmonic distortion in the a-c system voltage.

The principal emphasis of this chapter is to describe various methods of improving the inverter output voltage waveshape to meet the harmonic distortion requirements of practical systems. Various forms of a-c filters are used. However, several other circuit techniques may be employed to eliminate selected harmonics or otherwise reduce the total harmonic distortion. The L-C components used to provide commutation in the series and parallel inverters also provide filtering of the output voltage waveform. Pulse modulation, polyphase circuits, proper phase displacement of portions of complete inverter systems, and transformer tap changing are used with impulse-commutated inverters to reduce harmonic distortion.

The selection of the method for improving the inverter output voltage waveform, which includes the type of filter to use in a particular application, depends on a number of factors. Some of the most important of these are the inverter voltage and current rating, the range of loads, the operating frequency range, and the acceptable total harmonic content. All such factors must be carefully weighed to select the most advantageous approach for a given practical inverter system.

9.1   HARMONIC ANALYSIS OF WAVEFORMS

Fourier Analysis[1]

With the application of Fourier analysis the harmonic content of any waveform can be determined. A very brief discussion of such analysis is presented here as a refresher. Any periodic function can be represented by an infinite series of the form

$$f(x) = \tfrac{1}{2}a_0 + a_1 \cos x + b_1 \sin x + \cdots + a_n \cos nx \qquad (9.1)$$
$$+ b_n \sin nx + \cdots$$

where a_0 through a_n and b_1 through b_n are constants which can be determined by use of the expressions

$$a_n = \frac{1}{\pi} \int_{-\pi}^{\pi} f(x) \cos nx \, dx \quad (n = 0, 1, 2, \ldots) \qquad (9.2)$$

$$b_n = \frac{1}{\pi} \int_{-\pi}^{\pi} f(x) \sin nx \, dx \quad (n = 1, 2, 3, \ldots) \qquad (9.3)$$

When this analysis is applied to a voltage waveform such as $e(\omega t)$, expression (9.1) becomes

$$e(\omega t) = \tfrac{1}{2}a_0 + a_1 \cos \omega t + b_1 \sin \omega t + \cdots$$
$$+ a_n \cos n\omega t + b_n \sin n\omega t + \cdots$$

or

$$e(\omega t) = \frac{a_0}{2} + \sum_{n=1}^{\infty} (a_n \cos n\omega t + b_n \sin n\omega t) \qquad (9.4)$$

where the constants are the magnitudes of the nth harmonics, with a_0 the d-c component of the voltage waveform. These magnitudes are determined from the expressions

$$a_n = \frac{1}{\pi} \int_{-\pi}^{\pi} e(\omega t) \cos n\omega t \, d(\omega t) \quad (n = 0, 1, 2, 3, \ldots) \qquad (9.5)$$

$$b_n = \frac{1}{\pi} \int_{-\pi}^{\pi} e(\omega t) \sin n\omega t \, d(\omega t) \quad (n = 1, 2, 3, \ldots) \qquad (9.6)$$

Since inverter output voltages, which are basically square-wave, are of particular interest, a square wave will be used as an example to show how

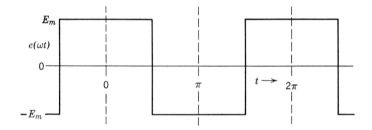

Figure 9.1   Square waveform.

the harmonics present in a waveform are determined. With $e(\omega t)$ a square wave (Figure 9.1), it is advantageous to select $t = 0$ at a particular point. If $t = 0$ is chosen so that $e(\omega t)$ is symmetrical about that point,

$$a_n = \frac{2}{\pi} \int_{0}^{\pi} e(\omega t) \cos n\omega t \, d(\omega t) \quad (n = 0, 1, 2, \ldots) \qquad (9.7)$$

$$b_n = 0 \quad (n = 1, 2, \ldots) \qquad (9.8)$$

The voltage function for the square wave of Figure 9.1 is given by

$$0 \le e(\omega t) \le \pi/2; \; e(\omega t) = E_m \qquad (9.9)$$

$$\pi/2 \le e(\omega t) \le \pi; \; e(\omega t) = -E_m \qquad (9.10)$$

Substituting these relationships into equation (9.7), the coefficients are found to be

$$a_0 = 0$$

$$a_1 = \frac{4E_m}{\pi}$$

$$a_2 = 0$$

$$a_3 = \frac{4E_m}{3\pi}$$

$$a_4 = 0$$

$$a_5 = \frac{4E_m}{5\pi}$$

$$a_6 = 0$$

$$a_7 = \frac{4E_m}{7\pi}$$

$$a_n = \frac{4E_m}{n\pi}$$

$$a_{n+1} = 0$$

where n is an odd integer.

In this case, the magnitude of the harmonics fits a pattern so that calculations are really not necessary beyond what is indicated. In the general case, however, the magnitudes must be individually calculated. The number of harmonics that must be determined depends on the required degree of accuracy in determining total harmonic content. Returning to the example chosen, and substituting the constants into equation (9.4), we get

$$e(\omega t) = \frac{4E_m}{\pi}\left(\cos \omega t + \frac{\cos 3\omega t}{3} + \frac{\cos 5\omega t}{5} + \cdots \frac{\cos n\omega t}{n}\right) \quad (9.11)$$

indicating that the fundamental has a peak value of $(4/\pi)E_m$. The only harmonics present in a square wave are odd harmonics whose magnitudes are those of the fundamental divided by the order of the harmonic; that is, the third harmonic appears with one third the magnitude of fundamental and so forth.

Simplified Graphic Fourier Analysis

It is fortunately true that for a number of waveforms, including many which exist in rectifiers and inverters, the harmonics can be determined by a simplified graphical approach to the Fourier analysis. This graphic approach is a very easy method for obtaining the harmonics for certain waveforms. In addition, it provides an excellent physical representation of the harmonics so that one can get a fairly good understanding of the effect of variations in the waveform on the harmonics which are present.

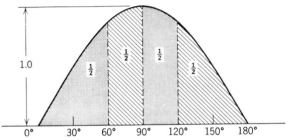

Figure 9.2   Incremental area of sine wave. Area of 0–30° increment = $1 - \cos 30° =$ $1 - \sqrt{3}/2 \approx 2/15$. Total area for half-cycle of unit sine wave = 2.0.

There are two fundamental facts which form the basis for the simplified method of obtaining the harmonics. The first of these is apparent from equation (9.5) or (9.6). The magnitude of a given harmonic is proportional to the integral of the instantaneous product of the waveform of interest and a unit sine wave (or cosine wave) at the harmonic frequency. This is expressed in equation form as

$$b_n \sim \int_{-\pi}^{\pi} e(\omega t) \sin n\omega t \, d(\omega t) \tag{9.12}$$

or

$$a_n \sim \int_{-\pi}^{\pi} e(\omega t) \cos n\omega t \, d(\omega t) \tag{9.13}$$

When the waveform of interest can be represented by constants during all intervals of time, the results of (9.12) or (9.13) can be easily determined if the integral of a sine wave is known for all increments of time. The integral or incremental area of a sine wave is

$$\int_{0}^{X_1} \sin x \, dx = -\cos x \Big|_{0}^{X_1} = 1 - \cos X_1 \tag{9.14}$$

This is shown in graphic form in Figure 9.2, for several increments of the

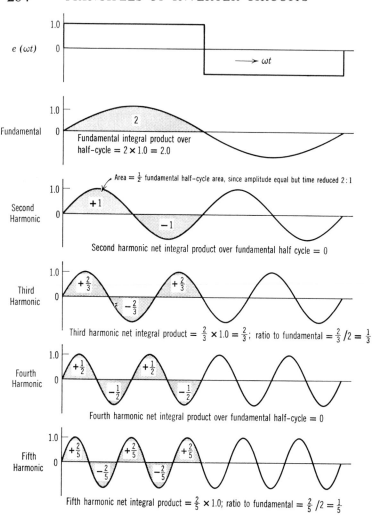

Figure 9.3   Square-wave harmonic analysis.

sine wave. The incremental area for a sine wave is the second basic fact of importance for this simplified method of harmonic analysis.

Generally, it is most important to know the ratio of a given harmonic to the fundamental. This can be determined quite simply by the use of this simplified graphic Fourier analysis, as will be illustrated for two specific cases. The procedure that is followed is indicated below.

(1) Draw the waveform of interest with unit amplitude.

(2) Draw a unit sine wave at fundamental frequency.

(3) Find the repeat distance net integral product for the fundamental.

(4) Draw a unit sine wave for the harmonic of interest.

(5) Find the repeat distance net integral product for the harmonic.

(6) Determine the ratio of the result of (5) to (3), and this is the ratio of the harmonic magnitude to the fundamental magnitude.

Figure 9.3 indicates the use of this method for a square wave. In this case, it is very simple, since the square wave form has a constant value over the full half-cycle repeat distance. Thus, the integral product is determined simply by finding the net area of the sine wave for the particular harmonic over a fundamental half-cycle. A waveform such as the simple square wave, which is symmetrical, except for sign, about the 180° point, contains no even harmonics. When such a wave is also symmetrical about the 90° points, the phase angles of the harmonics are as shown in Figure 9.3.[2] In the general case, when the phase angles of the harmonics are not known, it is necessary to carry through the procedure illustrated in Figure 9.3 twice for the fundamental and each harmonic. The magnitude of a particular harmonic is the square root of the sum of the squares of the two components resulting from using the procedure for two sine components 90° out of phase. The phase relation of the sine components, with respect to the waveform of interest, can be arbitrarily chosen to simplify the graphic analysis.

Figure 9.4 illustrates the use of the simplified method of Fourier analysis for a somewhat more complicated waveform, which often is encountered in inverter circuits. It should be noted here that it is not necessary to determine the area of certain increments of the sine wave when it is clear that they will cancel when finding the net integral product. It can be seen that the fourth harmonic is zero, by inspection, so that it is not necessary to know the area of the sine wave increments for this harmonic.

This simplified method of Fourier analysis is quite simple when applied to "stepped" waves of the form shown in Figures 9.3 and 9.4. In many cases, for more complicated waves, approximate values of the harmonics can be obtained relatively easily. If the waveform of interest is approximated by straight-line segments, the integral products may be determined for each time increment in the same manner as shown in Figures 9.3 and 9.4. The method is also very useful to determine the trend of the harmonics for certain variations in the waveform of interest. It is important to re-emphasize that this simplified method is strictly the Fourier analysis. However, it does not require the step-by-step mathematical solution of the equations for the harmonics, since a graphic incremental integration procedure is used.

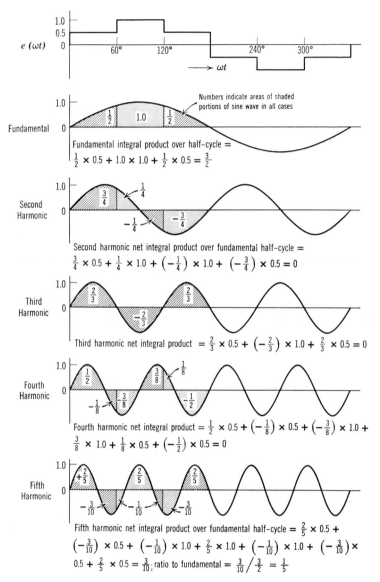

Figure 9.4   Stepped-wave harmonic analysis.

9.2 INVERTERS WITH INTERNAL FILTERING

The series capacitor-commutated inverter is capable of producing an output voltage which closely approaches a sine wave without requiring an external filter. The series capacitance and inductance, necessary for proper commutation, also provide filtering action. When the oscillating kva is high relative to the load power, the output voltage waveform is nearly sinusoidal. As discussed in Chapter 5, the output voltage waveshape can depart considerably from a sine wave when the circuit is not operated near its resonant frequency. With a relatively constant load and load power factor and over a limited frequency range, series capacitor-commutated inverters provide a very practical means of producing reasonably sinusoidal output voltage.

The filtering action actually occurs internal to the series capacitor-commutated inverter. The capacitor and inductor in the inverter form a series resonant circuit, which offers high impedance to harmonics and low impedance to the fundamental. While variations in operating frequency may not necessarily prevent proper operation of the circuit, such frequency variations will change the harmonic content in the output voltage.

A parallel capacitor-commutated inverter also may include appreciable filtering action within its circuit. When the commutating capacitor is large, this capacitor and the d-c reactor can provide sufficient filtering to produce nearly sinusoidal inverter output voltage. For these conditions, the inverter has a relatively large oscillating kva with respect to its power output.

9.3 FILTERS[3,4]

There are a wide variety of filters available with which to improve inverter output waveforms. In this section, filters, as applied to single-phase inverters with their conventional square-wave output voltages, are considered.

The objective of the filter applied to the output of an inverter is the reduction or attenuation of harmonics appearing at the load. The general approach is to provide a shunt path for the harmonic currents with a series impedance across which the harmonic voltages appear. This general arrangement is shown in Figure 9.5. The attenuation of any given harmonic depends on the ratio of the impedance of the parallel combination of the load and the shunt element to the total impedance at that frequency, that is

$$\frac{e_L}{e_I} = \frac{Z_0}{Z_1 + Z_0} \quad \text{where} \quad Z_0 = \frac{Z_2 Z_L}{Z_2 + Z_L} \quad (9.15)$$

The shunt filter element normally increases the total inverter output current. The series element produces a voltage drop, due to load current through it.

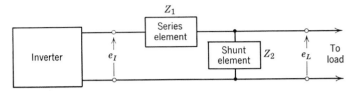

Figure 9.5   Basic configuration of external filter for inverter.

The basic considerations in designing a filter to attenuate certain harmonics adequately are as follows:

(a) The minimimizing of kva requirements of the inverter.

(b) The minimizing of variation in the load voltage, as the load is varied over its range (minimizing regulation).

(c) The minimizing of filter cost.

(d) The minimizing of filter size and weight.

Single L-C Filter

The simplest form of an efficient filter is the single section L-C filter indicated in Figure 9.6. In this filter the series element is an inductance.

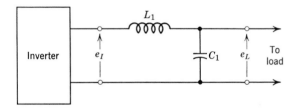

Figure 9.6   Single L-C filter.

The shunt element is a capacitor. The transfer function (using Laplace transforms) of this filter, unloaded or resistively loaded, has the form

$$\frac{e_L}{e_I}(s) = \frac{1}{\dfrac{s^2}{\omega_0{}^2} + \dfrac{2\delta}{\omega_0}s + 1} \tag{9.16}$$

where ω_0 is the resonant frequency of the filter, and δ is the damping ratio.

The general form of the frequency response can be found from this by replacing s by $j\omega$ and s^2 by $(j\omega)^2$ or $-\omega^2$ so that

$$\frac{e_L}{e_I}(j\omega) = \frac{1}{\dfrac{-\omega^2}{\omega_0{}^2} + \dfrac{j2\delta\omega}{\omega_0} + 1} \tag{9.17}$$

The substitution of u for ω/ω_0 in equation (9.17) gives the transfer function in normalized form:

$$\frac{e_L}{e_I} = \frac{1}{-u^2 + j2\delta u + 1} \tag{9.18}$$

The log of the magnitude of e_L/e_I from equation (9.18), plotted against u, the nondimensional frequency term, on a log scale, is the frequency

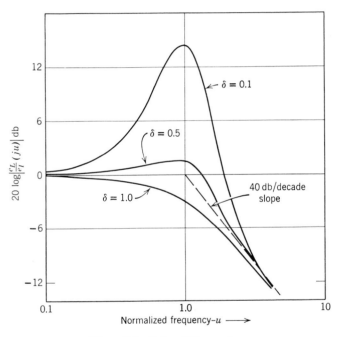

Figure 9.7   Gain of filter vs. frequency.

response curve. The frequency response with various values of damping ratio, δ, is shown in Figure 9.7.[5] Note that at low values of frequency, $u \ll 1$, the magnitude of the transfer function, is unity ($\log e_L/e_I(ju) = \log 1 = 0$). The rate of increase in attenuation with increasing frequency, above $u = 1.0$, approaches 40 db per decade or 12 db per octave. In other

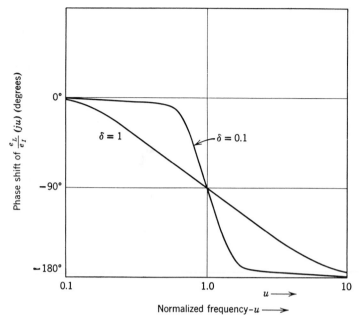

Figure 9.8   Phase shift of filter vs. frequency.

words, above $u = 1$, the attenuation of the filter will increase 4 to 1 when the frequency increases 2:1. As can be readily seen, there is considerable peaking at $u = 1$ with low values of δ.

The phase shift through the filter as a function of frequency is shown in Figure 9.8 in the same nondimensional form as used in Figure 9.7. There are two curves shown for different values of δ. The phase shift of the single section L-C filter is zero at very low frequencies, and approaches $180°$ at high frequencies.

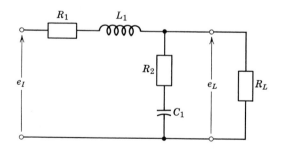

Figure 9.9   Generalized L-C filter.

The transfer function for the general form of the single section L-C filter will be developed.   Figure 9.9 shows this filter.   The losses associated with the reactor and capacitor are represented by resistors R_1 and R_2, respectively.   The load is assumed to be pure resistive.   Using Laplace transforms the transfer function is

$$\frac{e_L}{e_I}(s) = \frac{\dfrac{\left(R_2 + \dfrac{1}{sC_1}\right)R_L}{R_L + R_2 + \dfrac{1}{sC_1}}}{R_1 + sL_1 + \dfrac{\left(R_2 + \dfrac{1}{sC_1}\right)R_L}{R_L + R_2 + \dfrac{1}{sC_1}}} \tag{9.19}$$

$$= \frac{R_2C_1s + 1}{\left(L_1C_1 + \dfrac{R_2}{R_L}L_1C_1\right)s^2 + \left(R_1C_1 + R_2C_1 + \dfrac{R_2R_1}{R_L}C_1 + \dfrac{L_1}{R_L}\right)s + \dfrac{R_1}{R_L} + 1}$$

Since, for an efficient filter, $R_1/R_L \ll 1$, and since any reasonable filter capacitor has a very low effective series resistance, $R_2/R_L \ll 1$, this reduces to

$$\frac{e_L}{e_I}(s) = \frac{R_2C_1s + 1}{L_1C_1s^2 + \left(R_1C_1 + R_2C_1 + \dfrac{L_1}{R_L}\right)s + 1} \tag{9.20}$$

Comparing this equation to equation (9.16), it is seen that

$$\omega_0^2 = \frac{1}{L_1C_1} \tag{9.21}$$

$$\frac{2\delta}{\omega_0} = R_1C_1 + R_2C_1 + \frac{L_1}{R_L} \tag{9.22}$$

$$\delta = \frac{1}{2\sqrt{L_1C_1}}\left(R_1C_1 + R_2C_1 + \frac{L_1}{R_L}\right) \tag{9.23}$$

The transfer function as a function of frequency, $e_L/e_I(j\omega)$, is

$$\frac{e_L}{e_I}(j\omega) = \frac{j\dfrac{\omega}{\omega_2} + 1}{\dfrac{-\omega^2}{\omega_0{}^2} + j\left(\dfrac{\omega}{\omega_1} + \dfrac{\omega}{\omega_2} + \dfrac{\omega}{\omega_3}\right) + 1} \tag{9.24}$$

where

$$\omega_1 = \frac{1}{R_1 C_1}$$

$$\omega_2 = \frac{1}{R_2 C_2}$$

$$\omega_3 = \frac{R_L}{L_1}$$

The ω_2 term is usually very large compared to ω_0 or the other ω terms, since R_2 is very small in filter capacitors. The $[j(\omega/\omega_2) + 1]$ factor in the numerator modifies the general form of the frequency response and phase shift curves. This factor results in a reduced rate of attenuation per

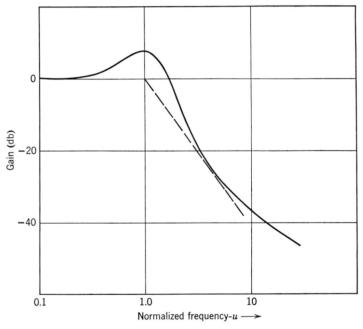

Figure 9.10   Gain vs. frequency for Figure 9.9.

decade to 20 db and phase-shift reduction to minus $90°$ at very high frequency when R_2 is much greater than X_C. This is illustrated in Figure 9.10. R_2 was included in the general form of the L-C filter to illustrate its effect. Usually, this effect is small enough so that it can be neglected.

With inductance in the load, there is an additional frequency-varying term added to the transfer function. This results in a third-order polynomial factor in the denominator of equation (9.20). Thus, the analysis of the L-C filter with inductive loads becomes quite complicated when approaching it in this fashion. The actual attenuation of particular sinusoidal frequency components may be quite easily calculated for most practical filters by using the impedances of each filter component, and then calculating the total circuit impedance ratio defined by equation (9.15)

Effects of L and C Value on a Filter

In general, when an L-C filter sufficiently attenuates the lowest harmonic present in the output of an inverter, the higher harmonics are automatically adequately reduced to a permissible level. This results because a practical filter is usually designed so that its resonant frequency ω_0 is below the lowest harmonic to be attenuated. Thus, the harmonics present occur in the frequency range where the filter attenuation curve has a 40 db/decade slope as shown in Figure 9.7. The form of the response curve indicates that the attenuation at a given frequency is determined by its relation to the resonant frequency of the filter.

It is relatively simple to determine the required L-C product to attenuate a harmonic a given amount. However, the values of L_1 and C_1 are still undetermined. Two important characteristics of the filter, which are affected by the particular values of L_1 and C_1 are given below.

(1) The fundamental regulation $|E_{L,1}/E_{I,1}|$.

(2) The value of inverter current $|I_{I,1}/I_{L,1}|$.

A large value of L_1 and small C_1 will give high regulation of fundamental voltage and a value of inverter current only slightly larger than the load current. On the other hand, a small value of L_1 and a large C_1 will result in low fundamental regulation but a large increase of inverter current over the load current value. In either case, the rating of the inverter is adversely effected. Thus, a trade-off is required. The power factor of the load has an important bearing on this trade-off.

As a guide in making the required trade-off, a study of the interrelation between the values of L and C and their affects on the inverter is required. Although the results may not provide a means of selecting the optimum values of L_1 and C_1 for a specific situation, they will indicate the trends present so that an intelligent choice may be made in any particular application.

First, consider the influence of the value of C on the inverter current. Figure 9.11 shows a simple L-C filter circuit with a load. An inductive load is shown, since this is the most general case. Series resistances in the reactor L_1 and capacitor C_1 are neglected.

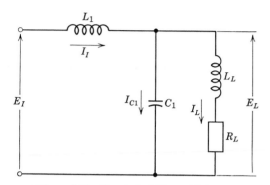

Figure 9.11   Simple L-C filter with load.

The term $I_{I,1}$, the fundamental component of I_I, is given by

$$I_{I,1} = I_{L,1} + I_{C1,1} = \frac{E_{L,1}}{Z_{0,1}} \tag{9.25}$$

where $Z_{0,1}$ is the equivalent impedance of the parallel combination of the filter capacitor and load to fundamental current.

$$Z_{0,1} = \frac{Z_{L,1} Z_{C1,1}}{Z_{L,1} + Z_{C1,1}} \tag{9.26}$$

$$Z_{L,1} = R_L + jX_{L(L),1} = |Z_{L,1}| (\cos \theta + j \sin \theta) \tag{9.27}$$

where θ = load power factor angle.

$$Z_{C1,1} = -jX_{C1,1} \tag{9.28}$$

Let

$$|X_{C1,1}| = K_1 |Z_{L,1}| \quad \text{or} \quad K_1 = \frac{(X_{C1,1})}{(Z_{L,1})} \tag{9.29}$$

then

$$Z_{C1,1} = -jK_1 |Z_{L,1}| \tag{9.30}$$

Substituting equations (9.27) and (9.30) in equation (9.26), the result is

$$Z_{0,1} = |Z_{L,1}| \frac{-jK_1(\cos \theta + j \sin \theta)}{\cos \theta + j(\sin \theta - K_1)} \tag{9.31}$$

Combining equations (9.31) and (9.25),

$$I_{I,1} = \frac{E_{L,1}}{|Z_{L,1}|} \frac{\cos \theta + j(\sin \theta - K_1)}{K_1(\sin \theta - j \cos \theta)} \tag{9.32}$$

Since

$$\frac{|E_{L,1}|}{|Z_{L,1}|} = |I_{L,1}| \tag{9.33}$$

$$\frac{|I_{I,1}|}{|I_{L,1}|} = \frac{\cos \theta + j(\sin \theta - K_1)}{K_1(\sin \theta - j \cos \theta)} \tag{9.34}$$

The term $I_{I,1}/I_{L,1}$ is the ratio of the fundamental component of inverter current to the corresponding component of load current. Figure 9.12 shows

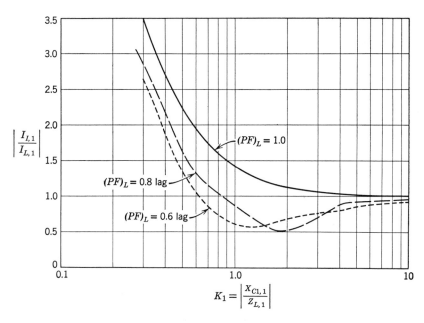

Figure 9.12 Trend of inverter current vs. value of filter capacitor at various load power factors [equation (9.34)].

this ratio plotted as a function of the constant K_1 for various load-power factors. Figure 9.12 shows that the ratio of inverter and load current can be less than one when the power factor of the load is lagging. The vector diagram of Figure 9.13 illustrates such a condition. In this case, the inverter current will be larger at no load than at the load condition shown in Figure 9.13.

The regulation of the fundamental component of voltage due to the filter can be expressed in general terms by the ratio of load voltage to inverter voltage. This ratio is affected by both the capacitor and the reactor values. The capacitor affects the fundamental current through the reactor, and this current multiplied by the reactor impedance determines the voltage drop across the reactor. The fundamental value of the load voltage is given by

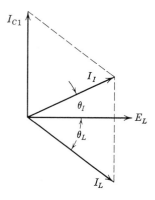

Figure 9.13   Vector diagram for a simple L-C filter with a load condition where the inverter current will increase as the magnitude of the load current is reduced. ($K_1 = 1.0$; $(PF)_L = 0.8$ lag.)

$$E_{L,1} = E_{I,1} - I_{I,1}Z_{L1,1} \Big\} \tag{9.35}$$
$$Z_{L1,1} = jX_{L1,1}$$

Let

$$|X_{L1,1}| = K_2 |Z_{L,1}| \quad \text{or} \quad K_2 = \frac{|X_{L1,1}|}{|Z_{L,1}|}$$

then
$$\tag{9.36}$$

$$\frac{E_{L,1}}{E_{I,1}} = 1 - \frac{I_{I,1}}{E_{I,1}} jK_2 |Z_{L,1}| \tag{9.37}$$

$$\frac{I_{I,1}}{E_{I,1}} = \frac{1}{Z_{I,1}} = \frac{1}{Z_{0,1} + jX_{L1,1}} \tag{9.38}$$

From equation (9.31),

$$Z_{0,1} = |Z_{L,1}| \frac{K_1(\sin\theta - j\cos\theta)}{\cos\theta + j(\sin\theta - K_1)} \tag{9.39}$$

Combining (9.37) through (9.39),

$$\frac{E_{L,1}}{E_{I,1}} = 1 - \frac{jK_2}{jK_2 + \dfrac{K_1(\sin\theta - j\cos\theta)}{\cos\theta + j(\sin\theta - K_1)}} \tag{9.40}$$

$$\frac{E_{L,1}}{E_{I,1}} = \frac{\dfrac{K_1(\sin\theta - j\cos\theta)}{\cos\theta + j(\sin\theta - K_1)}}{\dfrac{K_1(\sin\theta - j\cos\theta)}{\cos\theta + j(\sin\theta - K_1)} + jK_2} \tag{9.41}$$

In Figure 9.14 the ratio of the fundamental component of load voltage to inverter voltage is plotted as a function of K_2 for various values of K_1 and load power factor. Note that at small values of K_1 and K_2 the load voltage is appreciably higher than the inverter voltage. Such a condition is

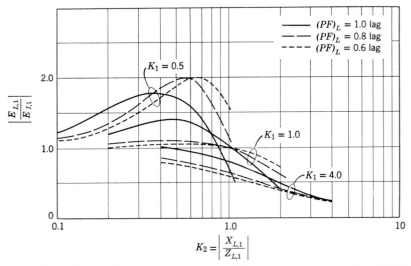

$$K_2 = \left| \frac{X_{L,1}}{Z_{L,1}} \right|$$

Figure 9.14   Trend of load voltage vs. value of filter reactor at various K_1 and $(PF)_L$ values [equation (9.41)].

indicated by the vector diagram shown in Figure 9.15(a). For the higher values of K_1 and K_2, the load voltage is less than the inverter voltage, as shown in Figure 9.15(b).

It is next desirable to consider the effects of the values of both L_1 and C_1 on attenuation of harmonics. This information is needed to determine the circuit relationships required to give a desired attenuation of a given

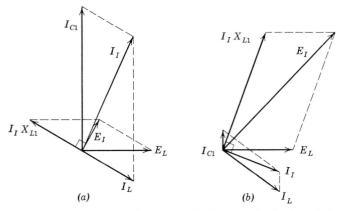

Figure 9.15   Vector diagrams for a simple L-C filter with a load condition where the inverter voltage becomes less than the load voltage when K_1 and K_2 are reduced. (a) $K_1 = 0.5$; $(PF)_L = 0.8$ lag; and $K_1 = 0.5$. (b) $K_1 = 4.0$; $(PF)_L = 0.8$ lag; and $K_2 = 2.0$.

frequency component. For the nth harmonic,

$$|X_{L1,n}| = n |X_{L1,1}| = nK_2 |Z_{L,1}| \tag{9.42}$$

$$|X_{C1,n}| = \frac{|X_{C1,1}|}{n} = \frac{K_1 |Z_{L,1}|}{n} \tag{9.43}$$

$$Z_{L,n} = R_L + j |X_{L(L),n}| = R_L + jn |X_{L(L),1}| \tag{9.44}$$

where

$$R_L = |Z_{L,1}| \cos \theta$$

$$j |X_{L(L),1}| = |Z_{L,1}| \sin \theta$$

$$Z_{L,n} = |Z_{L,1}| (\cos \theta + jn \sin \theta) \tag{9.45}$$

$$\frac{E_{L,n}}{E_{I,n}} = 1 - \frac{I_{I,n} |jX_{L1,n}|}{E_{I,n}} = 1 - \frac{j |X_{L1,n}|}{Z_{I,n}} \tag{9.46}$$

$$= 1 - \frac{j |X_{L1,n}|}{Z_{0,n} + j |X_{L1,n}|}$$

$$= 1 - \frac{jnK_2 |Z_{L,1}|}{Z_{0,n} + jnK_2 |Z_{L,1}|}$$

$$= 1 - \frac{jnK_2}{jnK_2 + \dfrac{Z_{0,n}}{|Z_{L,1}|}}$$

$$= \frac{jnK_2 + \dfrac{Z_{0,n}}{|Z_{L,1}|} - jnK_2}{jnK_2 + \dfrac{Z_{0,n}}{Z_{L,1}}} \tag{9.47}$$

$$jnK_2 + \frac{Z_{0,n}}{|Z_{L,1}|} = \frac{E_{I,n}}{E_{L,n}} \frac{Z_{0,n}}{|Z_{L,1}|} \tag{9.48}$$

$$K_2 = \frac{1}{jn} \frac{Z_{0,n}}{|Z_{L,1}|} \left(\frac{E_{I,n}}{E_{L,n}} - 1 \right) \tag{9.49}$$

Let

$$\frac{E_{I,n}}{E_{L,n}} = K_{A,n} \quad \substack{\text{(the attenuation required} \\ \text{of the } n\text{th harmonic)}} \tag{9.50}$$

$$Z_{0,n} = \frac{Z_{L,n}(-j\,|X_{C1,n}|)}{Z_{L,n} + (-j\,|X_{C1,n}|)} \tag{9.51}$$

$$= \frac{|Z_{L,1}|\,(\cos\theta + jn\sin\theta)\left(-j\dfrac{K_1}{n}\,|Z_{L,1}|\right)}{|Z_{L,1}|\,(\cos\theta + jn\sin\theta) + \left(-j\dfrac{K_1}{n}\,|Z_{L,1}|\right)}$$

$$\frac{Z_{0,n}}{|Z_{L,1}|} = \frac{-j\dfrac{K_1}{n}\,(\cos\theta + jn\sin\theta)}{(\cos\theta + jn\sin\theta) - j\dfrac{K_1}{n}} \tag{9.52}$$

$$\frac{1}{j}\frac{Z_{0,n}}{Z_{L,1}} = -\frac{\dfrac{K_1}{n}(\cos\theta + jn\sin\theta)}{\cos\theta + j\left(n\sin\theta - \dfrac{K_1}{n}\right)} \tag{9.53}$$

Substituting (9.53) and (9.50) in (9.49), and recognizing that K_2 is a real number,

$$K_2 = \frac{K_1}{n^2}\left|\frac{\cos\theta + jn\sin\theta}{\cos\theta + j\left(n\sin\theta - \dfrac{K_1}{n}\right)}\right|(K_{A,n} - 1) \tag{9.54}$$

Figure 9.16 is a plot of K_2 as a function of K_1 required for given amounts of attenuation of the third and fifth harmonics with different load-power factors. The general trend of these curves is such as to require higher values of K_2, as K_1 is increased.

The three sets of curves in Figures 9.12, 9.14, and 9.16 provide information on the effect of the important circuit parameters on the filter performance. There are several additional factors which should be considered in practical situations. For example, depending on the voltage and current ratings of the inverter involved and the ratings of the SCR's to be used, increased voltage may prove more desirable than increased current or vice versa; hence, effecting the trade-offs. Size and cost considerations are also important factors when selecting particular L_1 and C_1 values in an actual case. In many circuits, the inverter involves a full isolation transformer in which the primary-to-secondary leakage reactance may prove a convenient addition to the filter reactance required or, possibly, the full

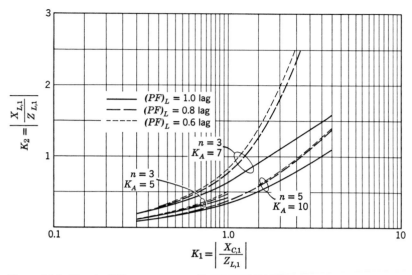

Figure 9.16   Trend of filter reactor vs. filter capacitor values for various attenuations and harmonics [equation (9.54)].

filter reactance required. In low voltage applications, the capacitor kva rating may be set by the minimum economically produced capacitor voltage rating. (Capacitors designed for low voltage require ultra-thin paper to allow for minimum size.) These and other considerations should not be overlooked in making the trade-offs necessary for a particular filter design.

Cascaded L-C Filter

Where higher values of harmonic attenuation are required, the use of cascaded *L-C* filters of the type discussed in the previous sections becomes attractive. A two-section *L-C* filter, with the same total kva rating of capacitance and inductance as in a single-stage filter, has considerably higher attenuation possible at values of total attenuation above a certain level. This general trend continues with an increasing number of stages of filters. In most practical inverters, however, the need for harmonic attenuation is satisfied with not more than a two stage filter. An approximate value for the attenuation where the two-stage filter is preferable to the single-stage filter is determined in this subsection.

Before entering into the detailed analysis of the two-stage filter, it is interesting to consider some factors which may influence the decision of whether to use a cascaded filter. First, servo-system stability considerations are important. When the filter involved is included in a closed-loop voltage regulating system, the presence of two L-C networks can cause a great deal of difficulty in stabilizing. This problem can be alleviated somewhat by separating the filter resonance frequencies by a factor of four or more.

Another consideration is the over-all size and construction cost of the filter. Although a cascaded filter may appear theoretically desirable at one value of attenuation to minimize total kva rating, the size and construction cost of the extra components may elevate the actual value of attenuation at which cascaded filters become advantageous.

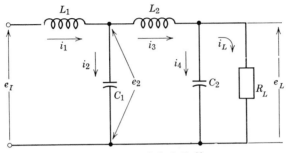

Figure 9.17   Casaded L-C filter.

The transfer function for the two-stage filter, illustrated in Figure 9.17, is

$$e_L(s) = i_L(s)R_L = i_4(s)\left(\frac{1}{sC_2}\right) \tag{9.55}$$

$$i_3(s) = i_L(s) + i_4(s) = \frac{e_L(s)}{R_L} + \frac{e_L(s)}{\dfrac{1}{sC_2}} = e_L(s)\left(\frac{1 + R_L sC_2}{R_L}\right) \tag{9.56}$$

Current i_3 is also given by

$$i_3(s) = \frac{e_2(s)}{sL_2 + \dfrac{R_L}{R_L sC_2 + 1}} \tag{9.57}$$

Substituting equation (9.56) into (9.57),

$$e_2(s) = e_L(s)\left(\frac{s^2 L_2 C_2 R_L + sL_2 + R_L}{R_L}\right) \tag{9.58}$$

$$\frac{e_L}{e_2}(s) = \frac{1}{s^2 L_2 C_2 + \dfrac{sL_2}{R_L} + 1} \tag{9.59}$$

This is, of course, the transfer function of a single-stage L-C filter, as we have seen previously.

Current i_2 is given by

$$i_2(s) = \frac{e_2(s)}{\dfrac{1}{sC_1}} = sC_1e_2(s) \tag{9.60}$$

Substituting equation (9.58) into equation (9.60),

$$i_2(s) = sC_1e_L(s)\left(s^2L_2C_2 + \frac{sL_2}{R_L} + 1\right) \tag{9.61}$$

The current i_1 is given by the expressions

$$i_1(s) = i_2(s) + i_3(s) \tag{9.62}$$

and

$$e_I(s) = sL_1i_1(s) + e_2(s) \tag{9.63}$$

Then, from equations (9.57) through (9.63),

$$e_I(s) = e_L(s)\left[(s^2C_1L_1)\left(s^2L_2C_2 + \frac{sL_2}{R_L} + 1\right)\right.$$

$$\left. + \frac{sL_1}{R_L} + s^2L_1C_2 + s^2L_2C_2 + \frac{sL_2}{R_L} + 1\right]$$

$$\frac{e_L}{e_I}(s) = \frac{1}{(s^2C_1L_1 + 1)\left(s^2L_2C_2 + \dfrac{sL_2}{R_L} + 1\right) + \dfrac{sL_1}{R_L} + s^2L_1C_2} \tag{9.64}$$

A few examples of attenuation calculations for single and cascaded filters point out an interesting consideration. The attenuation of the single L-C filter, as shown in Figure 9.6, may be determined as follows. Assuming the impedances below at the fifth harmonic,

$$X_{L1} = j1000$$

$$X_{C1} = -j50$$

$$\frac{e_L}{e_I}(j\omega) = \frac{-j50}{j1000 - j50} = \frac{-j50}{j950} = -\frac{1}{19}$$

The attenuation of a cascaded filter, as shown in Figure 9.17, having the same kva of filter components but divided into two identical cascaded L-C sections, may be calculated as follows. In this case, the impedances at the fifth harmonic are

$$X_{L1} = X_{L2} = j500$$
$$X_{C1} = X_{C2} = -j100$$

$$\frac{e_L}{e_I}(j\omega) = \left(\frac{-j100}{j500 - j100}\right)\left(\frac{\dfrac{[j400][-j100]}{j400 - j100}}{j500 + \dfrac{[j400][-j100]}{j400 - j100}}\right)$$

$$= \left(-\frac{1}{4}\right)\left(\frac{-j133}{j500 - j133}\right) = \left(\frac{1}{4}\right)\left(\frac{1}{2.75}\right) = \frac{1}{11}$$

Thus, the cascaded filter provides less attenuation with the same kva of components than the single-stage filter.

If the values of the components are changed to obtain impedance levels of

$$X_{L1} = j2000$$
$$X_{C1} = -j25$$

the attenuation for the single-stage filter is

$$\frac{e_L}{e_I}(j\omega) = \frac{-j25}{j2000 - j25} = -\frac{1}{79}$$

The corresponding component values for the cascaded filter are

$$X_{L1} = X_{L2} = j1000$$
$$X_{C1} = X_{C2} = -j50$$

and the attenuation is

$$\frac{e_L}{e_I}(j\omega) = \frac{-j50}{j1000 - j50}\left(\frac{\dfrac{[j950][-j50]}{j950 - j50}}{+j1000 + \dfrac{[j950][-j50]}{j950 - j50}}\right)$$

$$= -\frac{1}{19}\left(\frac{-j53}{j1000 - j53}\right) = \left(-\frac{1}{19}\right)\left(-\frac{1}{17.8}\right) = \frac{1}{340}$$

demonstrating that the cascaded filter at these attenuation levels is much superior to the single-stage filter.

Obviously, somewhere between these extremes, the single-stage filter and the cascaded filter are equivalent. Knowing the attenuation level at which the two are equivalent helps in deciding whether a cascaded filter is justified for a given application.

As an example of the type of calculation required to determine the attenuation at which the two-stage and single-stage filters are equivalent, assume that the cascaded filters, shown in Figure 9.17, have equal values of inductances and capacitances, each equal to half the values in the single section components.

$$L_1 = \text{inductance of single-stage filter}$$
$$C_1 = \text{capacitance of single-stage filter}$$
$$L_{11} = L_{22} \text{ inductance of two-stage filter}$$
$$C_{11} = C_{22} \text{ capacitance of two-stage filter}$$
$$L_{11} = L_{22} = L_1/2 \tag{9.65}$$
$$C_{11} = C_{22} = C_1/2 \tag{9.66}$$

The transfer function of a single L-C filter in the unloaded case is

$$\frac{e_L}{e_I}(s) = \frac{1}{s^2 L_1 C_1 + 1} \tag{9.67}$$

The transfer function of the cascaded L-C filter in the unloaded case, assuming the relationships of equations (9.65) and (9.66), is

$$\frac{e_L}{e_I}(s) = \frac{1}{\left(\dfrac{s^2 L_1 C_1}{4} + 2\right)\left(1 + \dfrac{s^2 L_1 C_1}{4}\right) - 1} \tag{9.68}$$

Expressing the transfer functions in (9.67) and (9.68) in terms of $j\omega$, and making use of the relationship between L_1, C_1 and the resonant frequency ω_0, we get

$$(\text{single})\ \frac{e_L}{e_I}(j\omega) = \frac{1}{-\dfrac{\omega^2}{\omega_0^2} + 1} \tag{9.69}$$

$$(\text{cascaded})\ \frac{e_L}{e_I}(j\omega) = \frac{1}{\dfrac{\omega^4}{16\omega_0^4} - \dfrac{3\omega^2}{4\omega_0^2} + 1} \tag{9.70}$$

To determine the value of attenuation at which the two filters are equivalent, the magnitudes of the two transfer functions are equated

$$\left| -\frac{\omega^2}{\omega_0^2} + 1 \right| = \left| \frac{\omega^4}{16\omega_0^4} - \frac{3}{4}\frac{\omega^2}{\omega_0^2} + 1 \right| \tag{9.71}$$

When ω is greater than ω_0, the left-hand term must be written $|\omega^2/\omega_0^2 - 1|$ so that both sides of the equation will produce positive numbers.

$$\frac{\omega^2}{\omega_0^2} - 1 = \frac{\omega^4}{16\omega_0^4} - \frac{3}{4}\frac{\omega^2}{\omega_0^2} + 1 \tag{9.72}$$

$$\frac{\omega^4}{16\omega_0^4} - \frac{7}{4}\frac{\omega^2}{\omega_0^2} + 2 = 0 \tag{9.73}$$

Solving for the ratio ω/ω_0, and substituting this into either equation (9.69) or (9.70), it is possible to determine the attenuation at which the two filters are equivalent.

$$\left(\frac{\omega}{\omega_0}\right)^4 - 28\left(\frac{\omega}{\omega_0}\right)^2 + 32 = 0$$

$$\left(\frac{\omega}{\omega_0}\right)^2 = \frac{28 \pm \sqrt{28^2 - 128}}{2}$$

$$\left(\frac{\omega}{\omega_0}\right)^2 = 14 \pm 12.8$$

$$\left(\frac{\omega}{\omega_0}\right)^2 = 26.8 \text{ or } 1.2 \tag{9.74}$$

This indicates that the two filters are equivalent at two points. Actually, the lower number arises because of equivalence in the resonant peaking region. The value of $(\omega/\omega_0)^2 = 26.8$ is the significant practical result. Substituting this into equation (9.69), we get

$$\frac{e_L}{e_I}(s) = \frac{1}{-26.8 + 1} = \frac{1}{-25.8}$$

$$\frac{e_L}{e_I}(s) = \frac{1}{25.8 \big/ 180°} \tag{9.75}$$

Equation (9.75) indicates that in filter applications where attenuations of up to approximately 26:1 are required, the single-stage filter is most desirable. In cases where attenuations greater than 26:1 are required, the cascaded filter should be considered. It is important to note that these conclusions resulted for this specific case where loading on the output of the filters was neglected and where the cascaded filter sections were assumed identical to each other. In addition, it should be re-emphasized that many practical considerations will affect the decision of whether single-section or cascaded filters should be used for particular applications.

Tuned Filters

The major disadvantages of simple $L\text{-}C$ filters are the added regulation in the series element and increased loading on the inverter because of the shunt element. Resonant networks in the series and shunt elements can partially overcome these disadvantages. For example, a series resonant circuit as shown in Figure 9.18(a), tuned to the fundamental frequency, minimizes the voltage drop to fundamental frequency current. This arrangement provides a series element with inductive impedance to harmonics but little regulation at the fundamental frequency.

Figure 9.18(*b*) shows a parallel resonant circuit used as the shunt element to reduce the added loading at the fundamental frequency. The parallel resonant circuit offers high impedance to fundamental, while

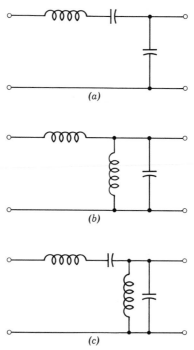

(*a*)

(*b*)

(*c*)

Figure 9.18   Tuned filters. (*a*) Series resonant series element. (*b*) Parallel resonant shunt element. (*c*) Series resonant series element and parallel resonant shunt element.

having low-capacitive impedance to higher harmonics. A combination of the two tuned circuits offers the advantages of each. Such a combination is shown in Figure 9.18(*c*).

The use of tuned filter circuits is limited to filtering applications which permit fixed or nearly fixed frequency operation.

9.4   POLYPHASE INVERTERS

Polyphase circuits provide the opportunity to minimize the harmonics in the output voltage from inverters. The square-wave output from the simple impulse-commutated circuits can be modified in polyphase arrangements, thereby reducing the external filter requirements. In general, polyphase schemes either provide cancellation of certain harmonics, or increase the fundamental component relative to certain harmonics. Since three-phase output is required from most high-power inverters, the third harmonics and multiples thereof are not present in the output voltage waveforms from these inverters. When a greater number of phases is employed, additional harmonics are eliminated.

An interesting parallel exists between the current waveshapes in rectifiers and the voltage waveshapes in impulse-commutated inverters. For this comparison, the commutation reactance in the rectifier and the commutation intervals in the inverter both must be negligible, and perfect transformers, rectifiers, and SCR's must be assumed. These assumptions are reasonable for most practical circuits. The square-wave voltage of the single-phase inverter is comparable to the square-wave current in a similar rectifier where the output current has negligible ripple. This comparison holds true in polyphase circuits, enabling one to make use of standard polyphase rectifier current waveforms to predict voltage waveforms for equivalent polyphase inverter circuits.

Polyphase inverter waveforms may be produced by the summation of square waves. Some of the harmonics of the square wave are canceled in this adding process. For example, consider the familiar three-phase inverter waveform as a summation of three single-phase inverter square waveforms. Figure 9.19 shows one possible way that a three-phase inverter waveform might be generated. There are three single-phase inverters with their output transformer secondaries wye-connected. If the gating signals for Inverter No. 2 are phased 120°, lagging those of Inverter No. 1, and the corresponding signals for Inverter No. 3 are phased 120°, lagging those of Inverter No. 2, the line-to-line voltage is as shown. The typical three-phase waveform can be the simple summation of square waves, that is

$$e_{AB} = e_{AN} + (-e_{BN}) \qquad (9.76)$$

$$e_{BC} = e_{BN} + (-e_{CN}) \qquad (9.77)$$

$$e_{CA} = e_{CN} + (-e_{CN}) \qquad (9.78)$$

With either Fourier analysis or the graphic method described in Section 9.1, it can be shown that the waveforms e_{AB}, e_{BC}, or e_{CA} have zero third harmonic, and multiples thereof, and the same other higher harmonics as a square wave. The fundamental components of the line-to-line voltages have the characteristic 120° phase relationships of three-phase systems. The filter required to obtain sinusoidal line-to-line

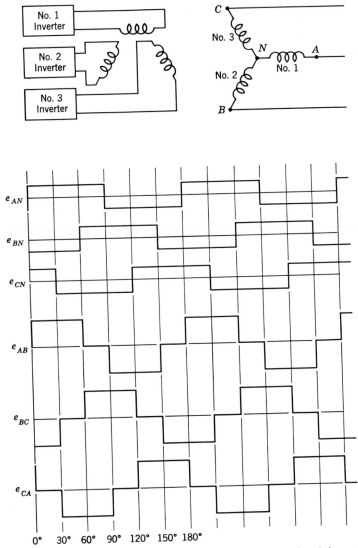

Figure 9.19   Addition of single-phase inverter outputs, producing 3ϕ waveforms.

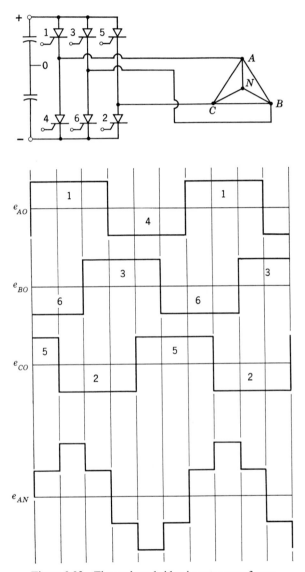

Figure 9.20   Three-phase bridge-inverter waveforms.

voltages is less of a problem, since the third harmonic is missing and need not be attenuated.

The process of adding square waves to form stepped waves can be carried further by combining the outputs from two three-phase circuits, each having waveforms similar to e_{AB}, e_{BC}, and e_{CA} in Figure 9.19. With a 30° phase relationship between two such three-phase circuits, the resulting output voltage waveforms contain no fifth and seventh harmonics. Such a system has been described in Chapter 8, where this process was used to obtain variable output voltage. This technique can be continued, thereby eliminating additional higher harmonics. Such techniques are of particular value in higher powered inverters, where filter component sizes and costs would otherwise be prohibitive and where multiple circuits or paralleled controller-rectifier devices are required to deliver the required power output.

The waveforms in Figure 9.19 show six evenly spaced commutations per fundamental cycle; that is, a commutation every 60°. Hence, the waveform is called a six-phase wave. Any circuit which has six evenly spaced commutations per cycle will have a six-phase waveform. The waveforms may take on a different appearance because of the phase relationship of the various harmonics, but they will have the same harmonic content. For example, consider the e_{AN} wave of Figure 9.20. Although it appears quite different from those of Figure 9.19, Fourier analysis proves it to have exactly the same harmonic content. Note that it has the same characteristic six evenly spaced commutations per cycle. The significant point to be made here is the relationship between the number of evenly spaced commutations per fundamental cycle and the harmonics present in the waveform. The lowest harmonics in these waveforms are the $n - 1$ and $n + 1$ harmonics, where n is the number of evenly spaced commutations per cycle. In the six-phase waveforms ($n = 6$), the fifth and seventh are the lowest harmonics. In twelve-phase waveforms ($n = 12$, as obtained by adding two six-phase waveforms at 30°), the lowest harmonics are the eleventh and thirteenth.

9.5   PULSE-WIDTH CONTROL

Single-Pulse Method

The stepped waveform of Figure 9.19 can be obtained in a single-phase inverter by pulse-width control. A basic circuit required to permit adjustment of pulse width is shown in Figure 9.21. This circuit has been stripped of all but the power-switching devices to convey most clearly this technique for waveform improvement. Figure 9.21 is a single-phase bridge

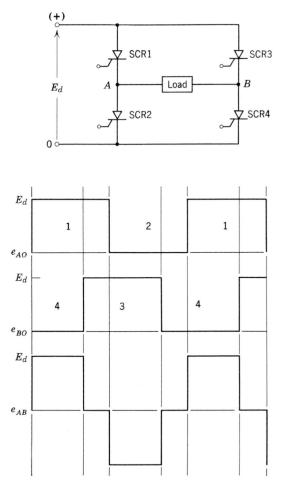

Figure 9.21   Three-phase waveform of Figure 9.19, produced from a single-phase inverter.

inverter. However, there is a difference in the SCR gating sequence. Unlike the simple bridge inverter, in which SCR1 and SCR4 are gated "on" together, the gating of SCR1 in Figure 9.21 lags that of SCR4 by 60°. Assuming 180° conduction intervals for each valve, the waveforms are as shown. The voltage waveforms e_{AO} and e_{BO} are the voltages of points A and B with reference to the negative of the d-c supply. The difference between these two voltages is the load voltage, e_{AB}, the stepped wave shown. In the arrangement of Figure 9.21, two square-waves with the proper phase relationship are added to produce stepped waves in which harmonics are canceled.

Pulse-width control is discussed in more detail in Chapter 8, where this type of control is used to vary the fundamental value of output voltage from inverters. Although the end results of the pulse-width control techniques discussed in Chapter 8 are different, the approaches described are directly applicable for reducing or eliminating selected harmonics in inverter output voltage waveforms.

Multiple-Pulse Methods

With inverter circuitry as described in Chapter 7, the conduction of an SCR can be started and stopped at the command of a control circuit. This

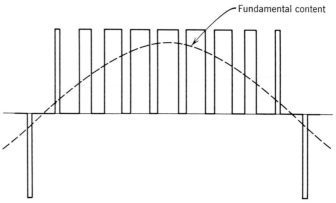

Figure 9.22   Multiple pulse-width control (the circuit is shown in Figure 9.21).

ability makes it possible to have several conduction intervals during a single half-cycle of the fundamental frequency. For example, suppose that SCR1 and SCR4 of Figure 9.21 are gated on, and then commutated off again nine times in a half-cycle of the inverter frequency, and SCR2 and SCR3 are gated on and off again a like number of times during the other half-cycle. If the conducting intervals of the SCR's are gradually increased and then decreased in a sinusoidal fashion, the load voltage varies in a sinusoidal fashion. This action is shown by the waveform in Figure 9.22. In this case, the lowest harmonic present in the output is the eighteenth harmonic or the repetition rate of the pulsing used. The filtering required to provide a sinusoidal output with acceptable harmonic content is considerably reduced as compared to that required for a square waveform. With the technique shown in Figure 9.22, the size and weight of the filter components are reduced at the cost of increased complexity in the control circuits.

Increasing the pulse repetition rate permits further reduction of filtering required. However, there is a definite practical limit to the repetition rate because of the fixed turn-off time of the SCR's used in the circuit. The shortest nonconducting interval must be greater than the specified turn-off time for the controlled rectifiers. In addition, the losses due to commutation are proportional to the number of commutations per second. As a result, the efficiency of the inverter is reduced as the repetition rate is increased.

Addition of Circuits with Pulse-Width Control

An inverter with much reduced harmonics results when several single-phase circuits have their outputs added, as shown in Figure 9.23. The

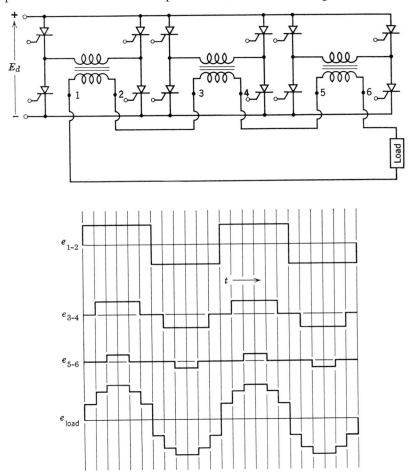

Figure 9.23 Pulse-width control with multiple circuits.

single-phase inverters are operated so that their outputs are as shown in the three waveforms e_{1-2}, e_{3-4}, and e_{5-6}. Pulse-width control is used in the inverters that deliver e_{3-4} and e_{5-6}. The summation of the three waveforms results in the load voltage waveform shown. All three of the single-phase inverters in Figure 9.23 can also be operated with square-wave outputs. The transformer ratios and inverter phase displacements are then adjusted to provide the desired output waveform.[6,7] The reduction of harmonics by adding the outputs from several single-phase circuits is of most practical significance when the inverter power rating is of sufficient magnitude to justify the multiple semiconductors involved.

9.6 TAP CHANGING IN THE INVERTER

Reduced harmonic content in the output of a single-phase inverter is possible by use of taps on the transformer used in the circuit. Figure 9.24 indicates a configuration required. Individual commutating circuits, are required for each pair of SCR's: *1-1'*, *2-2'*, and *3-3'*. The individual controlled rectifiers are turned on and off again in the sequence indicated in the waveform of the output voltage shown.

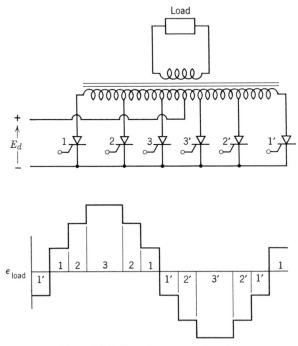

Figure 9.24   Transformer tap changing.

It is important to note that the waveform shown in Figure 9.24 results only with pure resistive load. When reactive load is present, controlled feedback valves are required in shunt with each of the controlled rectifiers to produce a waveform like that shown in the figure. Thus, this tap-changing arrangement requires relatively complex control to produce the most desirable output voltage waveform when supplying reactive loads.

REFERENCES

1. R. M. Kerchner and G. F. Corcoran, *Alternating Current Circuits*, John Wiley & Sons, New York, 1946, pp. 141–151.
2. Ibid., pp. 160–162.
3. Kerchner and Corcoran, op. cit., Chapter XIII.
4. R. V. Churchill, *Modern Operational Mathematics in Engineering*, McGraw-Hill, New York, 1944.
5. H. Chestnut and R. W. Mayer, *Servomechanisms and Regulating System Design*, Volume I, John Wiley & Sons, New York, 1951, pp. 310–314.
6. P. D. Corey, "Methods for Optimizing the Waveform of Stepped-Wave Static Inverters," AIEE Paper, CP 62–1147, Denver, June 17–22, 1962.
7. D. L. Anderson, A. E. Willis, and C. E. Winkler, "Advanced Static Inverter Utilizing Digital Techniques and Harmonic Cancellation," NASA Technical Note D–602, Washington, D.C. May, 1962.

Chapter Ten

D-C-to-D-C Power Conversion

by R. E. Morgan

There is much demand for changing from one d-c voltage to another. Some of the more important practical applications include power converters for armature voltage control of d-c motors, converting low- or high-battery source voltages to levels which best match practical load requirements, and controlling d-c power for a wide variety of industrial processes. For modern industrial and defense systems, d-c converters are required with ratings ranging from a few watts to thousands of kilowatts. The availability of the silicon-controlled rectifier makes the solid-state power converter practical for many of these applications. The SCR converter offers greater efficiency, faster response, lower maintenance, smaller size and, for many applications, it will offer lower cost than motor-generator set or gas-tube approaches.

Transistor d-c to d-c converters are attractive when the voltages are low: that is, 100 v or less. Thyraton and ignitron converters may be preferable for high voltages, above 5000 v. The SCR converter is suitable for applications at least covering the intermediate voltage range not best handled by the transistor, thyraton, or ignitron. With presently available SCR devices, this type d-c converter is applicable for voltages in the general range of 50 to 5000 v and currents between 1 and 1000 amp.

In controlled-rectifier d-c converters, commutating circuits are required, similar to those used in many inverters. Some of the most efficient d-c to d-c approaches involve the impulse-commutation techniques discussed in Chapter 7. This chapter covers two broad types of d-c to d-c SCR converters: the inverter-rectifier, and several forms of pulse-modulation control which we refer to as time ratio controls (TRC). The inverter-rectifier converter is very briefly discussed, since it is a simple and logical extension of the information in the previous chapters. The majority of

316

this chapter is concerned with time ratio control. The TRC is a relatively new form of control for d-c to d-c conversion. Generally, this control technique better utilizes the switching devices than is done in phase-controlled rectifiers. It is particularly advantageous where electric isolation is not required, where a wide voltage-control range is desired, and where the step-up or step-down voltage ratio is only several to one. With a TRC, it is possible to smoothly control the ratio of a d-c load voltage to the circuit input voltage, thereby accomplishing a function for d-c conversion similar to that which would be performed for ac by a continuously variable ratio transformer.

Four TRC techniques are discussed to illustrate the principles of this method of voltage conversion and control. Many possible SCR commutating techniques may be used in the TRC approaches, including those which have previously been discussed for inverter applications. Nonlinear magnetic components are used for the circuits in this chapter since this method of commutation has not been covered extensively in previous chapters, and since the combination of saturable reactors and the SCR often provides the most reliable, compact and economical TRC. The saturable reactor-capacitor commutating circuits discussed are another form of the impulse commutation techniques described in Chapter 7.

10.1 INVERTER-RECTIFIER

A block diagram of one inverter-rectifier type d-c to d-c converter is shown in Figure 10.1. Numerous types of inverters and rectifiers, including all those discussed in previous chapters, may be used in this converter. Square wave-type inverters are preferable when the simple transformer rectifier is used, as in Figure 10.1, since very little filtering is required to produce smooth d-c output voltage.

In its simplest form, the inverter-rectifier does not provide control of the output voltage. However, phase-controlled rectifiers and many other

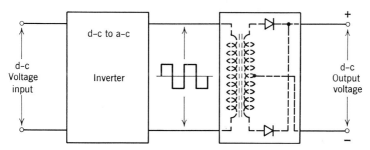

Figure 10.1   Inverter-rectifier.

a-c or d-c voltage controls may be used. A large voltage-control range may add considerably to the total equipment rating. With most forms of voltage control, there may be high output voltage ripple over a portion of the control range, which increases the filtering requirements.

The principal disadvantage of inverter-rectifier converters, when compared with TRC converters, is that the power is handled at least twice, once in the inverter and once again in the transformer-rectifier. A d-c to d-c converter with inverter-rectifiers is generally advantageous when the voltage ratio is large. For these applications, auto-transformer techniques do not afford appreciable reduction in kva rating compared to the full transformer. The full transformer can then be used without significantly adding to the total rating of the magnetic components required for a given circuit. The resulting d-c to d-c converter provides electric isolation and maximum flexibility in the selection of the output d-c voltage with respect to the input.

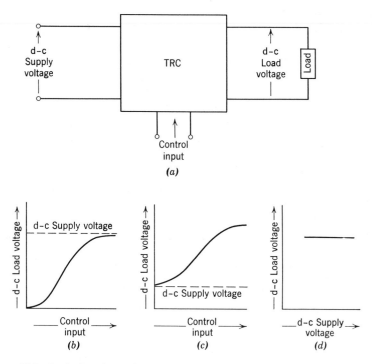

Figure 10.2   Basic functions of TRC converters. (*a*) Block diagram. (*b*) Step-down control. (*c*) Step-up control. (*d*) Constant output (includes closed-loop feedback, or similar automatic control, to hold load voltage nearly constant over wide supply-voltage range).

10.2 TIME-RATIO CONTROL—BASIC OPERATION

Many d-c to d-c converters require voltage control. A TRC can provide combination voltage conversion and control functions. The basic functions of such converters are indicated in Figure 10.2. The TRC performs a function for d-c to d-c conversion similar to that which would be performed for a-c to a-c conversion by a continuously variable turns-ratio a-c transformer. As shown in Figure 10.2, a TRC may deliver a proportionally controlled output voltage which is higher or lower than the d-c supply voltage. The step-down TRC circuits are capable of controlling the load voltage from a low value, which may be only a few per cent of the supply voltage, to a high value which may be over 95 per cent of the d-c supply. With a step-up TRC, the load voltage may be proportionally controlled over a range starting at the d-c supply voltage and extending to several times the supply voltage. Either of these techniques may be used in a closed-loop feedback control or with a similar form of automatic control to deliver an accurately regulated load voltage over a wide d-c input voltage range.

A time-ratio control involves the operation of a switch that is rapidly opened and closed. Figure 10.3 shows an elementary form of such a control to illustrate the switching action required. This circuit is only suitable for supplying resistive loads where smooth output current is not required, or for low power applications where efficient filtering is not essential. Appreciable power losses occur in filters containing linear circuit components when such filters are used to smooth the output voltage from the elementary circuit in Figure 10.3. Switch S_1 may be a solid-state device which is cyclically opened and closed 1000 times per second or more. The time that the switch is closed or opened is varied to control the load voltage. The average load voltage E_L over a repetitive time interval is related to the source voltage as follows.

$$E_L = E_d \frac{t_{\text{CLOSED}}}{t_{\text{CLOSED}} + t_{\text{OPEN}}} \tag{10.1}$$

With fast switching devices having low leakage current when open, and low voltage drop when closed, a pulsating d-c voltage may be proportionally controlled over a wide range with negligible power loss.

Figures 10.3(b), 10.3(c), and 10.3(d) show the load-voltage waveforms as the load voltage is reduced with a variable repetition rate or variable frequency mode of control. In these figures, the "on" time of the switch is constant. In many practical variable frequency TRC circuits, the "on" time may also be varied as the frequency is changed.

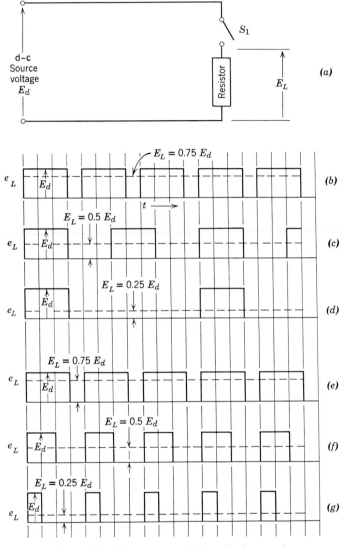

Figure 10.3   Elementary switching-circuit operation.

Figures 10.3(e), 10.3(f) and 10.3(g) show the load-voltage waveforms as the output voltage is controlled with a fixed switching frequency. For either fixed-frequency or variable-frequency controls, the output voltage is zero when the switch is open, and equal to the supply voltage when the switch is closed. Thus, assuming an ideal switching device, the average

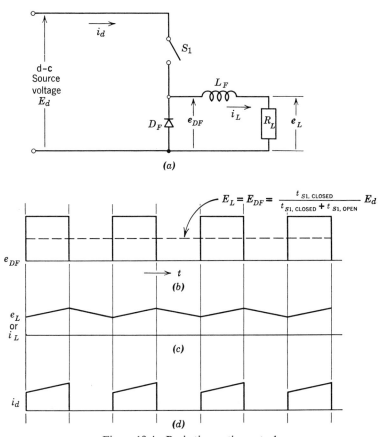

Figure 10.4   Basic time-ratio control.

value of the pulsating output voltage is proportionally controlled in a "lossless" manner.

Figure 10.4 illustrates an extremely important addition to the elementary switching circuit, which makes it possible to efficiently supply smooth dc to practical loads. The diode D_F provides a path for the load current when S_1 is open. This permits the use of a simple filter inductance L_F to provide sufficiently smooth d-c load current for many applications. When the switching frequency is in the kilocycle range, a relatively small inductance is often sufficient to reduce the ripple to a tolerable amount. For applications where only extremely low amplitude ripple is permissible, more complicated L-C filter networks are generally added to the basic arrangement in Figure 10.4. When the output is sufficient to require a number of switching elements in parallel, it is also possible to use a number of TRC circuits in parallel, time displaced from one another, to achieve lower ripple.

The diode D_F, in the circuit of Figure 10.4, permits the filter inductance L_F to provide the energy transformation for the d-c to d-c voltage conversion. Neglecting losses in switch S_1, and assuming that the load current is continuous, the average voltage across diode D_F is related to E_d as given by equation (10.1). Since there can be no average voltage across the ideal inductor L_F in steady state, the average load voltage is

$$E_L = E_{DF} = E_d \frac{t_{ON}}{t_{ON} + t_{OFF}} \qquad (10.1)$$

The circuit losses are assumed to be negligible, and the voltage across diode D_F is assumed to be zero during the entire "off" interval of switch S_1 for the above expression to be valid. This latter assumption, in general, means that the load current I_L is continuous. The efficiency of a TRC is, typically, 90 per cent or greater. When circuit losses are present, the load voltage E_L is reduced from that given in equation (10.1) for a given time ratio.

It is interesting to consider the transformation properties of the basic circuit in Figure 10.4 in somewhat more detail. This may be done by starting with the fact that the power output must equal the power input, assuming negligible losses. The power input is

$$P_{IN} = \frac{1}{T} \int_0^T E_d i_d \, dt \qquad (10.2)$$

or

$$P_{IN} = E_d I_d \qquad (10.3)$$

assuming negligible ripple in the source d-c voltage so that E_d is a constant.

The power output is

$$P_{OUT} = \frac{1}{T} \int_0^T e_L i_L \, dt \qquad (10.4)$$

When there is appreciable ripple in the load voltage and current, it is necessary to use some type of d-c wattmeter to measure the load power. When the output ripple is negligible, the integral of the instantaneous product of output voltage and current becomes simply the product of the average quantities

$$P_{OUT} = E_L I_L \qquad (10.5)$$

Thus, when there is negligible ripple in the source voltage and the load voltage or current, the following equation must be true.

$$E_d I_d = E_L I_L \qquad (10.6)$$

This equation quite clearly illustrates the transformerlike properties of the circuit in Figure 10.4. For example, when the average load voltage is one half the source voltage, the average load current must be twice the average source current. The step-down voltage ratio E_d/E_L must be the inverse of the corresponding current ratio I_d/I_L, as in the regular transformer.

In the TRC circuits discussed in the remainder of the chapter, switch S_1 is replaced with an SCR and its associated commutating circuit components. The TRC concept is possible with mechanical-type switches or similar relatively slow speed switching elements. However, it is most practical when devices which can switch in microseconds or less are employed. The filters required to achieve low output ripple are not objectionably large when only high-frequency ripple is present.

10.3 VARIABLE-FREQUENCY TRC

The circuit in Figure 10.5 is one of the more basic SCR time-ratio controls, which has been quite widely used.[1,2] This is one of the family in which the load voltage is controlled by controlling the TRC frequency. The combination of a saturable reactor and a capacitor provides a simple and reliable commutating circuit for the SCR, and the unijunction transistor oscillator (UTO) provides a simple means of controlling the

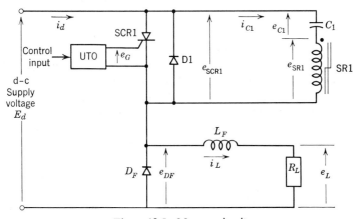

Figure 10.5 Morgan circuit.

TRC frequency. When SR1 in Figure 10.5 saturates, it acts as a switch to discharge the commutating capacitor. This capacitor discharges through the after saturation reactance of SR1 and SCR1 or D1 producing a form of impulse commutation of SCR1.

The circuit of Figure 10.5 operates in the same manner as the basic TRC circuit of Figure 10.4, described previously. SCR1 replaces switch S_1. The SCR is turned on by the UTO and turned off by the commutating circuit components, capacitor C_1 and saturable reactor SR1. For a given operating condition, once the SCR is turned on, the commutating circuit components C_1 and SR1 function to turn off the SCR after a fixed period of time. The frequency of the UTO is varied to control the load voltage E_L.

The load voltage E_L cannot exceed the supply voltage E_d in the circuit of Figure 10.5. The length of the SCR conducting interval, time interval t_{ON}, is determined principally by the design of saturable reactor SR1 and the d-c supply voltage. Modifications to the circuit of Figure 10.5 are discussed where the t_{ON} time interval is independent of the supply voltage. The UTO frequency f_0 must be somewhat less than $1/t_{ON}$ since, at this frequency, the SCR is gated on immediately after it has been turned off, and this may produce commutation failure.

Frequency Control

The frequency of the time-ratio control of Figure 10.5 may be controlled by a UTO, as shown in Figure 10.6. Many different controllable frequency oscillators may be used. The UTO is a very desirable approach, since it is a solid-state oscillator and the unijunction transistor is capable of delivering adequate gating pulses for firing a relatively wide range of SCR devices. The operation of a UTO has been described in Chapter 2, and in previous publications.[3-6] The frequency f_0 of the UTO is approximately $f_0 = 1/C_T(R_T + R_{Q1})$ where R_{Q1} is the equivalent resistance of the collector-emitter of transistor Q_1 in Figure 10.6. As the control input varies, the equivalent resistance of Q_1 varies. The frequency f_0 is thus controlled by the control input. The form of the transfer characteristic of Figure 10.5 with a UTO control is illustrated in Figure 10.6(b).

The maximum frequency of the UTO control must be limited to maintain proper operation of the SCR1-commutating circuit in Figure 10.5. The frequency of the UTO can be limited by resistor R_T. When the maximum load voltage need not exceed in the order of $E_L = 0.9 E_d$, R_T is a satisfactory frequency limit. However, when it is desired to operate with E_L as nearly equal to E_d as possible, the use of R_T to limit f_0 is not satisfactory. Variations in the UTO maximum frequency with temperature or variations in the length of time the SCR is on may be large enough so

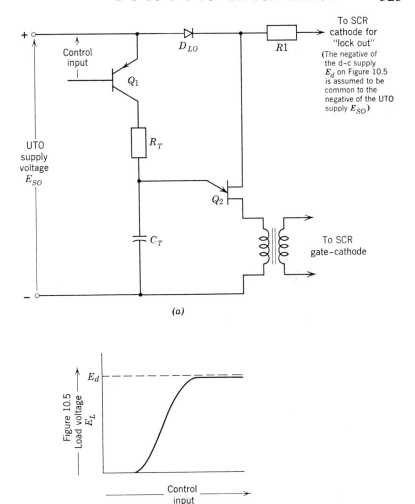

Figure 10.6 (a) Unijunction transistor control for the circuit in Figure 10.5. (b) Transfer characteristic of Figure 10.5, using *UTO* control.

that Figure 10.5 cannot be reliably operated where the load voltage closely approaches E_d. When it is desired to operate with E_L greater than 0.9 E_d, an automatic means of limiting the UTO frequency is used. The "lock-out" technique, indicated in Figure 10.6, is one means of providing such an automatic frequency limit. Resistor R1 connects the upper base lead of unijunction transistor Q_2 to the cathode of the SCR. When the SCR is on, the base-to-base voltage of the unijunction is raised to a value somewhat greater than twice the UTO supply voltage E_{SO}. The actual value of this voltage is determined by the relationship between R1 and the equivalent base-to-base resistance of the unijunction transistor. With its base-to-base voltage somewhat greater than $2E_{SO}$, the unijunction transistor firing will be delayed until the SCR turns off. After the SCR turns off, the UTO immediately turns the SCR back on, since the capacitor C_T would have been previously charged. This lock-out operation limits the maximum frequency to precisely match the frequency that corresponds to the time "on" established by the operation of the saturable reactor SR1.

Saturable Reactor-Capacitor Commutation[1,2]

In the circuit of Figure 10.5, the saturable reactor SR1 and the capacitor C_1 provide commutation of SCR1. For this form of commutation, the saturable reactor may be considered to operate as a switch. When it switches closed, it connects the commutating capacitor C_1 across SCR1, thus providing an alternate path for the load current and a means of momentarily reversing the SCR anode-cathode voltage. With this mode of commutation, for a given supply voltage E_d, the SCR conduction angle or "on" time interval is approximately constant as the UTO frequency is varied to control the load voltage E_L. The period of time that the SCR is on is determined principally by the time required for SR1 to move from positive saturation to negative saturation and return to positive saturation.

The waveforms of Figure 10.7 are believed to illustrate most clearly the saturable reactor-commutation principle. These waveforms assume ideal circuit components for the circuit of Figure 10.5, negligible losses, negligible ripple in the load current, and an SCR "on" time interval which is much greater than the commutation time interval required. At time t_0, it is assumed that capacitor C_1 has an initial positive voltage across it equal to E_d, and the core flux of SR1 is assumed to be at positive saturation, $+\Phi_s$. When SCR1 is gated on at t_0, the capacitor voltage appears across SR1, driving its core flux toward negative saturation. Assuming that negligible exciting current is required by SR1, the capacitor voltage will remain essentially constant from time t_0 to time t_{1-2}. During this time interval, SCR1 is conducting load current I_L from the d-c supply to the load. Diode D_F is not conducting as an inverse voltage of E_d appears across this diode.

Time t_{1-2} is reached when the core flux of SR1 reaches negative saturation, $-\Phi_s$. At this time, capacitor C_1 is discharged through SCR1 and the after-saturation reactance of SR1. This is actually a resonant discharge requiring a time of $\pi\sqrt{L_{SR1}C_1}$ seconds, assuming negligible losses; where L_{SR1} is the after-saturation inductance of SR1. However, the

resonant discharge time of capacitor C_1 is assumed very short relative
to the time interval from t_0 to $t_{1\text{-}2}$ for the waveforms in Figure 10.7.
Thus, the capacitor voltage drops very quickly, and is indicated by
a vertical line at time $t_{1\text{-}2}$ in Figure 10.7(a). At the conclusion of
the resonant discharge, the voltage across C_1 will again be equal to

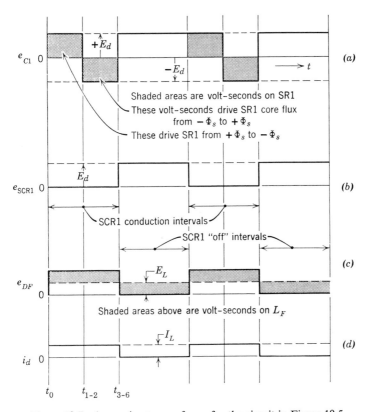

Figure 10.7   Approximate waveforms for the circuit in Figure 10.5.

E_d, but with the opposite polarity to that during the time interval from t_0
to $t_{1\text{-}2}$. It is assumed that there are negligible losses in the resonant
circuit containing C_1, L_{SR1} and SCR1.

From time $t_{1\text{-}2}$ to $t_{3\text{-}6}$, the capacitor voltage again remains constant
while the flux in SR1 is driven toward positive saturation, $+\Phi_s$. SCR1
continues to conduct load current I_L during the time interval from $t_{1\text{-}2}$ to
$t_{3\text{-}6}$. When time $t_{3\text{-}6}$ is reached, SR1 saturates in the positive direction.
Capacitor C_1 is again discharged in a resonant fashion, first through SCR1
and then through diode D1. After this second resonant discharge of C_1,

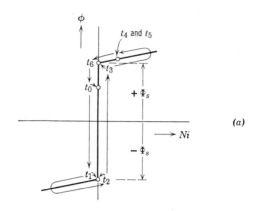

(a)

A_{2-3} = volt seconds to drive
SR1 from $-\Phi_s$ to $+\Phi_s$
Total reset volt seconds = $A_{1-0} + A_{6-0} = A_{2-3}$

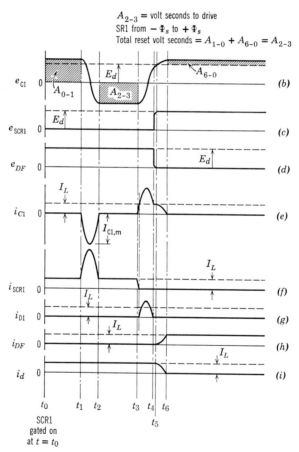

Figure 10.8 Waveforms for the circuit of Figure 10.5, showing detailed operation during capacitor C_1 charging and discharging intervals.

328

its voltage is returned to its original condition equal to positive E_d, since the losses have been assumed to be negligible. The resonant discharge time of L_{SR1} and C_1 provides the commutating time required for SCR1. SCR1 is thus commutated off at time t_{3-6}. The capacitor voltage remains constant, and the load current circulates through diode D_F until SCR1 is again gated on to start the next conduction interval.

The waveforms in Figure 10.7 are obtained in most practical circuits when the conduction interval of SCR1 is much longer than its commutating time requirement. These conditions usually exist when the maximum time-ratio control frequency is 1000 cps or less. The commutating-circuit losses in practical circuits are generally small enough so that capacitor C_1 does not lose appreciable charge during its resonant discharging with the after-saturation inductance of SR1.

The waveforms in Figure 10.8 illustrate the behavior of the circuit of Figure 10.5 when operating at higher time-ratio control frequencies. In this case, the commutating time of the SCR may be an appreciable fraction of its conduction interval. These waveforms show the detailed operation during resonant discharging of capacitor C_1. Saturable reactor SR1 is assumed to have a magnetization characteristic as shown in Figure 10.8(a). The load current is again assumed to have negligible ripple, and D1, D_F and SCR1 are again assumed to have negligible forward drop when conducting, and negligible leakage when off. In Figure 10.8, the losses in the commutating circuit are considered small but not negligible, as they were in Figure 10.7.

At time t_0, SCR1 is gated on to produce the waveforms in Figures 10.8(b) through 10.8(i). The core flux in SR1 is assumed to be as shown in Figure 10.8(a) at t_0. The initial charge on capacitor C_1 is slightly greater than the d-c supply voltage E_d. These initial conditions are chosen to produce steady-state operation over the cycle shown in Figures 10.8(b) through 10.8(i).

After SCR1 is turned on at time t_0, $e_{SR1} = e_{C1}$. The flux of SR1 is driven toward negative saturation. There is no significant change in the capacitor voltage since the magnetizing current of SR1 is negligible. From t_0 to t_1, SCR1 conducts load current I_L, the full-supply voltage appears at the output across diode D_F, and neither diode D1 or D_F is conducting. When the core flux of SR1 reaches negative saturation at time t_1, capacitor C_1 is discharged through SCR1 and the after-saturation inductance of SR1. Because of the losses in the resonant L-C circuit formed by C_1 and the after-saturation inductance of SR1, the capacitor voltage is somewhat less in magnitude at t_2 than it was at t_1.

As C_1 discharges between t_1 and t_2, the energy $\frac{1}{2}C_1 E_{C1,m}^2$ is transferred to L_{SR1} in the form of $\frac{1}{2}L_{SR1}I_{C1,m}^2$ where L_{SR1} is the saturated inductance of

SR1 and $I_{C1,m}$ is the peak value of the capacitor resonant discharge current as shown in Figure 10.8(e). All the energy, except for losses, initially stored in C_1 at time t_1 is transferred to L_{SR1} at the time between t_1 and t_2 when $e_{C1} = 0$ and $i_{C1} = I_{C1,m}$. As the current i_{C1} decreases again, the energy is returned to C_1, reduced by the losses. Thus at time t_2 the voltage across C_1 is slightly lower in magnitude and opposite in polarity to its value at time t_1. Starting at time t_2, the flux in SR1 is driven toward positive saturation, reaching $+\Phi_s$ at time t_3. Again, during the interval from t_2 to t_3, the capacitor voltage is constant, assuming negligible exciting current for SR1. The volt-second area A_{2-3} is that required to drive SR1 from $-\Phi_s$ to $+\Phi_s$.

At time t_3, a second resonant oscillation of capacitor C_1, with the after-saturation reactance of SR1, is initiated. When the capacitor current reaches a value equal to the load current I_L, SCR1 stops conducting, and diode D1 begins to conduct. The resonant oscillation of C_1 continues, otherwise, the same as during interval t_1–t_2, until time t_4. At this point, the capacitor current has reduced back to a value equal to the load current I_L. The diode D1 blocks; forward voltage is reapplied to SCR1; and the capacitor continues charging from the d-c supply with a constant current equal to the load current I_L during the interval from t_4 to t_5. When time t_5 reached, the capacitor has been charged to E_d. The reverse voltage across diode D_F reaches zero, and diode D_F begins conducting. The capacitor receives its final charge during interval t_5–t_6. This is a continuation of the resonant oscillation of C_1 and the after-saturation inductance of SR1 but, now, the capacitor current i_{C1} comes from the d-c supply, as both SCR1 and D1 are blocking. When the oscillation has proceeded to the point at which the current i_{C1} reaches zero, diode D_F is conducting the full load current I_L. The voltage across SR1 is again in the direction to drive its core flux toward negative saturation. The capacitor voltage remains approximately constant after t_6, resetting SR1 to the point t_0 in Figure 10.8(a) to begin the next cycle of operation.

The time interval t_0–t_1 is a function of the interval from t_6 to t_0, as SR1 is partially reset from $+\Phi_s$ between time t_6 and the beginning of the next conducting interval. The remainder of the resetting of SR1 to $-\Phi_s$ occurs between t_0 and t_1. For this reason, the SCR-conducting period is reduced somewhat as the UTO frequency is reduced. The variation in time interval t_0–t_1 is largely reduced by connecting a one- or two-turn bias winding on SR1 in series with diode D_F.

The time interval t_3–t_4 is a function of the load current I_L, as shown in Figure 10.8(e). Thus, as the load current is increased, the interval of time during which the supply voltage E_d is impressed on the load circuit is slightly reduced.

It is interesting to note that the circuit in Figure 10.5 is practically self-regulating to changes in supply voltage. When the supply voltage E_d is increased, SR1 is driven over its flux range in less time. Thus, the SCR-conduction interval is reduced when the d-c supply voltage E_d is increased. For a fixed UTO frequency, this means that there is a compensating effect for supply-voltage fluctuations.

Expressions for the time intervals indicated in Figure 10.8 are given in the appendix (pp.340–357).[7] The SCR-commutating time is approximately $\pi\sqrt{L_{SR1}C_1}$ at light load and assuming low damping in the $L_{SR1} - C_1$ resonant circuit. Several modifications to the circuit in Figure 10.5 are

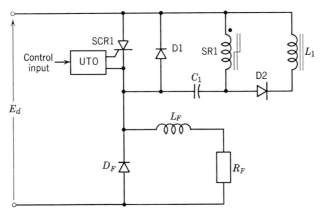

Figure 10.9   Morgan circuit with D2 and L_1 added.

also analyzed and discussed in the appendix. An additional modification of interest is shown in Figure 10.9. Diode D2 and inductance L_1 have been added to the circuit of Figure 10.5. This addition makes the first oscillatory reversal of capacitor C_1 occupy essentially the full interval of time from t_0 to t_3 in Figure 10.8. This prevents saturation of SR1 between t_0 and t_3 and the resulting high peak of current through SCR1, SR1, and C_1. The conduction interval for SCR1 is approximately $\pi\sqrt{L_1C_1}$ seconds. In a practical case, the inductance of L_1 is many times the after-saturation inductance of SR1.

The operation of the modified circuit in Figure 10.9 during the commutating interval from time t_3 to t_4 and during the interval from t_4 to t_5 is the same as the operation of Figure 10.5, since diode D2 is blocking. During the interval from t_5 to t_6, diode D2 is forward-biased, since the voltage across capacitor C_1 is greater than the supply voltage E_d. Diode D_F is conducting from t_5 to t_6, thereby producing essentially zero potential difference between the negative of the supply and the common connection of SCR1, D1, and C_1. The operation of the circuit of Figure 10.9 from t_5 to t_6 is, nevertheless, approximately the same as without D2 and L_1, since the after-saturation inductance of SR1 is many times less than the inductance L_1. Thus, when SR1 is saturated, it provides the principal path for current flow from C_1, even though diode D2 may be forward-biased to permit conduction. The operation from t_6 to t_0 is similar to that shown in Figure 10.8. The principal effects of the addition of diode D2 and inductance L_1 are to reduce the variation in SCR "on" time with supply voltage changes, reduce the pulse current or RMS rating of SCR1 and SR1 since L_1 is considerably larger than L_{SR1}, and to reduce the volt-second rating of SR1 for a given SCR "on" time. As a result, SR1 of Figure 10.9 is less than one half the size and volt-ampere rating of SR1 of Figure 10.5. Usually L_1 is approximately one half the size of SR1 of Figure 10.9.

10.4   CONSTANT-FREQUENCY TRC

Many d-c to d-c conversion systems require high speed of response to control or regulate the load voltage. In other applications, it is desirable to minimize load-voltage ripple with a minimum size of filter components. For these requirements a time-ratio control that operates at a constant frequency is generally preferred.

A constant-frequency TRC circuit controls the output voltage by vary-ing the length of time that the SCR is on. This type of TRC cannot generally control the load voltage to zero and, usually, the control circuit of the constant-frequency TRC is somewhat more complicated than the corresponding control for a variable-frequency circuit.

Saturable Reactor Control of SCR "on" Interval

The TRC shown in Figure 10.10 controls the load voltage by con-trolling the "on" time interval of SCR1. This SCR is turned on at a constant repetition rate by the UTO. Saturable reactor SR2 reduces the flux excursions of SR1 during the charging and discharging periods of capacitor C_1, thereby decreasing the SCR "on" time interval.

SR2 is reset, each cycle of operation, to a flux level determined by con-trol current I_c. The amount of flux change required to swing the core flux of SR2 from its reset flux level to positive saturation determines the length of time that SCR1 is on.

To illustrate the operation of SR2 in Figure 10.10, first consider the case with a very large d-c current I_c. This allows SR1 to operate the same as previously shown in Figure 10.8. The core flux of SR2 is reset by control current I_c to negative saturation during each period of time that diode D2 blocks. Reactor SR2 is designed to have approximately the same volt-second rating as SR1. Starting at the instant in time when SCR1 is gated on, first the capacitor voltage appears across SR1 to produce a negative voltage e_{SR1}. This voltage drives the flux of SR1 toward negative saturation, as previously shown in Figure 10.8. In addition, this same voltage, e_{SR1}, appears on SR2 to drive the flux in SR2 toward positive

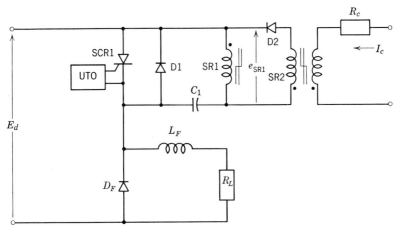

Figure 10.10   Constant-frequency time-ratio control.

saturation. Since SR2 was previously reset a large amount by control current I_c, it is assumed that SR2 does not reach positive saturation until SR1 saturates. This occurs at time t_1 in Figure 10.8. The capacitor voltage then reverses, and voltage e_{SR1} becomes positive. During the interval of time from t_2 to t_3, as previously shown in Figure 10.8, diode D2 blocks, and SR2 is reset by I_c in the same manner as in a conventional half-wave magnetic amplifier.[8]

There will be some voltage across SR2 beginning at the point after time t_3, Figure 10.8(b), when the capacitor voltage e_{C1} has discharged to zero. In addition, there will be some voltage across SR2 during the "off" interval of SCR1, as indicated in Figure 10.8(b). However, the principal positive flux change in SR2 will take place during the time interval from t_0 to t_1 immediately after SCR1 is gated on. For the condition where the control current I_c is large, the circuit of Figure 10.10 operates the same as without the components SR2 and D2.

When control current I_c is reduced to a small value, the flux of SR2 resets a lesser amount during the diode D2 blocking interval. Thus, after SCR1 is gated on, SR2 will reach saturation before SR1. This essentially reduces time interval t_0–t_1 in Figure 10.8. When SR2 saturates, capacitor C_2 is discharged in a resonant oscillation with the after-saturation inductance of SR2. Thus, the voltage across capacitor C_1 is reversed in a similar manner to that when SR1 saturated at time t_1 in Figure 10.8. After the capacitor voltage is reversed, e_{SR1} is again positive, diode D2 blocks, and the circuit operates as previously shown in Figure 10.8, starting at time t_2. The time interval from t_2 to t_3 is also reduced, since SR1 will reach positive saturation sooner, as it was not driven as far toward negative saturation during the reduced t_0–t_1 time interval. SCR1 is on for a shorter period of time when the control current I_c is reduced and, therefore, the load voltage in Figure 10.10 is controlled by varying the d-c current, I_c.

The gate supply to turn on the controlled rectifier in Figure 10.10 can be any oscillator circuit that supplies the required voltage pulses. The unijunction transistor oscillator of Figure 10.6, with transistor Q_1 removed, can reliably drive SCR1 in Figure 10.10, and provide reasonably constant frequency for practical TRC circuits.

Usually the frequency of the gate supply is held within 10 per cent of a fixed value. The time for a cycle of the gate-supply oscillator must remain greater than the SCR "on" interval. The lock-out technique of Figure 10.6 again may be used to prevent gating the SCR, until after commutation is complete.

10.5   AUXILIARY SCR USED IN COMMUTATION CIRCUIT

An SCR-diode circuit can be used in place of SR1 in Figures 10.5, 10.9, and 10.10. Such a circuit arrangement is shown in Figure 10.11. The firing of SCR2 is delayed from the firing of SCR1 to determine the conduction interval of SCR1. In this circuit, capacitor C_1 is reversed rapidly, as during the t_1–t_2 interval in Figure 10.8, starting at the instant SCR2 is turned on. The interval from t_2 to t_3, in Figure 10.8, no longer exists. The commutation interval for SCR1 begins immediately after C_1 has completed its first reversal. Diode D2 provides the path for capacitor current flow during commutation of SCR1 and during the charging interval of C_1.

Diode D2 and controlled-rectifier SCR2 can be reversed in Figure 10.11. The operation is similar, except the voltage of C_1 reverses after SCR1 is turned on, and SCR1 is commutated when SCR2 is turned on.

Figure 10.12 shows an alternative version of Figure 10.11, in which the current rating of SCR2 is considerably reduced. The commutating circuit in Figure 10.12 may be more economical, and is generally more compact than the previous circuits discussed in this chapter.

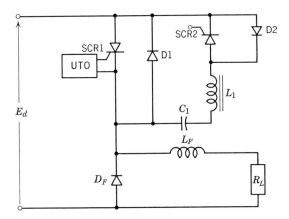

Figure 10.11   TRC with an auxiliary SCR used to commutate the main SCR.

When SCR1 in Figure 10.12 is gated on, capacitor C_1 remains charged, since diode D3 prevents circulating current flow in the forward direction of SCR1. When SCR2 is turned on, C_1 is reversed in an oscillatory manner through the circuit containing L_1, D2, C_1, and SCR2. The inductance of L_1 is large enough so that this first reversal of the charge on C_1 occurs slowly, generally in the order of ten times as long as the commutating time of SCR1. This permits SCR2 to have a low current rating. In practical circuits, SCR2 is 10 per cent the current rating of SCR1. As soon as the voltage on C_1 has reversed so that e_{C1} is negative, a commutation process starts, involving SR1 and C_1 the same as for the circuit in Figure 10.9.

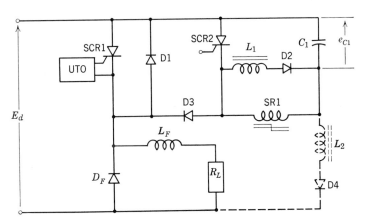

Figure 10.12   Alternate version of Figure 10.11 to reduce the rating of SCR2.

With proper control of SCR1 and SCR2 in Figures 10.11 and 10.12, these circuits can provide constant frequency or variable frequency time-ratio control. The circuit in Figure 10.12 may include two additional components which provide commutation even during load transients where the load current drops to zero. Inductor L_2 and diode D4 provide a path for supplying commutating circuit losses when the load current is very low or zero.

10.6 CONVERTER FOR STEPPING UP VOLTAGE

This section describes d-c to d-c converters in which the output voltage exceeds the source voltage. Their operation is similar to the circuits previously discussed in this chapter. However, TRC step-up converters permit controlling the load voltage over a range from the supply voltage to several times the supply voltage.

Voltage Step-Up Technique

The principle of the technique used to step up the voltage is shown in Figure 10.13. Switch S_1 will later be replaced by an SCR. The operation is most clearly understood by considering the case when L_F is sufficiently large so that there is negligible ripple in the source current, C_F is large enough so that there is negligible ripple in the load voltage, and the losses in the circuit are negligible.

When switch S_1 is rapidly opened and closed in a cyclical fashion, the load voltage will be greater than the source voltage. When switch S_1 is closed, additional energy is stored in inductance L_F. When switch S_1 is opened, energy is transferred from inductance L_F to the filter capacitor C_F and the load. For example, if S_1 is closed, for the same amount of time it is open, the load voltage will be twice the supply voltage. The additional energy stored in L_F when S_1 is closed is $E_d I_d t_{\text{CLOSED}}$ and the energy transferred from L_F to C_F and the load when the switch is open is $(E_L - E_d)I_d t_{\text{OPEN}}$. The load voltage as a function of the time ratio is determined as follows.

$$E_d I_d t_{\text{CLOSED}} = (E_L - E_d)I_d t_{\text{OPEN}} \tag{10.7}$$

$$E_L t_{\text{OPEN}} = E_d t_{\text{CLOSED}} + E_d t_{\text{OPEN}}$$

$$E_L = E_d \frac{t_{\text{CLOSED}} + t_{\text{OPEN}}}{t_{\text{OPEN}}} \tag{10.8}$$

The load voltage E_L is controlled by time-ratio control, as shown in equation (10.8). This method of stepping up and controlling a d-c voltage is generally more advantageous than inverter-rectifier approaches when considerable control range is required with a step-up ratio of the order of two to one or less.

Losses in the circuit result in the load voltage being slightly less than given by equation (10.8). The value of inductance L_F in Figure 10.13 is determined principally by the input ripple current desired and the operating frequency. In practical circuits, approximately 20 per cent ripple is

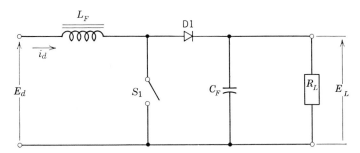

Figure 10.13   Circuit technique for step-up TRC converter.

allowed at full load so that the input current remains continuous down to a fairly low load. It is generally desirable to minimize the input ripple both from source considerations and to minimize the peak current rating of switch S_1.

SCR Constant-Frequency Time-Ratio Control

Figure 10.14 shows a practical circuit using the principle shown in the previous figure. Switch S_1 is replaced by SCR1. The UTO provides constant frequency gating pulses for the SCR. A commutating and control circuit similar to that shown previously in Figure 10.10 is used. This circuit is a constant-frequency approach, where the SCR "on" time is varied by changing the control current I_c.

Control-to-Load Voltage Same as Source Voltage

Figure 10.15 shows a variable-frequency step-up approach. This circuit is particularly advantageous when it is desired to control to the point where the load voltage is approximately equal to the source voltage. The SCR commutating and control circuits are similar to those previously shown in Figures 10.5 and 10.6. The unijunction transistor provides a variable frequency oscillator to deliver variable repetition rate gating

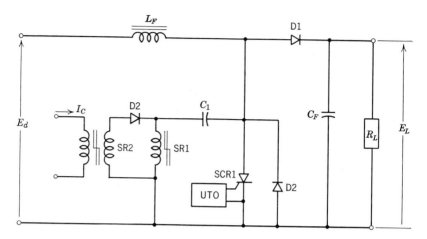

Figure 10.14   Constant-frequency TRC step-up converter circuit.

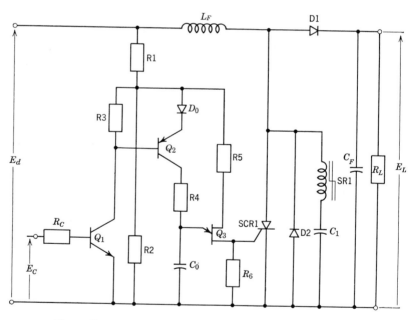

Figure 10.15   Variable-frequency TRC step-up converter circuit.

pulses to SCR1. The frequency of the unijunction transistor oscillator can be reduced to zero. At this point, the SCR is "off" continuously, so that $E_L = E_d$, neglecting the d-c drop in L_F and D1.

Load Voltage Much Greater Than the Source Voltage

The use of a tap on inductor L_F of Figure 10.16 provides a practical circuit for controlling the load voltage E_L to several times the supply

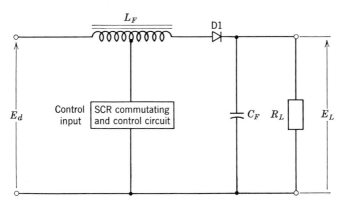

Figure 10.16   TRC converter with increased step-up ratio.

voltage. This combined reactor and autotransformer circuit provides a means of stepping up the supply voltage above the voltage rating of the SCR and its associated commutating circuit. The maximum SCR voltage is less than E_L, but the peak current of the SCR is increased accordingly.

In all of the TRC converters discussed in this chapter, it is possible to use a full isolation transformer in the load circuit. As a consequence, the total magnetic component rating of a given circuit is increased. In addition, such transformers may require air gaps or special core flux resetting circuits to handle the pulsating d-c voltage produced in TRC converters.

REFERENCES

1. R. E. Morgan, "Magnetic Silicon–Controlled Rectifier Power Amplifier," U.S. Patent 3,019,355, January 30, 1962.
2. R. E. Morgan, "A New Magnetic–Controlled Rectifier Power Amplifier with a Saturable Reactor Controlling on Time," *AIEE Transactions*, Volume 80, Part I, 1961, pp. 152–155.
3. F. W. Gutzwiller *et al.*, *Silicon–Controlled Rectifier Manual*, Second Edition, General Electric Company, Auburn, N.Y., 1961, pp. 44–65.

4. I. A. Lesk, "Nonlinear Resistance Device," U.S. Patent 2,769,926, November 6, 1956.
5. R.W. Aldrich *et al.*, "Semiconductor Network," U.S. Patent 2,780,752, February 5, 1957.
6. V. P. Mathis, "Sawtooth Wave Generator," U.S. Patent 2,792,499, May 14, 1957.
7. W. McMurray, "SCR D–C to D–C Power Converters," IEEE International Conference on Nonlinear Magnetics, Washington, D.C., April 17–19, 1963.
8. H. F. Storm, *Magnetic Amplifiers*, John Wiley & Sons, New York, 1955, pp 309–319.

APPENDIX.   SCR D-C–TO–D-C POWER CONVERTERS

W. MCMURRAY*

This appendix deals with chopper circuits that control or regulate the flow of power from a d-c source to a d-c load, which have a common terminal. That is, the inverter-transformer-rectifier type of converter is excluded. Also, the scope is limited to the voltage step-down arrangement, where the average load voltage can approach, but not exceed, the supply voltage.

The principle of operation is "time-ratio control." Suppose a switch is connected in series with the d-c source and the load. The ratio of the average load voltage to the source voltage will equal the fraction of time for which the switch is closed. By closing and opening the switch at a high frequency, the size of filter required to obtain smooth output voltage is minimized. The load voltage may be regulated by controlling the ratio of "on" time to "off" time of the switch.

The silicon-controlled rectifier is a suitable form of static switch for this type of application. It may easily be turned on by a gate signal. However, it requires auxiliary circuitry to turn it off. The voltage on the controlled rectifier must be reversed for a certain turn-off time in order to recover its ability to block forward voltage. The necessary inverse voltage can be obtained by means of a lightly damped *L-C* circuit which can be employed in a number of circuit configurations. The "on" time of the controlled rectifier can be extended and controlled by means of an auxiliary controlled rectifier or a saturating reactor. Gating the auxiliary controlled rectifier or saturation of the reactor initiates the commutating oscillation by the *L-C* circuit.

The purpose of this analysis is not to describe the design of a chopper for any particular application, but to analyze the basic mode of operation and discuss the merits of a number of circuit variations. Emphasis will be placed upon the waveforms produced by the circuits. For simplicity, the following assumptions are made.

* A paper that was presented at the IEEE International Conference on Nonlinear Magnetics, Washington D.C., April 17–19, 1963.

(1) The static and switching characteristics of the rectifiers are ideal, except for the turn-off time required by the controlled rectifiers. Forward voltage drop, inverse leakage current, inverse recovery current, and turn-on time are neglected. The components that may be necessary to limit voltage overshoots, dv/dt and di/dt, are not considered.

(2) The saturating reactors have negligible magnetizing current when unsaturated, and a constant after-saturation inductance.

(3) The d-c supply voltage E_d is stiff. An input filter capacitor is generally necessary.

(4) The output filter choke, plus any load inductance, is large enough to maintain the load current constant in steady-state operation.

Basic Chopper Circuit

A simple chopper circuit that uses a capacitor C and a linear inductance L to turn off the SCR is shown in Figure 10.17. During the "off" intervals, the load current I_L, maintained by the choke L_F, coasts through the rectifier D_F. The point Z is at the potential of the negative d-c line, so the capacitor C is charged to the supply voltage E_d, assuming that previous oscillations have died out.

When the SCR is gated on, point Z rises to the potential of the positive d-c line, rectifier D_F blocks the voltage E_d, and the load current I_L transfers to the path through SCR, drawing power from the supply. Also, capacitor C discharges through SCR and inductance L in an oscillatory manner. After one half-cycle of the natural frequency, the capacitor has reversed its charge. With light damping, the half-period is approximately $\pi\sqrt{LC}$ seconds. The peak magnitude of the reversed capacitor voltage is a fraction β of the initial voltage E_d, where β is the half-cycle decrement factor:

$$\beta = \epsilon^{-(\pi/2Q)} \tag{10.9}$$

As the oscillatory cycle continues, the capacitor current reverses, reducing the current flowing through the controlled rectifier. When the capacitor current i_c equals the load current I_L, the SCR blocks, and the remaining reverse capacitor voltage appears as inverse voltage on the SCR. The magnitude E_r of this peak inverse voltage is approximately

$$E_r \approx E_d\sqrt{\beta^2 - \lambda^2} \tag{10.10}$$

where λ is the normalized load current factor

$$\lambda = \frac{I_L}{E_d} \sqrt{\frac{L}{C}} \tag{10.11}$$

The load current I_L continues to flow from the d-c supply through the capacitor C and inductance L until the capacitor has charged back to the d-c line voltage E_d. Since the load current has been assumed constant, no

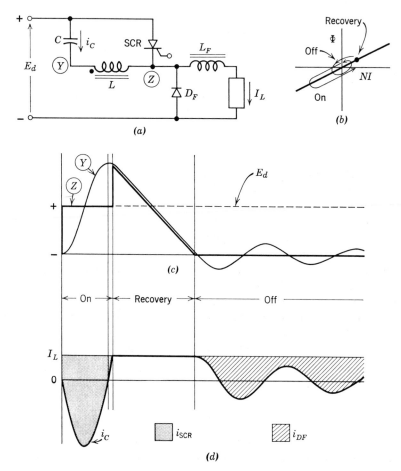

Figure 10.17   Basic chopper. (*a*) Circuit diagram. (*b*) Flux-ampere-turn diagram for inductance L. (*c*) Voltage waveforms. (*d*) Current waveforms.

voltage appears across the inductance L during this interval. The initial part of the interval, during which the capacitor and SCR voltage remain negative, is the time t_0 available for turn-off of the controlled rectifier

$$t_0 = \frac{CE_r}{I_L} \tag{10.12}$$

When the capacitor voltage reaches the value E_d, rectifier D_F unblocks and clamps the potential of point Z to the negative d-c line. However, the inductance L is still carrying the load current I_L, which forces the capacitor to overcharge and ring. The peak value of the first overshoot in capacitor voltage is

$$I_L \sqrt{\frac{L}{C}} \, \epsilon^{-(\pi/4Q)} = \lambda \sqrt{\beta} \, E_d \qquad (10.13)$$

Eventually, the oscillations decay and dissipate the energy $\frac{1}{2}LI_L^2$ that was trapped in the inductance. If the controlled rectifier SCR is turned on again before the ringing has stopped, the initial conditions for the next pulse are different. Thus, the waveforms become dependent upon repetition rate as well as load current.

The waveforms shown in Figure 10.17 are drawn for the conditions $\lambda = 0.33$, $Q = 10$, $\beta = 0.855$. The voltage of point Z, relative to the negative d-c line, is the voltage across rectifier D_F or the unfiltered load voltage. The voltage across the SCR can be seen in the sketch as the voltage of point Z relative to the positive d-c line at the level E_d. Similarly, the voltage on capacitor C is the potential of point Y relative to the positive line level, while the voltage across inductance L appears in the figure as the difference between the traces of Y and Z. The current waveform sketches show the load current I_L and the capacitor current i_c directly. The currents in the SCR and the coasting rectifier D_F are indicated by cross-hatched areas. It is believed that this method of presenting waveform sketches aids in comprehending the mode of operation, and the technique is followed throughout this appendix.

The voltage oscillograms in Figure 10.18 were obtained with a breadboard circuit having $E_d = 120$ v, $C = 4\ \mu f$, $L = 17\ \mu H$. In Figure 10.18(a), the load current I_L is 20 amp and $\lambda \approx 0.33$, corresponding to the sketch in Figure 10.17 while Figure 10.18(b) is for 10 amp load current. The small oscillations in the traces for point Z during the "off" interval are caused by a small reactor inserted in series with the coasting rectifier D_F to limit its recovery current when the controlled rectifier is turned on. The recovery current of the controlled rectifier at the beginning of the turn-off interval also causes a minor departure from the ideal waveform. Both the controlled rectifier SCR and rectifier D_F are shunted by small R-C filters to limit voltage spikes.

In the arrangement of Figure 10.17, the trailing or recovery end of the unfiltered load voltage pulse, after the controlled rectifier has been turned off, is a major portion of the total pulse width, and is inversely proportional to the load current (see Figure 10.18). This necessitates a wide range in the repetition frequency in order to control the circuit when the load current varies. If a wide range in load voltage is also desired, the frequency problem is compounded. Another disadvantage of this basic circuit is the poor form factor of the current in the controlled rectifier. The conducting period of the SCR is comparable to the available turn-off time, which

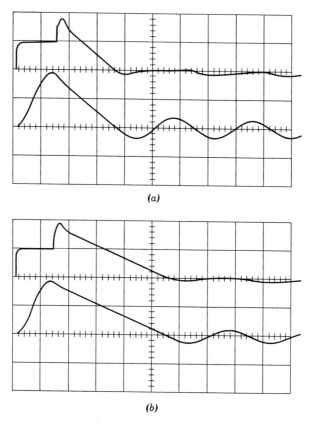

(a)

(b)

Figure 10.18   Oscillograms of Figure 10.17 circuit. (a) 20-amp load. (b) 10-amp load. (Voltage of point Z-upper traces; voltage of point Y-lower traces; scales are 120 volts/division and 20 μsec/division.)

results in either high switching losses or oversized commutating components. Alleviation of these difficulties is the major objective of the circuits to follow.

Saturating Reactor to Increase "on" Time

In Figure 10.19, a saturating reactor SR replaces the linear inductance L of Figure 10.17. This arrangement was first described by Morgan.[1] The after-saturation inductance of SR is equal to the linear inductance L.

During the "off" intervals, the load current I_L coasts through the rectifier D_F as before, but unsaturation of SR stops the ringing of the capacitor voltage after the first quarter-cycle. The first peak overshoot voltage,

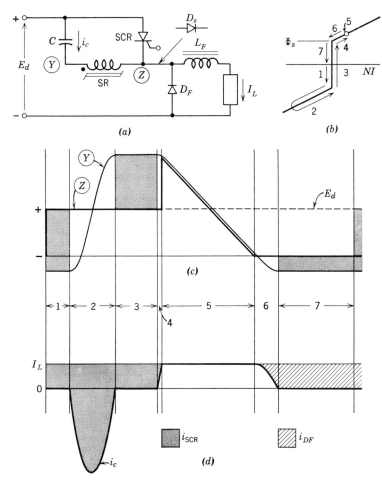

Figure 10.19   Chopper with saturating reactor. (*a*) Circuit diagram. (*b*) Flux-ampere-turn diagram for saturating reactor SR. (*c*) Voltage waveforms. (*d*) Current waveforms.

equation (10.13), is held off by the saturating reactor and resets it to some flux level Φ_0 by the start of the next "on" interval. A cycle of operation can be divided into a number of intervals that are bounded by changes in the state of the controlled rectifier or coasting rectifier (conducting or blocking) or in the state of the reactor (saturated or unsaturated). These intervals are numbered in sequence on the waveforms in Figure 10.19.

Interval 1. The controlled rectifier SCR is gated on and the load current I_L switches from the coasting rectifier D_F to the SCR and power source. Reactor SR holds off the total capacitor voltage $(1 + \lambda\sqrt{\beta})E_d$ and is reset from the initial flux level Φ_0 to negative saturation, $-\Phi_s$. The duration t_1 of Interval 1 is

$$t_1 = \frac{N(\Phi_0 + \overline{\Phi}_s)}{(1 + \lambda\sqrt{\beta})E_d} \qquad (10.14)$$

where N is the number of turns on the reactor SR.

Interval 2. Reactor SR saturates, and capacitor C reverses its charge in one half cycle of oscillation with the after-saturation inductance L. The charge reversal current pulse is superimposed upon the load current in controlled rectifier SCR. This is similar to the initial interval in the circuit of Figure 10.17, and its duration t_2 is

$$t_2 = \pi\sqrt{LC} \qquad (10.15)$$

Interval 3. Reactor SR comes out of saturation and holds off the reversed capacitor voltage, which suffers a magnitude decrement as a result of losses during Interval 2. The flux is set from negative saturation to positive saturation in the time t_3:

$$t_3 = \frac{2N\Phi_s}{\beta(1 + \lambda\sqrt{\beta})E_d} \qquad (10.16)$$

Interval 4. The oscillatory cycle, begun in Interval 2 and interrupted by the flux setting Interval 3, continues for a brief period until the load current is commutated out of the controlled rectifier. The peak inverse voltage E_r that is now applied to the SCR is approximately

$$E_r \approx E_d\sqrt{\beta^2(1 + \lambda\sqrt{\beta})^2 - \lambda^2} \qquad (10.17)$$

The duration t_4 of Interval 4 is approximately

$$t_4 \approx \sqrt{LC}\,\sin^{-1}\frac{\lambda}{\beta(1 + \lambda\sqrt{\beta})} \qquad (10.18)$$

Interval 5. As in the circuit of Figure 10.17, the load current I_L now charges the capacitor C from the voltage $-E_r$ to the d-c supply voltage E_d in a time t_5:

$$t_5 = \frac{C(E_r + E_d)}{I_L} \qquad (10.19)$$

Interval 6. Rectifier D_F starts to conduct, while capacitor C overshoots to the same voltage given by equation (10.13), as discussed previously.

$$t_6 = \frac{\pi}{2}\sqrt{LC} \tag{10.20}$$

Interval 7. This is the "off" interval during which the load current coasts through D_F and reactor SR is reset by the capacitor over-voltage $\lambda\sqrt{\beta}E_d$ from positive saturation to the original lower flux level Φ_0 in a time t_7:

$$t_7 = \frac{N(\Phi_s - \Phi_0)}{\lambda\sqrt{\beta}E_d} \tag{10.21}$$

Note that t_1 and t_7 are interdependent, so that the "on" time is a function of the repetition frequency as well as the load current. It is assumed that the flux Φ_0 does not reach negative saturation, otherwise the ferroresonant equivalent of the ringing seen in Figures 10.17 and 10.18 ensues. Thus, the above analysis implies minimum and maximum frequency limits. The maximum period is obtained by setting $\Phi_0 = -\Phi_s$ in equations (10.14) and (10.15) and summing t_1 through t_7. Similarly, $\Phi_0 = +\Phi_s$ gives the minimum period. Operation outside of these limits is possible, but there may be a discontinuity in the mode.

This limitation can be avoided by inserting a rectifier D_S at the location indicated in Figure 10.19. Rectifier D_S, rather than reactor SR, now holds off the overvoltage on the capacitor during the "off" Interval 7, so that $\Phi_0 = +\Phi_s$ and t_7 is unrestricted. If Q and λ are large, the total capacitor voltage can rise to several times the supply voltage E_d. The commutating capability of the circuit can be designed to increase as the load current that has to be commutated increases. However, a higher voltage controlled rectifier is required.

The waveform sketches in Figure 10.19 are drawn for the same conditions $\lambda = 0.33$, $Q = 10$, $\beta = 0.855$ as in Figure 10.17, and for a hypothetical reactor that has $2N\Phi_s/E_d = \pi\sqrt{LC}$. The particular condition $\Phi_0 = 0$ is shown. Also, the oscillograms in Figure 10.20 have the same parameters as Figure 10.18, but the reactor has $2N\Phi_s/E_d \approx 8\pi\sqrt{LC}$, a condition more typical than that in Figure 10.19. It has an Arnold Engineering Co. 5320-D1 core, with $N = 75$ turns.

Comparing Figures 10.18 and 10.20, the saturating reactor greatly extends "on" time, without increasing the size of the circuit components. However, the pulse width decreases as the load current increases, as a result of both variable capacitor overvoltage during Intervals 1, 3, and 7 and the current-sensitive trailing end of the pulse, Interval 5.

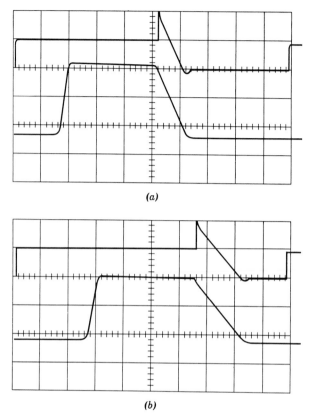

(a)

(b)

Figure 10.20   Oscillograms of Figure 10.19 circuit. (a) 20-amp load. (b) 10-amp load. (Voltage of point Z-upper traces; voltage of point Y-lower traces; scales are 120 volts/division, and 50 μsec/division.)

Feedback Rectifier to Reduce Trailing End of Pulse

A rectifier D connected in inverse-parallel with the controlled rectifier SCR (Figure 10.21) greatly reduces the length of Interval 5. Intervals 1, 2, 3, 6, and 7 are the same as in Figure 10.19. However, the commutating oscillation (Interval 4) continues through rectifier D after SCR blocks reverse current. The forward voltage drop of rectifier D, while it is conducting, provides the necessary inverse voltage across SCR for the required turn-off time. Since the voltage is low and the rate of change of current is rapid, it is very important to minimize lead inductance in making connections to rectifier D, but the inductance through D should be greater than the inductance through SCR to prevent circulating current in the loop SCR-D. That is, the wiring layout should conform to the diagram in Figure 10.21, and not have the positions of SCR and D interchanged.

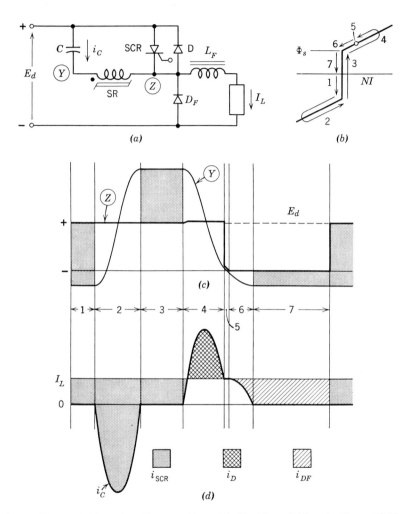

Figure 10.21   Addition of feedback rectifier. (*a*), (*b*), (*c*), and (*d*) as in Figure 10.19.

Interval 4 ends when the capacitor current i_c falls back to equal the load current I_L again, and rectifier D blocks. At this time, most of the energy stored in the capacitor C at the start of Interval 4 has been fed back through rectifier D into the capacitor again but with the opposite voltage

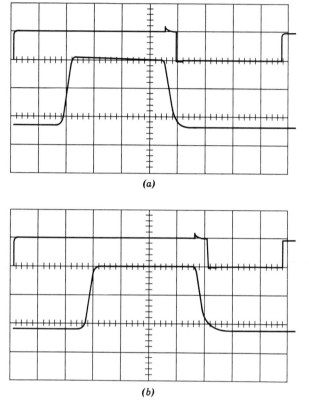

(a)

(b)

Figure 10.22   Oscillograms of Figure 10.21 circuit. Conditions as in Figure 10.20.

polarity. The capacitor voltage now appears as suddenly reapplied forward voltage E_f across the controlled rectifier.

$$E_f \approx E_d\sqrt{\beta^4(1 + \lambda\sqrt{\beta})^2 - \lambda^2} \qquad (10.22)$$

The duration t_4 of Interval 4 is approximately

$$t_4 \approx \sqrt{LC}\left[1 - \sin^{-1}\frac{\lambda}{\beta^2(1 + \lambda\sqrt{\beta})}\right] \qquad (10.23)$$

In Interval 5, the load current I_L charges capacitor C from the voltage E_f to line voltage E_d in time t_5:

$$t_5 = \frac{C(E_d - E_f)}{I_L} \qquad (10.24)$$

If the load parameter λ is large, E_f may be greater than E_d, in which case Interval 5 is absent.

The waveform sketches and oscillograms (Figures 10.21 and 10.22) correspond to those in Figures 10.19 and 10.20 with the addition of the feedback rectifier D. Note that the contribution of Interval 5 to the average load voltage becomes insignificant. However, with very light load or high repetition rate, a problem still arises when the SCR is refired before the capacitor C is fully charged to E_d. The high rate of rise of reapplied forward voltage (dv/dt) on the controlled rectifier is a disadvantage of the circuit. Extra components to limit dv/dt may be needed. The difficulty is increased by the recovery current of rectifier D, which should therefore be a fast recovery device.

Relocation of Coasting Rectifier to Clamp Capacitor Voltage

If the cathode of coasting rectifier D_F is connected to point Y instead of point Z, (Figure 10.23) the capacitor C is prevented from overcharging above line voltage E_d. During the "off" interval, the load current I_L coasts through D_F and reactor SR, holding it saturated. When the controlled rectifier SCR is gated on, the load current I_L commutates from D_F and SR to SCR linearly in a time t_{1A}:

$$t_{1A} = \frac{LI_L}{E_d} \tag{10.25}$$

This interval is designated $1A$ to preserve correspondence of main interval numeration in Figures 10.19, 10.21, and 10.23.

Intervals 1 through 5 now follow, and are similar to those in Figure 10.21. However, the fixed initial capacitor voltage and reactor flux render the pulse width independent of load current and frequency.

$$t_1 = \frac{2N\Phi_s}{E_d} \tag{10.26}$$

$$t_3 = \frac{2N\Phi_s}{\beta E_d} \tag{10.27}$$

The duration of Interval 5 is still given by equation (10.24), but the value of E_f becomes

$$E_f \approx E_d\sqrt{\beta^4 - \lambda^2} \tag{10.28}$$

Rectifier D_F immediately conducts the whole load current I_L at the end of Interval 5 when the capacitor voltage reaches E_d.

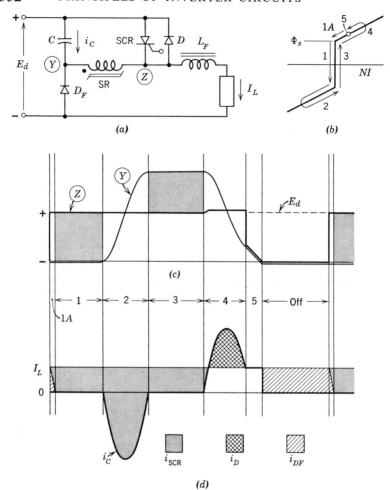

(a)

(b)

(c)

(d)

Figure 10.23   Relocation of coasting rectifier. (a), (b), (c), and (d) as in Figure 10.19.

The insensitivity of this circuit to load changes is apparent in the oscillograms of Figure 10.24. It is particularly suited to voltage regulator applications, since an increase in supply voltage E_d is compensated by shrinkage of Intervals 1 and 3. Thus, only a small variation in the pulse repetition frequency is required to regulate the load voltage against wide variations in supply voltage and load current.

Current Transformation to Modify Pulse Width

If an additional nN turns are wound on reactor SR and placed in series with the load, the "on" time again becomes a function of the load current,

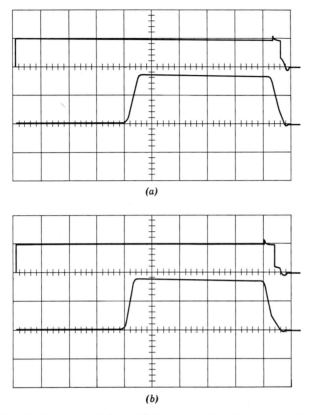

(a)

(b)

Figure 10.24 Oscillograms of Figure 10.23 circuit. Conditions as in Figure 10.20.

in a manner that is useful in certain applications.[1] This circuit variation is shown in Figure 10.25 in combination with a relocated coasting rectifier D_F but without feedback rectifier D. Other combinations of the circuit elements are possible, of course. The waveforms in Figure 10.25 are drawn for the same conditions $\lambda = 0.33$ and $2N\Phi_s/E_d = \pi\sqrt{LC}$ as the previous sketches, with the fractional turns ratio $n = 0.55$.

A cycle begins with an interval (1A) in which the load current I_L commutates from the coasting path through D_F and SR to the power path through SCR according to equation (10.25). Another short Interval (1B) follows as the capacitor current i_c rises to $-nI_L$, balancing the ampere-turns on reactor SR as it unsaturates. Main Interval 1 now commences, with SR being reset by the capacitor voltage which decreases linearly as C is discharged by the transformed load current. If the capacitor voltage reaches zero before the reactor reaches negative saturation, the duration

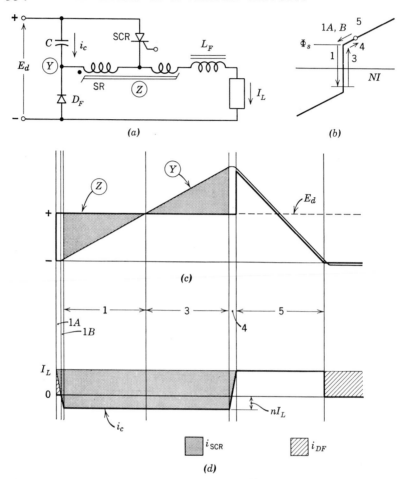

Figure 10.25   Use of current transformation. (a), (b), (c), and (d) as in Figure 10.19.

t_1 of Interval 1 is

$$t_1 = \frac{CE_d}{nI_L} = \frac{\sqrt{LC}}{n\lambda}$$ (10.29)

Since SR does not saturate, Interval 2 is absent, and Interval 3 is a smooth continuation of Interval 1. The capacitor voltage rises linearly to $-E_d$ and SR is set back into positive saturation in a time t_3 that is equal to t_1, equation (10.29). Controlled rectifier SCR is now turned off in the same manner as in Figures 10.17 and 10.19, and the duration t_5 of Interval 5 is given by equation (10.19), where E_r is now

$$E_r \approx E_d\sqrt{1 - \lambda^2(1 - n^2)}$$ (10.30)

Operation of the breadboard circuit in this mode is shown in Figure 10.26(a), where $\lambda = 0.33$ and $n = 1/15$ so that SR is reset to $-0.75\Phi_s$ at the end of Interval 1, which is the same per unit flux reset depicted in Figure 10.25. The critical load condition where the reactor SR just reaches negative saturation is shown in Figure 10.26(b). The parameter λ has the value λ_c:

$$\lambda_c = \frac{E_d\sqrt{LC}}{4nN\Phi_s}$$ (10.31)

Also, the maximum "on" time is attained:

$$(t_1 + t_3)_{\max} = \frac{8N\Phi_s}{E_d}$$ (10.32)

When $\lambda > \lambda_c$, the "on" time is inversely proportional to λ, or to load current I_L.

When $\lambda < \lambda_c$, reactor SR goes into negative saturation, Interval 2 reappears, and the "on" time decreases [see Figure 10.26(c)]. In this region, the "on" time increases as the load current increases, reversing the usual trend. This is desirable in power amplifier applications, since a smaller frequency range can control a given range of load voltage. Also, the form factor of the current in the controlled rectifier SCR, capacitor C, and reactor SR improves with loading, raising the efficiency. When $\lambda \geq \lambda_c$, the charge reversal current of the capacitor is spread out uniformly over the whole "on" time, and not concentrated in a short, high, lossy pulse.

Discussion

For high-frequency operation, the saturating reactor should have a low-loss tapewound toroidal core, of material such as grain-oriented 50-50 nickel iron. If the current is high, insulated, stranded "Litz" wire may be needed for the windings. Generally, the field intensity after saturation is so high that the incremental permeability is negligible and the inductance L can be calculated from the geometry of the windings as if the core were not present. As an approximation, the dimensions of the mean turn can be used in one of the formulae for the inductance of a toroid.

Design of a reactor for a specified flux linkage $N\Phi_s$ and after-saturation inductance L restricts the selection of core size. Often, a rather wide tolerance in $N\Phi_s$ or L is permissible, and a suitable compromise core can be found. If an otherwise satisfactory design has a too-small value of L, extra linear inductance can be placed in series, or the core can be stacked with a dummy former to increase the area enclosed by the turns.

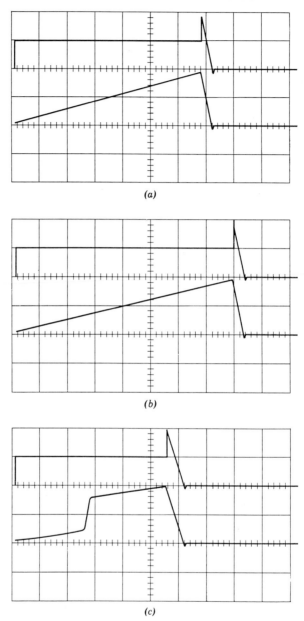

(a)

(b)

(c)

Figure 10.26 Oscillograms of Figure 10.25 circuit. (a) 20-amp load. (b) 17.5-amp load. (c) 10-amp load. (Voltage of point Z-upper traces; voltage of point Y-lower traces; scales are 120 volts/division, and 100 μsec/division.)

All of the circuits described above do not have a directly controllable "on" time, but rely upon variable repetition frequency for control. It is often desired to operate at a fixed frequency and achieve control by varying the "on" time. For this purpose, an auxiliary saturable reactor and rectifier can control the reset of reactor SR in the manner of a magnetic amplifier.[1] Alternatively, a back-to-back combination of an auxiliary controlled rectifier and an ordinary rectifier can be placed in series with the linear inductance L of Figure 10.17. The "on" time is then controlled by varying the delay between gating the main and auxiliary controlled rectifiers. This arrangement can be used in combination with either location of the coasting rectifier D_F, and with or without the feedback rectifier D. The waveforms are very similar to the corresponding cases with the saturating reactor.

The operation of a basic chopper circuit has been described in detail, and some of the most significant and useful modifications have been discussed. However, many other circuit variations, both major and minor, are possible and useful in particular applications.

Abstract

The operation of a d-c chopper circuit using a silicon-controlled rectifier is analyzed. The use of a saturating reactor and capacitor to extend the conducting interval of the controlled rectifier and then to turn it off is described. Some useful variations in the basic circuit are presented. Emphasis is placed upon the waveforms produced, and typical oscillograms are shown. The specific circuits analyzed employ variable repetition frequency as the means of load voltage control, but modification to operate at a fixed frequency with variable "on" time is mentioned.

REFERENCE

1. R. E. Morgan, "A new Magnetic-Controlled Rectifier Power Amplifier with a Saturable Reactor Controlling On Time," *AIEE Transactions*, Part I (Communications and Electronics), Vol. 80, May 1961, pp. 152–155.

Chapter Eleven

Applications

by J. D. Harnden, Jr.

The application of solid-state devices for electric power control and conversion necessarily started at the low end of the power spectrum in December of 1957 when SCR's were first announced.[1] During the ensuing period, equipment power ratings have steadily increased to nearly 1000 kw at present, as shown in Figure 11.1. At first, SCR's were employed in modified thyratron-type circuits. This does not necessarily take advantage of the unique capabilities of the device, such as small size, short turn-off time, and low forward-voltage drop. The preceding chapters have delineated newer circuit developments which have resulted from an attempt to make use of the advantageous properties of solid-state power-switching devices.

The need for reliable battery-supplied standby power sources, precision frequency controlled a-c motors, efficient and lightweight converters for military applications, particularly where d-c power sources are employed, and a wide range of other industrial and commercial products (including lamp dimmers, heat controls, low cost motor controls, and the like) all have contributed to the growth of a vast new solid-state electric power control and conversion technology. The high degree of interest in unconventional power sources, such as, for instance, fuel cells, thermoelectric generators, thermionic converters, and photovoltaic cells, also has been a factor in accelerating the development of solid-state power-conversion circuits. These new power sources are actually somewhat dependent on conversion circuit developments for their own success. Since the basic output from present unconventional sources is dc, inversion to ac with regulation is often required, in addition to d-c transformation and regulation.

The reliability of solid-state converters will only be proved after years

358

of operation, but indications are that high reliability is possible in properly designed circuits, particularly when suitable attention is given to transient voltage problems. The circuit and application aspects are important but, alone, cannot insure high reliability. In certain applications involving a large number of signal level semiconductor devices, such as in many

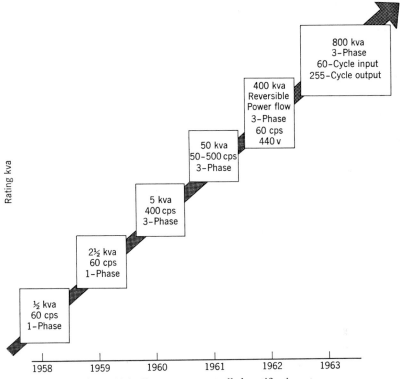

Figure 11.1   D-c–to–a-c controlled-rectifier inverters.

military and industrial computers, the devices are no longer the weak link but, instead, the interconnections between circuit components are proving to be the more frequent cause of failure. This has generated considerable interest in and development activity on integrated and microelectronic circuits.

This chapter includes a description of a number of practical circuit applications of the techniques presented earlier in this book. The particular applications selected illustrate some of the present equipment possibilities which have resulted from the exploitation of the rapidly expanding solid-state power control and conversion technology.

11.1 PHASE-CONTROLLED APPLICATIONS

Simple Power Controls

The use of SCR phase-controlled circuits for a vast range of applications could well be the subject of a separate book. Not only have these powerful semiconductors become widely accepted in industrial and military equipment, but they are fast becoming common-place in consumer products for incandescent and fluorescent lamp dimming[2] and low-cost motor speed controls.[3] This market is developing rapidly because of the high efficiency, small size, instant operation, and low price, as compared with other approaches which may have been used previously.

Incandescent lamp dimmers are an important application where controlled-rectifier devices are used to provide phase-controlled a-c power. The lamp-dimmer ratings start around 300 watts, and range up to many kw for professional studio use. One of the severe problems with a lamp dimmer is the inrush resulting from the cold-to-hot resistance change of the tungsten filament. Some of the more de luxe circuits incorporate a current inrush control to keep the semiconductor junction temperature within safe limits. Similar circuits are very useful for the control of resistance heaters. This is perhaps the easiest application from all aspects, since the load is not usually waveform-sensitive, has unity power factor, and normally does not change more than 10 per cent with temperature changes. Half-wave phase-controlled circuits requiring only one switching element or circuits including the recently developed semiconductor a-c switch are used for the simplest and lowest cost power controls. When smoother d-c output is desired, simplified bridge circuits may be used where only part of the rectifying devices are controlled.[4] Gating circuits for phase-controlled applications are fairly sharply divided between the all-magnetic amplifier type and the all-semiconductor type. Isolation is somewhat easier to incorporate in the former, but the magnetic approaches may be more expensive. In simple power controls, pulse gating is often used while inverters generally require continuous gating signals during the full conduction interval for the particular SCR.

Regulated Power Supplies

SCR's are also becoming widely applied to both d-c and a-c power supplies. Perhaps the simplest of these is the former, since only controlled rectification and filtering are basically involved. Figure 11.2 shows an early SCR assembly rated at 75 kw, 250 v d-c, adjustable output. This assembly uses twelve SCR's, connected in a three-phase bridge arrangement. Each leg of the bridge contains two 400-v SCR's, connected in series, thereby providing considerable voltage rating to allow for line voltage transients. Higher kw ratings can be easily obtained by connecting

Figure 11.2   Silicon-controlled rectifier power assembly (top view);   75 kw, 250 v d-c adjustable output.

assemblies in parallel or by using higher rated SCR's.   Figure 11.3 shows a large industrial application involving a number of the SCR assemblies shown in Figure 11.2.   The main advantages of SCR approaches in applications where controlled d-c output is required are size, weight, cost, and speed of response.

The a-c-to-a-c constant voltage regulator provided an opportunity for innovation, based on the use of the SCR.   Figure 11.4 shows a typical unit designed for rack mounting.[5]   Most of these supplies provide an output a-c with less than 3 per cent total rms harmonic content.   In general, the SCR equipment can be operated over a wider frequency

Figure 11.3   SCR adjustable-voltage d-c power-supply assembly.

Figure 11.4   SCR voltage stabilizer, rated 57 to 63 cycles; 1.0 kva, 95–130/190–260 v input, 118 v output.

range than is practical with most magnetic voltage stabilizers. The basic power circuit diagram is shown in Figure 11.5.[6,7] It is expected that a number of new a-c–to–a-c regulated power supplies will be developed around inversion and waveshaping concepts—which have been set forth

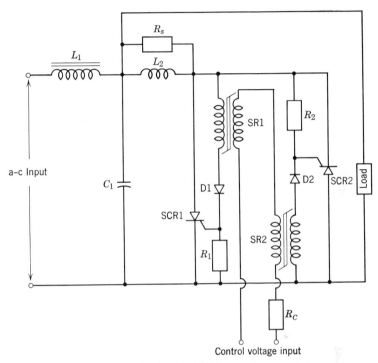

Figure 11.5   Basic circuit of SCR a-c voltage regulator.

in this book. In fact, the full exploitation of these techniques should provide supplies with greatly improved performance and features such as lighter weight, smaller size, and higher efficiency.

11.2   TIME-RATIO CONTROL

General

When only d-c power is available, a time ratio control (TRC) circuit is the most practical means to provide an efficient, compact, and reliable d-c voltage control. Even when a-c power is available, it may be preferable to use a simple rectifier followed by a TRC to provide adjustable d-c voltage. This approach will provide better input a-c power factor, less ripple for a given rating of filter components, faster response, better utilization of controlled rectifier devices, and lower cost in certain applications than a phase-controlled rectifier. In a number of inverter systems the d-c input voltage to the inverter is controlled by a TRC.

Static Power Switching

Static switches, contactors, and circuit breakers[8] commutate on a very low duty cycle; that is, the SCR is commutated only when the switching function is required. A d-c static contactor is shown in Figure 11.6, and

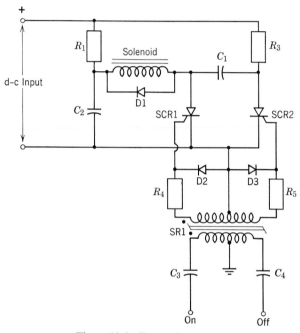

Figure 11.6   D-c static contactor.

a photograph of the actual equipment is shown in Figure 11.7. Another whole area of usage where the commutation duty cycle is lower than in most time-ratio controls is in the area of pulsing, as for high-power pulse modulators.[9] In addition, a highly reliable and high-performance solid-state ignition system can be built for operation from a d-c source.[4]

Static Exciters

One of the first applications of time-ratio control, using SCR's, was for reliable and accurate static regulation of d-c aircraft generators operating at 120°C ambient.[10] A photograph of a packaged unit is shown in Figure 11.8. The steady-state rating at this temperature is 7 amp and increases to around 10 amp for lower ambients. The over-all weight is about 6¼ lb. The output is maintained at 28 v ± 0.75 v over the speed, load, and temperature extremes. The block diagram for this circuit

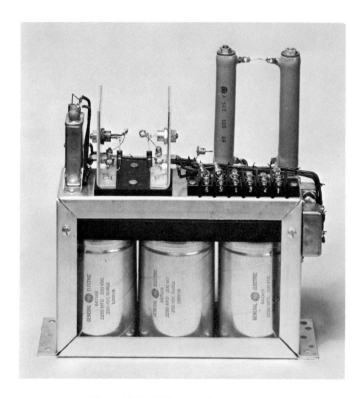

Figure 11.7   SCR static d-c contactor.

approach is shown in Figure 11.9. The power circuit uses an SCR basically in a "Morgan" circuit arrangement using a saturable transformer for switching the commutating capacitor, similar to the circuits discussed in Chapter 10.

A number of TRC's have been applied to other field regulators, some where the field voltage is reversed to force the current to decay rapidly. The transfer characteristic for a single-direction TRC field regulator circuit is shown in Figure 11.10, which indicates the wide control range. These amplifiers are self-contained, normally operating from d-c sources or rectifier supplies. They have a predictable and repeatable voltage

Figure 11.8   Aircraft d-c generator regulator.

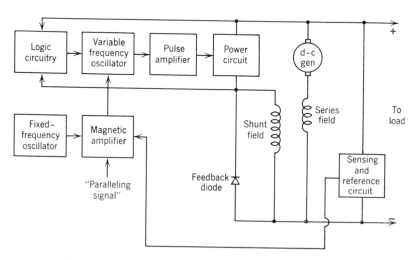

Figure 11.9   Block diagram of aircraft d-c generator regulator.

output as a function of the control signal. In the example shown, a magnetic amplifier is used as the input device, so that the control input is in terms of ampere-turns. As a result of the high switching speeds incorporated, the TRC amplifier time constants are so short that they can be ignored. By controlling the ratio of the SCR "on" to "off" times,

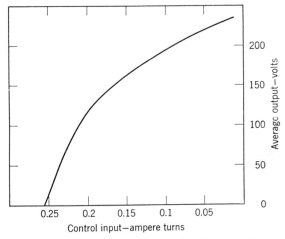

Figure 11.10   TRC field regulator transfer characteristic.

an average field voltage theoretically covering the range between zero and the d-c supply voltage can be obtained. In the particular unit for which Figure 11.10 applies, the static excitation panel is composed of three basic sections, which are listed below.

1. A 200 cps square-wave signal inverter.
2. A firing amplistat and load SCR.
3. A commutating SCR and associated circuits.

An elementary diagram of this system is shown in Figure 11.11. Several pertinent waveforms are shown in Figure 11.12. This is a constant-frequency variable pulse-width TRC where an external impulse coupled from an auxiliary SCR provides commutation of SCR1, the load SCR. The amplistat gate determines the firing point of the load SCR which is turned off at a fixed repetition rate. SCR2 is turned on to generate a pulse which is coupled to SCR1 by transformer $T1$ to commutate SCR1. Although this is a somewhat more complex arrangement than the simple TRC using a variable pulse rate, the filtering problems, noise problems, and transient response problems are lessened. The signal inverter frequency chosen is a matter of systems needs, such as response, size, and gain, for

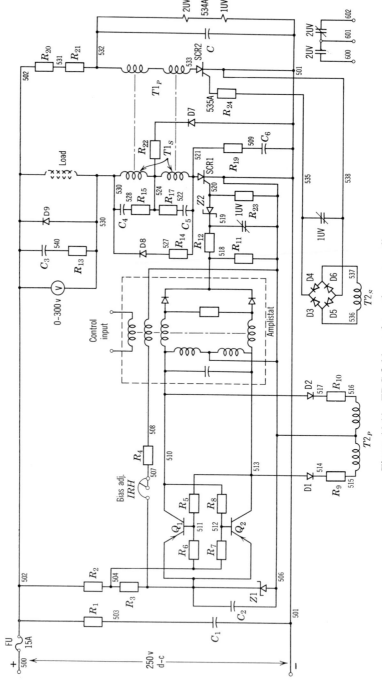

Figure 11.11   TRC field regulator elementary diagram.

instance, and usually is a compromise based on circuit-efficiency considerations. The response time of this TRC is governed principally by the amplistat time constant and delay times. Figure 11.13 shows an installation of these TRC field regulators, each rated for 220 v d-c input, and 12 amp output.

Regulated Power Supplies

Figure 11.14 shows a 2-kw, fixed-output, step-down voltage converter. This equipment is designed for use as a battery charger and for supplying control power at 37.5 v, as required for certain locomotive and rapid transit car controls when the available supply is nominally 75 v d-c. In addition to providing the transformation, output voltage variations caused by supply voltage changes are essentially eliminated. Figure 11.15 shows a block diagram of this equipment. It is designed to operate with input variations from 45 to 80 v, and holds the output within 2 v of the nominal 37.5-v output. The circuit is self-protecting against overloads, and is designed to operate from −20 to +85°C. At the lower temperature extremes, increasing output ripple from 0.1 to 0.5 v occurs. SCR's with 16-amp average current rating are used in the power circuit, while 7-amp average SCR's are used for commutation.

A 5-kw unit has also been designed, which is very similar to the above, except that the input is 600 v d-c. Several SCR's are used in series because of the high-voltage and transient-voltage problems. The over-all efficiency is about 85 per cent, which is 15 to 20 per cent higher than for the rotating auxiliary power supply which it is designed to replace. A current limit circuit is included to provide self-protection in the case of overloads.

TRC principles are becoming widely used in high-performance power supplies of all types for such reasons as, for example, response, size, weight, and cost.[11] This is particularly true in the field of high-accuracy high-performance computer supplies.[12] Figure 11.16 shows a computer power supply rated for 1.6 kw which occupies a volume of 1 cu ft. The total weight for all three modules is 110 lb. The over-all efficiency from a-c input to regulated d-c output is 65 per cent. The switching frequency for this power supply is 3000 cps, thus providing very fast response. The basic power circuit is shown in Figure 11.17. This circuit combines two half-wave TRC systems to provide greater output, with less inherent ripple. Each TRC contains its own output transformer. The complete closed loop regulated supply includes SCR gating and commutating circuits, circuits to aid resetting of each transformer, interlocking and synchronization controls for proper co-ordination of the two half-wave systems, and closed loop voltage regulating controls in addition to the

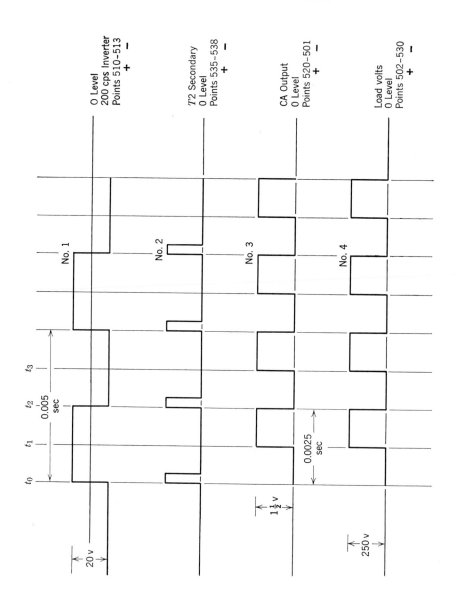

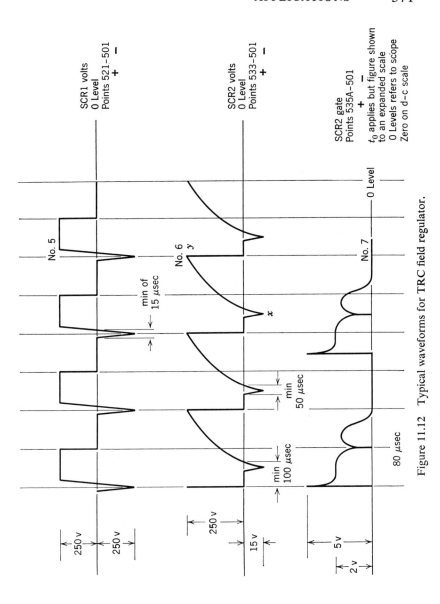

Figure 11.12   Typical waveforms for TRC field regulator.

Figure 11.13   TRC field regulators installed in paper mill.

basic power circuits of Figure 11.17. A constant-frequency TRC approach is used with auxiliary SCR's to commutate SCR1 and SCR2. Since the switching frequency of 3000 cps is substantially greater than the highest frequency disturbance to which the output filter may respond without considerable attenuation, the response of the system is not limited by the TRC power control. A closed loop stability analysis can neglect the time

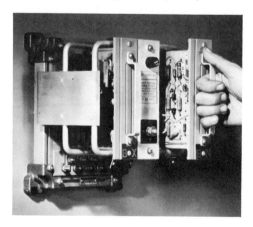

Figure 11.14   Static voltage regulator for railroad applications.

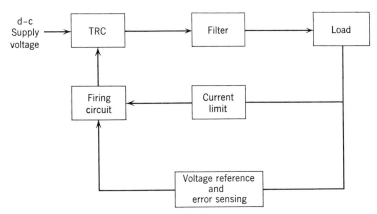

Figure 11.15 Block diagram of static regulator for railroad applications.

Figure 11.16 1.6 kw Computer power supply.

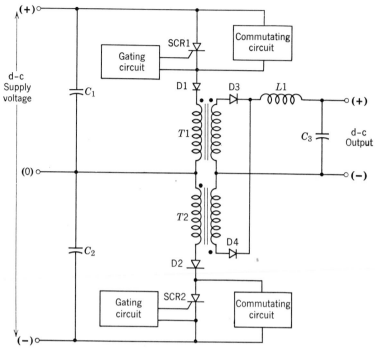

Figure 11.17   Computer power-supply circuit.

constants or delay time of the TRC when a high switching frequency is employed, as there are usually much longer time constants in the output filter or load circuit.

D-C Motor Drives

The application of TRC techniques to electric propulsion drives is very promising. This allows for smooth speed variation over a very wide range without the energy loss normally associated with resistor-type control. A basic circuit concept is shown in Figure 11.18. This particular circuit is a variable frequency TRC. Ideally, the circuit should be designed so that the load current is continuous at all times with relatively low ripple. Under these conditions, the circuit operates as a "variable ratio d-c transformer," as discussed in Chapter 10. The ratio of the average motor current to the input average current is the inverse of the ratio of the average motor voltage to the d-c supply voltage. However, the "switch" must handle the peak load current, as must the commutation

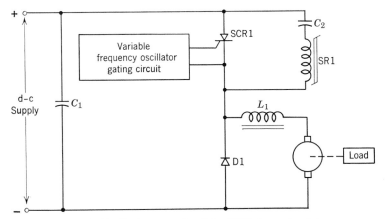

Figure 11.18 Basic power circuit for TRC d-c motor drive.

circuit. Time-ratio control circuits are being developed which provide a given transformer ratio with less peak current requirements for both the semiconductor switch and the commutation components for a given load current.

Figure 11.19 shows a 30-hp series motor control, which is designed for operation from a sea-water battery.[13] The battery voltage varies from 100 to 150 v, so that by using a TRC unit similar to that shown in Figure

Figure 11.19 30-hp d-c Torpedo motor drive (over-all view).

11.18, speed control independent of battery voltage is provided. Also, the battery energy can be "metered" out to the motor to provide increased operating time.

The actual circuit uses five 50-amp SCR's in parallel to handle the load current, the highest available device rating at the time of the design of this equipment. The motor speed is varied over a 5–to–1 range, which requires an output voltage range of approximately 10 to 1.

Figure 11.20   TRC power controller for electric vehicle.

This same circuit concept may have advantages for motor drives[14] and power supplies when preceded by a rectifier to allow operation from a-c sources. The TRC circuit provides better SCR utilization, reduced SCR reverse voltage, and simpler control circuits than the polyphase phase-controlled rectifier which requires at least three SCR's to deliver adjustable d-c output. In some cases, an advantage may also exist when compared with a single-phase circuit. In addition, if the TRC is operated at high frequency, faster response results than is possible with a phase-controlled rectifier.

There is, currently, renewed interest in electric automobiles, battery trucks, and small golf carts. This market is expected to grow to vast numbers, particularly as battery improvements continue to be made. Figure 11.20 shows a picture of a TRC power controller which is utilized in the vehicle shown in Figure 11.21 to provide a stepless and "lossless-type speed control." The efficiency of power conversion in these applications is extremely important, since the stored energy in the battery is limited.

Figure 11.21 Electric vehicle with TRC control of Figure 11.20.

If four of the basic circuits of Figure 11.18 are incorporated into a bridge arrangement as shown in Figure 11.22, a reversible drive results. A practical circuit of this type to drive a 30-hp motor has been demonstrated in the laboratory, but widespread application awaits further development to reduce the complexity of the associated control circuits.

A split field d-c motor-reversing drive, using time-ratio control, is shown in Figure 11.23.[15] Maximum speed in a given direction is achieved by operating with the associated SCR turned continuously on, while its counterpart remains turned off. For intermediate speeds, the two SCR's may share differing proportions of "on" time, or one may be turned on for repetitive short conduction intervals while the other remains continuously off. Speed control is a matter of establishing the proper degree of intermittent on-off switching of that particular SCR which gives the correct direction of rotation. The circuit outlined employs a method of external impulse commutation which is independent of the turn-on of the SCR's. Fixed-frequency turnoff pulses are provided, with the turn-on

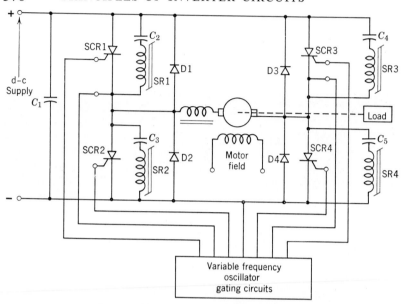

Figure 11.22   TRC reversible d-c motor drive.

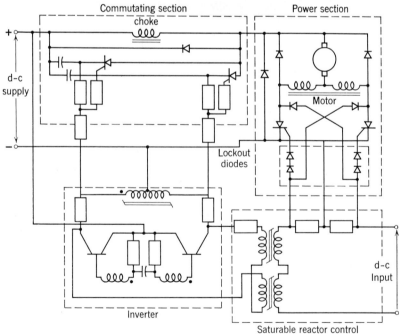

Figure 11.23   Split series field d-c motor control.

being variable. The transistor oscillator, which generates the turnoff impulses, also provides a voltage source for the control saturable reactor, thus allowing multiple isolated control windings. The frequency of operation is sufficiently high so that the response time is mainly determined by the motor and its load. Operating as a closed loop speed control, a step function input changed the speed from 2500 rpm in one direction to the same value in the opposite direction in $\frac{1}{20}$ of a second, which is the mechanical time constant of the motor used. When operating as a closed loop position control, the system responds to a position error of four revolutions of the motor in $\frac{1}{6}$ of a second with a quarter revolution peak overshoot. This overshoot is corrected and steady state is reached in a total time of $\frac{1}{4}$ of a second.

Voltage Step-Up Converters

There are several ways in which TRC concepts can be utilized to provide voltage step-up rather than step-down. Figure 11.24 shows the

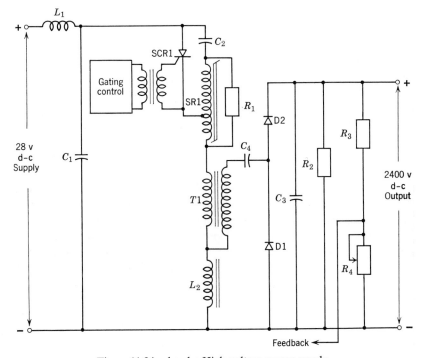

Figure 11.24 d-c–d-c High-voltage power supply.

power circuit for a voltage step-up from 28 to 2400 v.[16] The block diagram
for the complete unit is shown in Figure 11.25. This step-up circuit
utilizes saturable reactor-capacitor commutation.

It is important to note that the TRC circuit in Figure 11.24, with an
output transformer, is actually an inverter with time ratio control of the
a-c output voltage. This is also true of the step-down circuit, previously
shown in Figure 11.17, and similar TRC circuits where an output transformer is used to transform the a-c component of the unidirectional

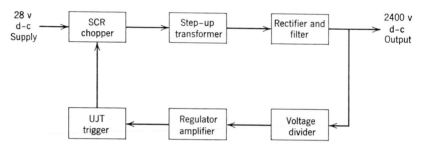

Figure 11.25   Block diagram of d-c–d-c high-voltage supply.

pulsating d-c output from a TRC. The inverters of Figures 11.17 and
11.24 both operate with nearly a unity power factor load since the input
to a rectifier is approximately unity power factor.

Figure 11.26 shows the block diagram of another d-c–to–d-c converter,
similar in principle to the circuit in Figure 11.17. However, in the system
of Figure 11.26, a single output transformer is used with a center-tapped
primary winding and multiple secondary windings. The upper SCR
chopper, during its "on" interval, energizes the upper half of the transformer primary winding. When this chopper is turned off, there is an
interval when neither TRC is "on." The lower TRC chopper energizes
the lower half of the transformer primary winding to deliver the next
half-cycle of transformer output voltage. In this case, the output transformer operates with symmetrical positive and negative flux changes as
in a conventional a-c transformer. This requires that alternate SCR
chopper "on" intervals are equal. The SCR choppers of Figure 11.26
use TRC circuits with saturable reactor-capacitor commutation similar
to that shown in Figure 10.5 (Chapter 10). The output voltage is controlled by controlling the frequency of the square-wave oscillator to vary
the frequency of the SCR choppers. Therefore, this system includes an
inverter whose frequency is controlled to regulate the output voltage.
A photograph of a typical unit is shown in Figure 11.27.

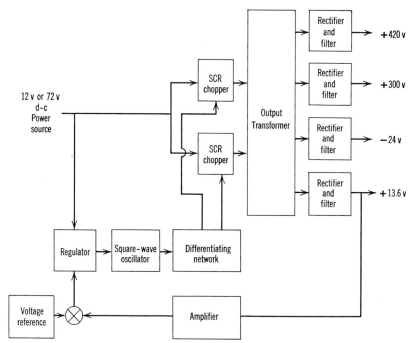

Figure 11.26   d-c–d-c Power supply with multiple outputs.

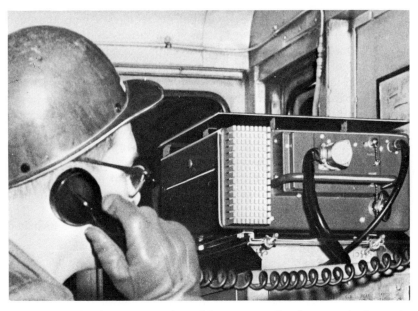

Figure 11.27   d-c–d-c Power supply used in two-way railroad communication equipment.

381

11.3 INVERTERS

The power ratings of inverters that are being developed and marketed have closely paralleled the basic capacity of the SCR's that are available. Many applications are already in operation up to 400 kw, while megawatt units are now in the design stages. Semiconductor inverters are used to provide emergency ac for microwave relay equipment, computers, lighting, and the like; acoustically quiet and highly reliable d-c–to–a-c inverters for shipboard use; and small, light weight, and efficient equipments to convert battery power to ac on missiles and space vehicles, for example. In applications requiring a highly versatile power supply, such as changing dc to 50, 60, or 400 cps in one equipment, the static inverter is an increasingly competitive and attractive means of providing the solution.[17] In time, it is expected that these static equipments will be lower in cost for many applications than other methods presently being used.

The adjustable voltage and frequency characteristics of SCR inverters make them ideal for use with variable speed a-c motor drives.[14] A frequency accuracy of 0.1 per cent is easily provided with solid-state components, and very fine adjustment of the frequency can be provided by small multiturn potentiometers. Precision frequency stability is obtained by using crystal-controlled gating circuits.

The design of static inversion equipments requires a most thorough consideration of the complete system specifications and requirements. An optimum arrangement is most closely approached when the unique capabilities of static inverter hardware are thoroughly considered at the point when the total system requirements are established. Inverter cost is directly effected by the performance complexity required. In the power range from 1 to 10 kva, single-phase, square-wave output may be the most economical while, for large ratings, polyphase has many advantages. Polyphase arrangements often allow the use of many SCR's in a system without actual series or parallel connection and, if properly applied, voltage control and harmonic reduction of significant magnitude can be achieved. One of the major factors in inverter design concerns the amount of overload and unbalanced load which is to be handled. The peak currents to be commutated by the SCR's must be anticipated beforehand to assure reliable operation.

Parallel Capacitor Commutated Inverter

An early SCR inverter, rated to deliver approximately 500 watts, is shown in Figure 11.28. The schematic circuit diagram is shown in Figure 11.29. Emphasis in the design was placed on the use of standard parts to minimize construction time. The inverter is basically a parallel capacitor-commutated inverter.[18] Several modifications to the basic circuit were made necessary by the nature of the practical load. In this application the inverter supplied a number of solenoids and magnetic amplifiers so that there was considerable inductive load which varied over a wide range. When the load impedance is high, the commutating capacitor

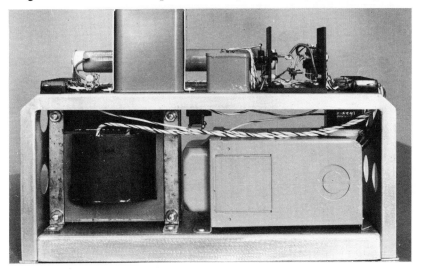

Figure 11.28 A 500-watt parallel capacitor-commutated inverter.

may oscillate to an excessively high voltage. Feedback inductance L_1 and diode $D3$ provide a means of limiting the amplitude of the load voltage and restoring energy to the d-c source when the inverter load impedance is increased. The resistance R_3, in parallel with d-c inductance L_3, limits the self-induced voltage which results from sudden load changes. A fixed dummy load resistance R_4 is used to prevent excessive SCR voltage when the normal load is removed. This inverter was designed to operate from 120 v d-c and to supply 130 v 60 cps. The output voltage waveform was not critical, and contained appreciable harmonics, which varied considerably over the load range. The circuit used a pair of 300-v SCR's.

A separate external transistor voltage regulator was used to control the output voltage by adjusting the d-c input to the inverter. The small saturating pulse transformer was added in the gate drive circuit to eliminate false triggering. This transformer remains saturated between gate voltage pulses and provides a very low impedance when it is not delivering triggering pulses.

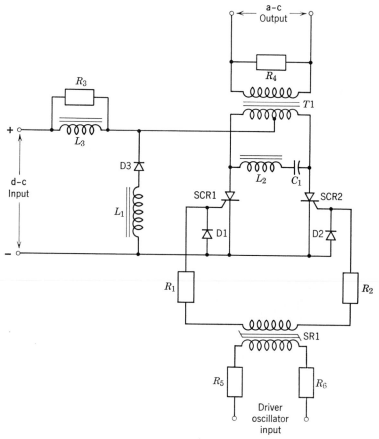

Figure 11.29 Schematic circuit diagram of 500-watt parallel capacitor-commutated inverter.

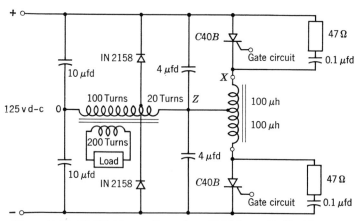

Figure 11.30 Half-bridge configuration of McMurray-Bedford inverter.

384

Impulse-Commutated Inverters

An improved inverter circuit is shown in Figure 11.30. This is a half-bridge configuration of the McMurray-Bedford inverter discussed in Chapter 7. As stated in Chapter 7, there are several advantages associated with this method of commutation that allow much greater load flexibility than the circuit of Figure 11.29.[19] Table 11.1 summarizes the performance operating with a resistive load; the circuit operates equally well over a wide range of inductive loads.

Table 11.1   Performance of inverter in Figure 11.30 with resistive load

	120	400	1000
Output frequency, cps	120	400	1000
Primary Voltage (OZ) Rms volts {no load	74.5	75.0	78.0
Primary Voltage (OZ) Rms volts {maximum load	66.0	66.2	66.9
Maximum load power (watts)	655	585	300
Efficiency at maximum load (per cent)	96	92	67
Power input at no load (watts)	18	35	64

A block diagram of a 3.5-kva inverter system is shown in Figure 11.31. This system contains a bridge version of the McMurray-Bedford inverter. In applications where this equipment is to be used for emergency power, a static SCR switch automatically connects the standby battery to provide uninterrupted power output.[20] A picture of the packaged unit is shown

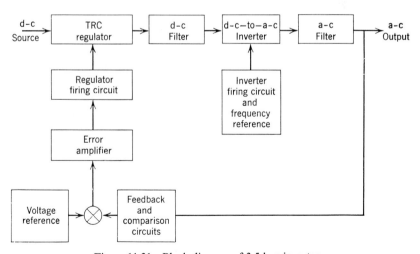

Figure 11.31   Block diagram of 3.5-kva inverter.

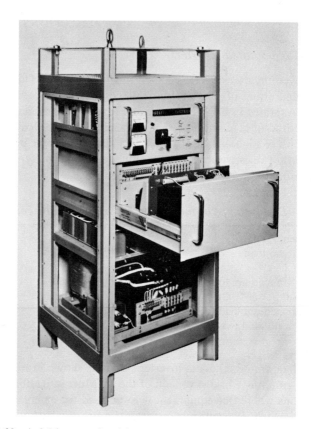

Figure 11.32 A 3.5-kva regulated inverter; 140–105 v battery input; 115 v ± 5%, 60 cps ± 2 cps, 3.5 kva single-phase output at 0.7 pf lagging with 5% total harmonic distortion.

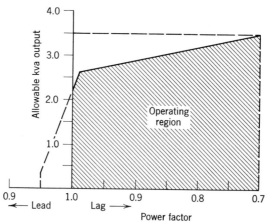

Figure 11.33 Rating vs. power factor for inverter of Figure 11.32.

386

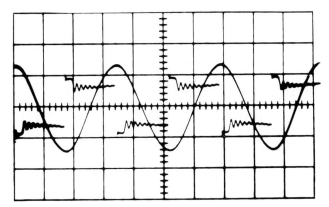

Figure 11.34 Output voltage waveshape of the inverter in Figure 11.32 before and after *L-C* filter.

in Figure 11.32. The equipment is designed to operate over a wide range of load power factors with a slight derating as shown in Figure 11.33. The output voltage is 115 v, ±5 per cent, single phase. The frequency is derived from a local unijunction oscillator, and is nominally 60 ± 2 cps.

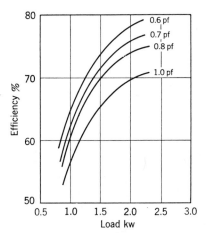

Figure 11.35 Efficiency vs. load at various power factors—inverter of Figure 11.32.

The output is filtered so that the total rms content is less than 5 per cent. Figure 11.34 shows the output voltage waveshape before and after filtering. The efficiency for various power factors as a function of load is shown in Figure 11.35, and the transient response to step loads is shown in Figure 11.36.

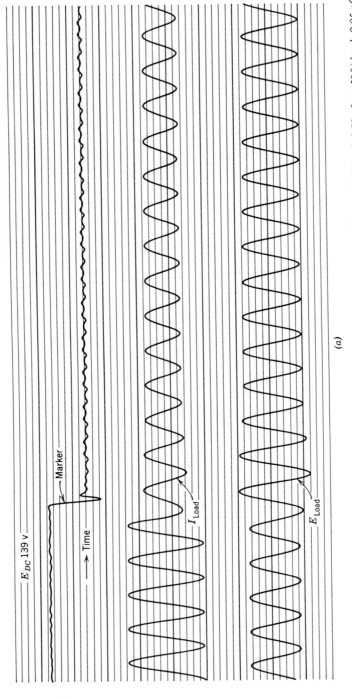

Figure 11.36(a)   Transient recording of load voltage and load current for a step change in load from 100% load, 0.72 pf, to 50% load, 0.95 pf, with $E_d = 139$ v.

(a)

388

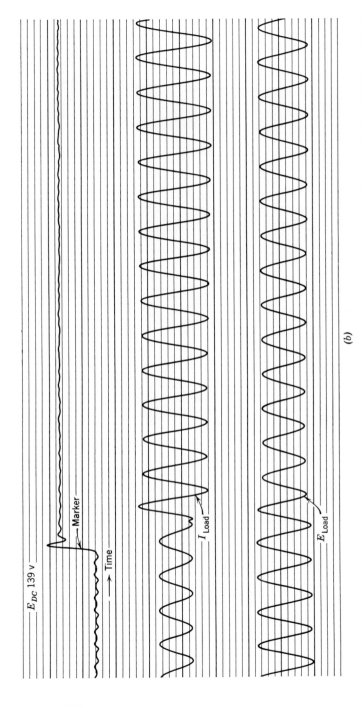

389

Figure 11.36(b) Transient recording of load voltage and load current for a step change in load from 50% load, 0.95 pf, to 100% load, 0.72 pf, with $E_d = 139$ v.

Figure 11.37   A 115 v, 725 va—two minutes, 400 cps, single phase aircraft inverter.

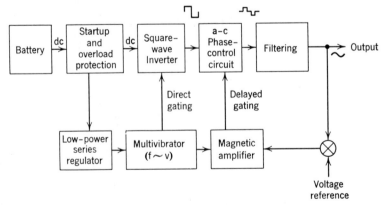

Figure 11.38   Block diagram of the inverter shown in Figure 11.37.

Figure 11.37 shows an aircraft inverter, using a form of McMurray-Bedford impulse commutation similar to that shown in Figure 11.30 and discussed in Chapter 7. A block diagram of this system is shown in Figure 11.38. The following is a summary of the specifications.

Input:	22 to 28 v d-c with transients up to 45 v.
Output:	725 volt-amperes at 0.5 to 0.7 pf for 2 minutes with 60 minutes off between load cycles.
	Steady-state output voltage within limits of 103-120 v rms and 380–420 cps with maximum no-load voltage of not more than 136 v rms.
	Crest factor of output voltage between 1.0 and 1.55.
Environment:	−40°F to +130°F (maximum storage temperature + 185°F). Withstand aircraft vibration, shock, and acceleration. Operate in 100 per cent relative humidity and sand, dust, salt spray, and fungus conditions.
Short circuit:	Inverter protected against output short circuit.
Size and weight:	10.2 in. × 6 in. × 9.06 in.
	20 lb.

Figure 11.39 shows a 50-kva inverter.[21] This equipment incorporates a three-phase version of the McMurray inverter, discussed in Chapter 7.

Figure 11.39   A 120/240 v adjustable output, 50-kva, 50–500 cps, three-phase static inverter.

The block diagram of Figure 11.40 reveals that the power circuit is composed of four three-phase inverters, involving a total of twenty-four 50-amp, 200-v SCR's. A like number of 16-amp units are used to provide commutation. The advantages of the multiphase approach to provide voltage control and reasonably good output waveform have been covered in Chapters 8 and 9. The basic input is 125 v d-c, suggesting that greater output power could be readily achieved by using higher current and

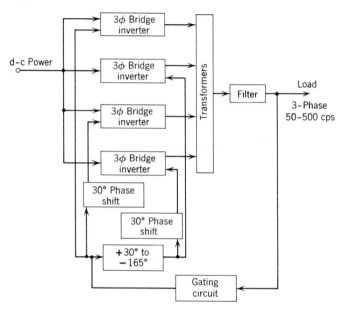

Figure 11.40   Block diagram of the 50-kva inverter shown in Figure 11.39.

higher voltage SCR's with a higher d-c input voltage. The three-phase output voltage is isolated by transformers, and is regulated to ±1 per cent from 0 to 100 per cent load, with ±10 per cent input voltage variations. The output voltage is adjustable by a small multiturn potentiometer in the control circuits from 20 to 240 v with a total harmonic content less than 5 per cent. A 2:1 step-down autotransformer in the inverter output circuit is included to provide a similar voltage control range with 120 v rated output. An *L-C* filter arrangement is contained in the output circuit to achieve good waveform. The output can be adjusted below 20 v, but the harmonic content is then greater than 5 per cent. The frequency is variable from 50 to 500 cps by a small multiturn potentiometer, adjusting the frequency of the reference oscillator which drives the gating circuit. The frequency accuracy is ±½ per cent, which is achieved with a simple unijunction oscillator. This "universal" power supply will provide accurately controlled adjustable a-c output from unity to 100 per cent lagging or leading power factor in ambients up to 50°C.

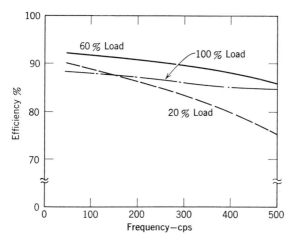

Figure 11.41   Efficiency vs. frequency at various loads—50-kva inverter of Figure 11.39.

The efficiency under different operating conditions of load and frequency is shown in Figure 11.41. It should be pointed out that the efficiency at full load is less than at 60 per cent loading, partly because of excessive source regulation during these tests near full load, resulting in heavier currents and, hence, more losses for a given output power. The scope traces shown in Figure 11.42 emphasize the "stiffness" of static inverters. The inverter output was unregulated and filtered for this test. The upper trace is motor line current, and the lower trace is motor voltage during an across-the-line start of the specified synchronous motor. The motor inrush on starting is 26.5 kva. Expanded time-scale waveforms show that the frequency is also unaffected by starting transients of this sort. Therefore, static inverters are ideal sources for synchronous motors in precise speed control applications.

An inverter unit, rated for 322 kva, is shown in Figure 11.43. This unit also includes the transformer rectifier equipment, so that the inverter is directly operable from three-phase a-c power. Modular construction is used for the power circuitry as well as for the control circuits. This particular equipment uses a polyphase configuration of the inverter shown in Figure 7.28 (Chapter 7).

The concept of harmonic reduction by means of stepped output wave-shape is very important for large inverters. In smaller inverters, this

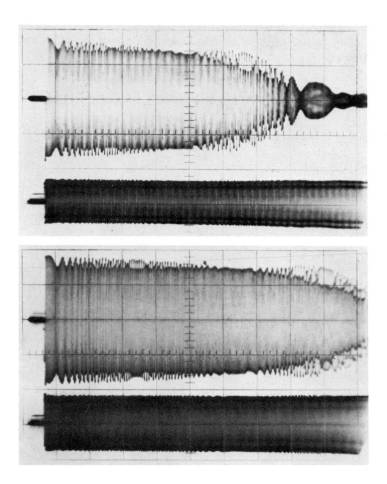

Figure 11.42   The 50-kva inverter of Figure 11.39—waveforms during across-the-line start of synchronous motor. Oscillograms showing motor line current (*upper wave form*) and line voltage (*lower wave form*) for 50 cps start (*upper photo*) and 65 cps start. Both photographs were taken at $\frac{1}{2}$ sec/div sweep, 50 amp/div, and approximately 420 v/div. Inverter output was unregulated but filtered. Motor starting voltage (across-the-line starts at 50 cps) was 185 and, at 65 cps, 240. The motor was a 0.1/0.54 hp, 360/1950 rpm, 55/240 v, 12/65 cps, 3ϕ machine with an $11\frac{1}{4}$ in. diameter $\times$ $3\frac{1}{2}$ long solid steel flywheel. Starting kva at 65 cps was 26.5 kva.

method of harmonic reduction may also be used when it is necessary to meet stringent waveform requirements.[22] It is important in any design to consider the trade-offs involved in choosing just how many "steps" to generate. Figure 11.44 shows a 2-kva three-phase 400-cycle unit designed for airborne use. The lowest harmonic frequency in this inverter output waveform is 11 times the fundamental frequency so that excellent sine-wave outputs may be obtained with a relatively small a-c filter.

Figure 11.43   A 322-kva three-phase 210 or 270 cps inverter.

A complete inverter system, which was developed for a military application, is shown in Figure 11.45. It is packaged, as shown in Figure 11.46, to meet high shock specifications, in an ambient of +20 to +40°C with convection cooling. The input d-c is regulated by a TRC circuit to allow operation with an input voltage range of 200 to 355 v. The performance of the regulator for various line and load conditions is summarized in Table 11.2.

The over-all efficiency corresponding to the 250-v input and full load on the inverter is 73 per cent, where Table 11.2 indicates that the regulator portion has 94.8 per cent efficiency. The regulated dc is filtered and applied to a three-phase bridge inverter using McMurray-Bedford commutation. The step-wave output of the bridge inverter is transformer-coupled to a monocyclic network filter, which provides a sine-wave output and load-current limit for the inverter. The maximum total harmonic content is 10 per cent. The no-load and full-load output

Figure 11.44(*a*)   A 2-kva, 3ϕ, 400 cps stepped-wave inverter for airborne applications.

voltage waveforms are shown in Figure 11.47. The output voltage and current are sensed and applied to the regulator gating circuit controlling the preregulator. The inverter gating circuit controlling the firing of the three-phase bridge inverter obtains a frequency signal from a master oscillator which establishes the 400 cps output. The nominal output is 120 v, line-to-line, three-phase, with a maximum of 3 per cent voltage unbalance, and is adjustable plus 7 per cent and −3 per cent from the nominal. Output voltage waveforms are shown in Figure 11.48 for step

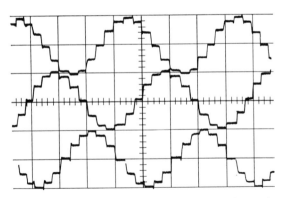

Figure 11.44(*b*)   Output voltage waveforms of stepped-wave inverter.

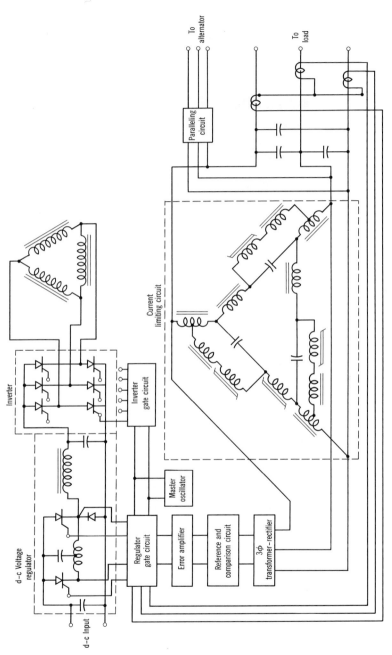

Figure 11.45   Basic power circuit of 5-kva inverter system.

load transients from 25 per cent to full load, and no load to full load. The inverter is designed to be momentarily paralleled with a motor generator to allow the set to be removed from the a-c bus without loss of output voltage.

Figure 11.46   A 5-kva inverter for military application.

The next few years will provide many new equipment concepts which are direct outgrowths of the new power conversion and control flexibility which power semiconductors make possible. Figure 11.49 shows an interesting example where a specific waveshape and frequency are required

Table 11.2   Performance data for d-c voltage regulator

Input			Output				Ambient Temperature (°C)	Load
Voltage (volts)	Current (amps)	Power (kw)	Voltage (volts)	Current (amps)	Power (kw)	Efficiency (per cent)		Load on d-c regulator
355	17.2	6.12	160	33.8	5.41	88.5	30	Resistive
200	27.2	5.45	154	33.8	5.20	95.5	30	Resistive
250	21.7	5.43	154	33.4	5.15	94.8	50	Inverter with rated load
355	16.7	5.73	154	34.5	5.31	89.6	30	Inverter with rated load
200	27.7	5.45	154	34.5	5.31	97.5	30	Inverter with rated load
355	3.75	1.33	136	6.37	0.867	65.2	30	Inverter with no load
200	5.25	1.05	136	6.37	0.867	82.5	30	Inverter with no load
175	5.62	0.985	136	6.37	0.867	88.0	30	Inverter with no load

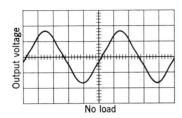 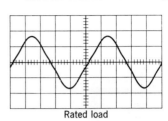

No load   Rated load

Figure 11.47   Output voltage waveforms for the inverter of Figure 11.46. Vertical: 100 v/division. Horizontal: 500 μsec/division.

for the testing of a large electrical apparatus. This new method of high-potential testing offers the advantages of 60 cps a-c testing, yet requires smaller equipment. A block diagram of the system is shown in Figure 11.50. The equipment uses a 2500 cycle per second inverter in order to minimize the size of transformation equipment involved, and to achieve maximum portability. The rated electrical input is approximately 4 kva, and the end output of this low frequency test set is 50 kv rms, 0.1 cps into a ½ microfarad capacitor load.

6A–20A

(*a*) Transients applied with approximately 25% fixed load

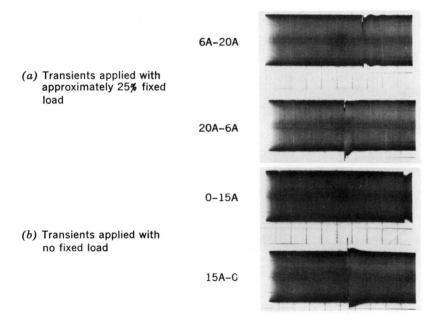

20A–6A

0–15A

(*b*) Transients applied with no fixed load

15A–0

Figure 11.48   Transient response of the inverter in Figure 11.46. (*a*) Transients applied with approximately 25% fixed load. (*b*) Transients applied with no fixed load. Vertical: 100 v/division. Horizontal: 50 μsec/division.

Figure 11.49 Low frequency a-c high potential test set. Rating 50 kv rms, 0.5 μf, 0.1 cps.

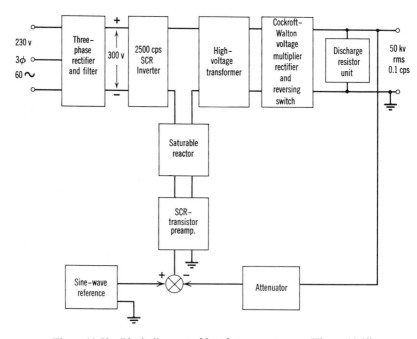

Figure 11.50 Block diagram of low frequency test set (Figure 11.49).

Frequency Converters

There are several additional areas of interest where a-c power is available, but for various reasons it is not suitable for the final desired output. An example is shown in Figure 11.51. Constant output frequency is desired from a higher frequency input generator which is coupled to a variable speed prime mover.[23] The modulator provides the proper firing for the switching devices. The exciter-regulator acts to control the

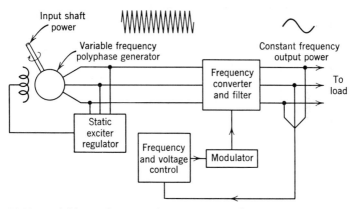

Figure 11.51   Variable-speed constant-frequency system, frequency converter approach.

excitation of the generator in such a way that the proper voltage is available to the input of the frequency changer. The frequency and voltage control block contains an oscillator, which determines the system frequency, and a phase splitter, which determines the phase displacement and sequence. The basic power circuit of the converter is shown in Figure 11.52. An SCR is connected from each input phase to each output phase in both directions. The proper SCR's are consecutively fired for a large number of input cycles, corresponding to the period of one positive output half-cycle and, then, alternate SCR's are fired to produce the negative output, such that the output frequency is independent of the input frequency. If each SCR is turned on at the same phase angle of the incoming supply voltage, for the entire output half-cycle, the unfiltered output voltage will contain appreciable high, low, and subharmonics of the desired frequency. When the firing angle is varied in a sinusoidal manner, the low and subharmonics are considerably reduced.

The circuit of Figure 11.52 is generally operated as a line-commutated inverter where SCR commutation is accomplished by the a-c supply voltage. Thus, it operates very much like the circuits discussed in Chapter

3, and it may feed power in either direction, assuming the proper gating circuits.

For the system of Figure 11.51, the SCR's are gated to produce an alternating output voltage at a lower frequency than the a-c supply frequency. A similar circuit may also be used to produce a reversible

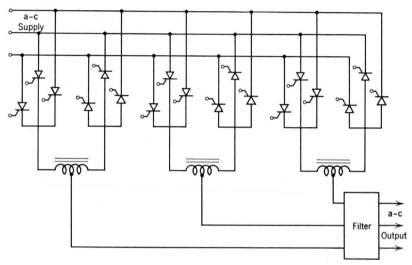

Figure 11.52   Basic frequency converter power circuit.

phase-controlled d-c output. In other applications, the input supply frequency may actually be fixed, but the output will be varied, or converted to a substantially lower frequency. If the control circuit is properly designed it is possible to provide a reversible output so that positioning systems of very slow speed and high accuracy are possible. One of the major advantages of line-commutated circuits, similar to that shown in Figure 11.52, is that reliable commutation is maintained even when the output voltage is reduced to zero.

Another inverter is shown in Figure 11.53 in which there is no apparent d-c link, since the system is basically self-rectifying.[24] The circuit is characterized by providing more than one chance for commutation to occur. Thus, if a commutation failure should occur as a result of a momentary transient disturbance, the output waveform will be momentarily distorted. However, normal operation will be resumed at a later time in the supply voltage cycle if the transient disturbance disappears. The lower over-all utility of components detracts from widespread use of this approach. With the parallel capacitor commutation shown, the circuit cannot be operated over wide load ranges nor open circuited, as is

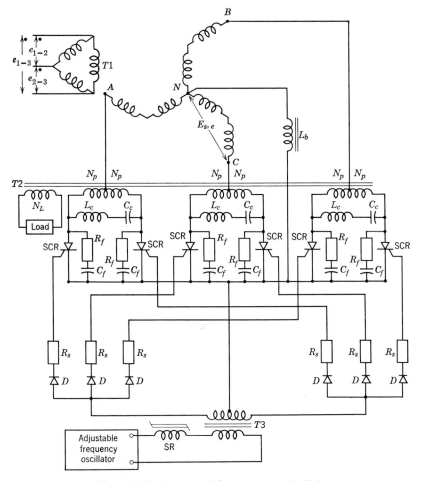

Figure 11.53 Inverter with no apparent d-c link.

characteristic of the conventional parallel inverter. Many other basic circuit arrangements are possible, including the incorporation of other commutating techniques. However, in general, these circuits, as well as the arrangement of Figure 11.53, require more semiconductor component rating per kva of output than inverters with a d-c link.

A-C Motor Drives

Static inverters are finding wide acceptance as power sources for a-c motor control where ease and accuracy of control are required. In general, all of the inverters which have been discussed may be used to provide variable frequency power to control the speed of a-c motors. Several additional motor control approaches are described in this section which differ from the basic variable frequency inverter. A simple circuit operating from a fixed frequency a-c source is shown schematically in

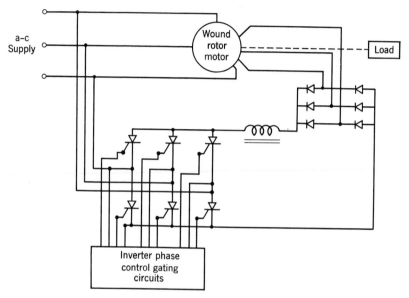

Figure 11.54   Wound rotor-motor control.

Figure 11.54. This system involves rectifying the rotor power and feeding it back into the a-c source through an a-c line-commutated inverter circuit. An amount of energy proportional to the slip is returned to the a-c supply rather than dissipated in resistance, as in the conventional wound rotor-motor speed control. The motor speed is controlled by adjusting the inverter d-c voltage to control the amount of energy returned to the a-c system. Adjustment of the phase angle of the firing pulses to the inverter adjusts the inverter d-c voltage level as discussed in Chapter 3. Figure 11.55 shows a picture of a 15-hp prototype machine, using this method of speed control.

In many applications it is desirable to eliminate slip rings so that considerable interest exists in "solid-state commutator" motors. These are often referred to as "commutatorless" motors, since the conventional d-c machine commutator is not used. Figure 11.56 illustrates one possible scheme suitable for relatively small motors where dissipation of a fraction

Figure 11.55   A 15-hp wound rotor motor with the speed control of Figure 11.54.

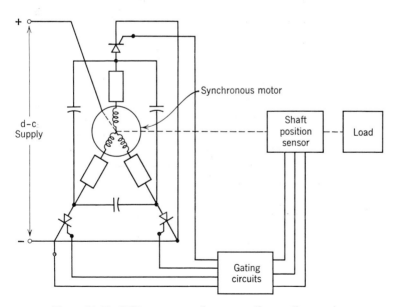

Figure 11.56   SCR commutatorless motor (for small motors).

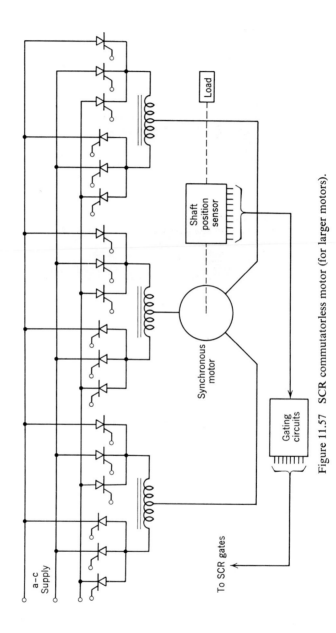

Figure 11.57   SCR commutatorless motor (for larger motors).

of the motor power in resistors is not a major disadvantage. The function of the commutator is accomplished by the shaft position sensor controlling the SCR's. Since current is only flowing in each winding for one third of the time, the utility of the motor winding is poor which restricts the use of this approach to small machines.

Figure 11.58   A 15-hp commutatorless motor.

The system shown in Figure 11.57 is similar to the previous circuit, but is more practical for larger motors. A picture of a 15-hp commutatorless motor, using this basic circuit, is shown in Figure 11.58. The basic power circuit is similar to that developed many years ago for use with gas thyratrons.[25] There are a number of SCR's used but, when the same number of paralleled units would be required anyway as a result of the required power rating, no great disadvantage results, and the utility of the motor is very good. In this system, the motor starts with the circuit operating as a step-down frequency converter with commutation provided by the a-c supply. When the motor runs at rated speed, the circuit is also line-commutated, but with commutation provided by the motor back emf. Regenerative breaking can be provided by including the proper firing circuits for each SCR, but these firing circuits are fairly complex, since the full range from inverter action to rectifier action must be covered.

In both of the "commutatorless" motor schemes discussed, some form of shaft position sensing is required to provide proper operation. This information can be generated, using variable reluctance devices, semiconductor devices such as hall crystals, light beam devices, or various mechanical devices.

11.4 SUMMARY

The silicon-controlled rectifier has had a revolutionary impact on the power conversion technology. Figure 11.59 shows the range of SCR device ratings which have become available and which are predicted for

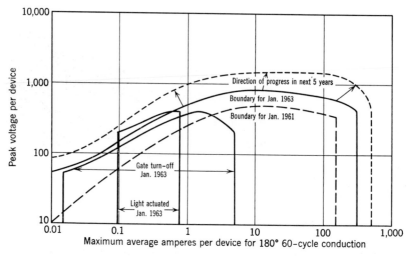

Figure 11.59 Maximum SCR device ratings (available and predicted).

the near future. It is presently anticipated that major breakthroughs will be required to radically and economically push beyond the maximum voltage and current indicated in Figure 11.59.

There is expected to be a continuing effort to produce devices capable of higher frequency operation. At the present time, SCR inverters are limited to around 10,000 cps depending on the type of circuit and the characteristics of the devices used. Three main factors currently govern high frequency performance: (1) turn-off time, (2) allowable dv/dt, and (3) di/dt.

The availability of faster semiconductor power-switching devices will open whole new areas of application, including modulators, sonar and radar pulsers, solid-state transmitters, induction heating supplies, and the like. It is also important to emphasize that higher frequency operation can provide improvements which are desirable in other applications, such as reduced acoustical noise, faster response, improved flexibility in terms of waveshaping, wider control range, and smaller size and lighter weight. These improvements will be beneficial for both TRC and inverter equipments.

Improvements in semiconductors have occurred at an unbelievable rate, which has offered new challenges for conventional components as well. Since the process of forced commutation is very rapid, adequately designed capacitors and inductors are critical to the success of the system. Capacitors with low internal impedance are needed for both energy storage and commutating circuits. Reactor core losses and copper losses may make the use of Litz wire and special core materials imperative if maximum performance is to result. The problems of interconnection have required the design of new low-inductance methods of wiring between components, even in equipment operating at low frequency such as 60 cps.

One of the important factors, which requires serious consideration in the proper design of static equipment, concerns the question of system stability. In some cases, large filters are required to provide acceptable waveshape for the output, or energy storage, to ensure proper commutation. It is important to devise stabilization techniques which achieve the extremely fast transient response inherent in solid-state power converters.

The advent of semiconductor devices for power conversion and control has made possible new levels of performance not heretofore obtainable. As device ratings continue to increase, as dynamic characteristics become even more favorable and, as totally new devices become available, the solid state power conversion technology will continue its phenomenal growth with ever expanding horizons.

REFERENCES

1. B. D. Bedford, D. A. Paynter, and J. D. Harnden, Jr., "Solid State Power Inversion Techniques," *Semiconductor Products*, Part I, March 1960, pp. 51–56; Part II, April 1960, pp. 50–55.
2. E. E. Von Zastrow, "Semiconductor Dimmers in Architectural Lighting," *Lighting*, April 1962.
3. F. W. Gutzwiller, "A Plug–In Speed Control for Standard Portable Tools and Appliances," Rectifier Components Department Application Note 201.1, General Electric Company, Auburn, N.Y.
4. F. W. Gutzwiller *et al.*, *Silicon Controlled Rectifier Manual*, Second Edition, General Electric Company, Auburn, N.Y., 1961.

5. G. M. Bell, "A Frequency Compensated Magnetic Voltage Stabilizer," AIEE District Paper DP 62-613, Fort Wayne, Ind., April 25–27, 1962.
6. E. W. Manteuffel, "Magnetic Voltage Stabilizer Employing Controlled Silicon Rectifiers," U.S. Patent 3,076,924, February 5, 1963.
7. E. W. Manteuffel and T. A. Phillips, "The Shunt Loaded Magnetic Amplifier, Part I and Part II," *IEEE Transactions on Communications and Electronics*, January 1963, pp. 440–457.
8. B. D. Bedford and L. J. Goldberg, "Direct Current Static Electric Switch," U.S. Patent 3,042,838, July 3, 1962.
9. E. W. Manteuffel and R. E. Cooper, "D-C Charged Magnetic Pulse Modulator," *AIEE Transactions*, Vol. 78, Part I, 1960, pp. 843–850.
10. A. L. Wellford, "A Controlled Rectifier Regulator for Aircraft DC Generators in 120°C Applications," *AIEE Transactions*, Vol. 79, Part II, 1960, pp. 411–416.
11. J. L. Fink and T. W. Macie, "The SCR as a Building Block in Power Conversion Systems," AIEE Conference Paper CP60-1293, Chicago, October 9–14, 1960.
12. T. W. Macie and E. F. Chandler, "Computer Power Supply Utilizing SCR Conversion Techniques," AIEE Conference Paper CP62-419, New York, January 28–February 2, 1962.
13. F. G. Turnbull, "Controlled Rectifier DC to DC 30 HP Motor Drive," *IEEE Transactions on Communications and Electronics*, January 1963, pp. 458–462.
14. R. B. Jones and A. R. Olds, Jr., "Static Power Supplies for Adjustable Speed Drives," *Electrical Engineering*, March 1962, pp. 178–186.
15. J. B. McFerran, "A Controlled Rectifier DC Servo Drive," AIEE District Paper DP62-1006, Erie, Pa., May 14–16, 1962.
16. G. E. Snyder, "SCR High Voltage Power Supply," Rectifier Components Department, Application Note 610.5, General Electric Company, Auburn, N.Y.
17. J. L. Fink, "The Application of Static Inverters for Emergency Power and Frequency Changing," *Distribution*, January, 1963, pp. 10–11.
18. F. W. Gutzwiller, "Silicon Controlled Rectifier Circuit Including a Variable Frequency Oscillator," U.S. Patent 3,040,270, June 19, 1962.
19. W. McMurray and D. P. Shattuck, "A Silicon Controlled Rectifier Inverter With Improved Commutation," *AIEE Transactions*, Vol. 80, Part I, 1961, pp. 531–542.
20. J. A. Laukaitis, W. D. Modern, and J. L. Fink, "A Static Approach to the Continuous Power Problem," AIEE Conference Paper, CP61-1169, Detroit, October 16, 1961.
21. C. W. Flairty, "A 50 KVA Adjustable Frequency, 24 Phase Controlled Rectifier Inverter," *Direct Current*, December 1961, pp. 278–282.
22. P. D. Corey, "Methods for Optimizing the Waveform of Stepped-Wave Static Inverters," AIEE Conference Paper CP62-1147, Denver, June 17–22, 1962.
23. S. C. Caldwell, L. R. Peaslee, and D. L. Plette, "The Frequency Converter Approach to a Variable Speed, Constant Frequency System," AIEE Conference Paper CP60-1076, San Diego, August 8–12, 1960.
24. C. W. Flairty and J. D. Harnden, Jr. "Controlled Rectifier Frequency Multipliers Using Inverter Principles," AIEE Conference Paper CP60-321, New York, January 31–February 5, 1960.
25. E. F. W. Alexanderson and A. H. Mittag, "Thyratron Motor," *Electrical Engineering*. Vol. 53, November, 1934, pp. 1517–1523.

Index

411